D1162865

THE COMPLETE GUIDE TO THE HAZARDOUS WASTE REGULATIONS

Other books by Travis Wagner:

Hazardous Waste Identification and Classification Manual

The Hazardous Waste Q&A (Revised Edition)

In Our Backyard: A Guide to Understanding Pollution and Its Effects (also available in Spanish)

THE COMPLETE GUIDE TO THE HAZARDOUS WASTE REGULATIONS

RCRA, TSCA, HMTA, OSHA, AND SUPERFUND

Third Edition

TRAVIS P. WAGNER

JOHN WILEY & SONS, INC.
New York · Chichester · Weinheim · Brisbane · Singapore · Toronto

This book is printed on acid-free paper.

Library of Congress Cataloging-in-Publication Data:

Wagner, Travis.
The complete guide to the hazardous waste regulations : RCRA, TSCA, HMTA, OSHA, and Superfund / Travis P. Wagner.—3rd ed.
p. cm.
Includes bibliographical references and index.
ISBN 0-471-29248-6 (cloth : alk. paper)
1. Hazardous wastes—Law and legislation—United States.
I. Title.
KF3946.W34 1998
344.73'04622—dc21 98-29064

0-471-29248-6

Printed in the United States of America

10 9 8 7 6 5 4 3 2 1

Nature never did betray the heart that loved her.

—William Wordsworth (1798)

CONTENTS

PREFACE

This is the third edition of *The Complete Guide to the Hazardous Waste Regulations*. Thankfully, since 1991, the year the second edition was released, the pace at which major regulations have been promulgated has slowed significantly. The slowdown has been due in part to a reduced reliance on command-and-control as the primary means of controlling hazardous waste and to a general maturation of the hazardous waste program. Nonetheless, there have been significant changes since 1991, and this third edition includes all of them. In addition to these regulatory updates, the third edition of *The Complete Guide to the Hazardous Waste Regulations* includes the following new elements for increased utility and understandability:

- A summary of state-specific hazardous waste requirements.
- Additional, detailed compliance checklists.
- Different regulatory requirements are combined into text and tables wherever possible.
- A comprehensive spill reporting chapter dealing with multiple federal laws.
- Relevant information on the Emergency Planning and Community Right-to-Know Act (EPCRA).
- A modified overall format and organization to improve readability.
- Additional flowcharts, check lists, and tables.
- An appendix that lists the Superfund reportable quantity for each RCRA hazardous waste.
- Detailed information on resources available on the Internet, including each state environmental program.

My primary purpose in writing the first edition of *The Complete Guide to Hazardous Waste Regulations* was to help readers understand the federal hazardous waste regulatory program. This original purpose has not changed. In my experience, I have heard many wrong interpretations, as well as basic misunderstandings, of the regulations, which can be attributed to the sheer complexity and nebulousness of the regulatory language. *The Complete Guide to the Hazardous Waste Regulations* alleviates this problem by aggregating the information and presenting it in a straightforward and lucid manner.

The Complete Guide to the Hazardous Waste Regulations goes far beyond the typical inclusion of a copy of the various laws along with a brief, canned summary. It is an in-depth, step-by-step guide to the regulations that contain compliance guidance, as well as EPA interpretations, which are not readily communicated. My experience as a technical information specialist on both the RCRA/Superfund and TSCA hotlines required research and answers to thousands of questions on this subject, and, combined with my 15 years of experience as an environmental consultant, has enabled me to present this comprehensive collection of hazardous waste information. Responding to those thousands of questions concerning regulations required the presentation of very technical information in general, understandable language to persons of varying backgrounds. I have written *The Complete Guide to the Hazardous Waste Regulations* in that same manner.

To understand the federal hazardous waste management program and the interrelationships among the statutes, it is important to review a brief history of the program.

Hazardous waste regulation is a relatively recent initiative. The pollution scare and environmental movement in the 1960s brought about the first major attempts to rectify the adverse effects of environmental pollution. These first attempts were focused on the most visible media, air and surface water, with the passage of the Clean Air and Clean Water Acts. Because these new acts prevented disposal of pollutants into the air or water, industry relied more heavily on land disposal for large amounts of wastes. Because hazardous waste disposal practices at that time were inadequate and environmentally unsound, and because knowledge of the behavior and effects of pollutants in the environment was limited, the nation became aware of a new pollution problem: hazardous and toxic substances contaminating groundwater and tainting drinking water wells. In response, Congress enacted the Resource Conservation and Recovery Act (RCRA) in 1976 specifically to control the generation, transportation, and management of solid and hazardous wastes The Toxic Substances Control Act (TSCA) also was enacted in 1976 to regulate the use and management of toxic substances, specifically polychlorinated biphenyls (PCBs).

In 1978, Love Canal and similar incidents emphatically demonstrated the need for legislation to ensure the cleanup of past hazardous waste disposal sites. To meet this need, in 1980, Congress passed the Comprehensive Environmental Response, Compensation, and Liability Act (CERCLA, also called "Superfund"). During the early 1980s, perceiving shortcomings in both RCRA and CERCLA, as well as a reluctance by the Reagan administration to enforce and administer the nation's environmental laws; Congress began to implement sweeping changes that reduced the reliance on federal government action during the reauthorizations of both RCRA and CERCLA. The reauthorization of RCRA added the Hazardous and Solid Waste Amendments of 1984 (HSWA), which includes a mandate to halt reliance on the traditional practice of land disposal. The reauthorization of CERCLA in 1986 added the Superfund Amendments and Reauthorization Act (SARA), which established the Emergency Planning and Community Right-to Know Act (EPCRA) and includes a mandate to speed the cleanup of former waste management sites. Although the Hazardous Materials Transportation Act (HMTA) was established in 1974, increased information on the potential health and the environmental impacts from transportation-related incidents have brought about sweeping changes in this hazardous materials transportation program as well.

This array of statutes, their subsequent reauthorizations, and their respective implementing regulations constitute the crux of the nation's hazardous waste regulatory program.

Travis P. Wagner
September 1998

ACKNOWLEDGMENTS

I would like to thank the following persons for their assistance and contributions in preparing the third edition of this book:

Jim Byrne, U.S. Environmental Protection Agency, Region I*
Patricia Overmeyer, Science Applications International Corporation
Bob Stewart, Science Applications International Corporation
Kristin Tensuan, Science Applications International Corporation

Special thanks to Chuck Tobler for editorial assistance, Clark Whittington for production of the figures, Paddy Fitz for the cover concept, and Julie Whittington for production of the tables.

I would also like to express my gratitude to those who assisted in the previous editions of this book: Leigh Benson, Karen Burchard, Kathie Gabriel, Saskia Mooney, Suzanne Parent, Ingrid Rosencrantz, and Lance Traves.

* The review by this person in no way represents endorsement of this book or its approval by the U.S. Environmental Protection Agency.

ACKNOWLEDGMENTS

A GUIDE TO READERS

This book has been written and structured in the most straightforward manner possible, given the diversity of its audience and the nature of its subject. Some minor regulations, or regulations that affect a very small portion of the regulated community, have been purposely omitted.

The Complete Guide to the Hazardous Waste Regulations: Third Edition is separated into three parts, each relating to one of the three major statutes (RCRA, CERCLA, and TSCA) that control "hazardous" waste. Other statutes are addressed throughout the book where appropriate. Regulatory compliance checklists have been included at the end of appropriate chapters.

Although regulations are written with the intent to be clear-cut, the results rarely achieve this lofty goal. There are many factors that can affect their applicability, interpretation, or intent. One way to improve the understanding of regulations is to pay close attention to the regulatory definitions of terms used throughout this book, which appear as the first chapter in each part. One may tend to use the commonly implied meaning of a term instead of its strict regulatory meaning. Misunderstanding or ignorance of definitional usage can easily lead to noncompliance or "over"-compliance.

This book follows the federal regulatory program as established in the *Code of Federal Regulations* (CFR). A major thrust of the Resource Conservation and Recovery Act (RCRA), the principal law addressing hazardous waste, is to have states operate the program in lieu of the federal government. As a result, some state hazardous waste management programs may differ from the federal program. (State programs may be more stringent or broader in scope, but never less stringent than the federal program. In fact, most state programs are identical to the federal program.) To help the reader understand the differences between state programs and federal programs, Chapter 10 contains a detailed table that summarizes the major differences between the individual state programs and the federal program. The other statutes addressed in this book, Superfund and the Toxic Substances Control Act, do not have a specific state authorization provision; a discussion of differences is unnecessary.

Obviously, this book cannot provide individualized legal advice, and it is not intended to do so. However, it does provide the information needed to understand various regulatory situations and, thus, the necessary tools to make a prudent decision. I have tried to be as accurate and objective as possible and have not included judgments about the need, value, or practicality of the implementation of these regulations.

Knowledge of how regulations are developed and published at the federal level is helpful in understanding the hazardous waste regulatory program. (See Figure P-1.)

The process must be initiated through the enactment of a law, which is passed by the Congress. Although the law may stand alone, regulations are typically required

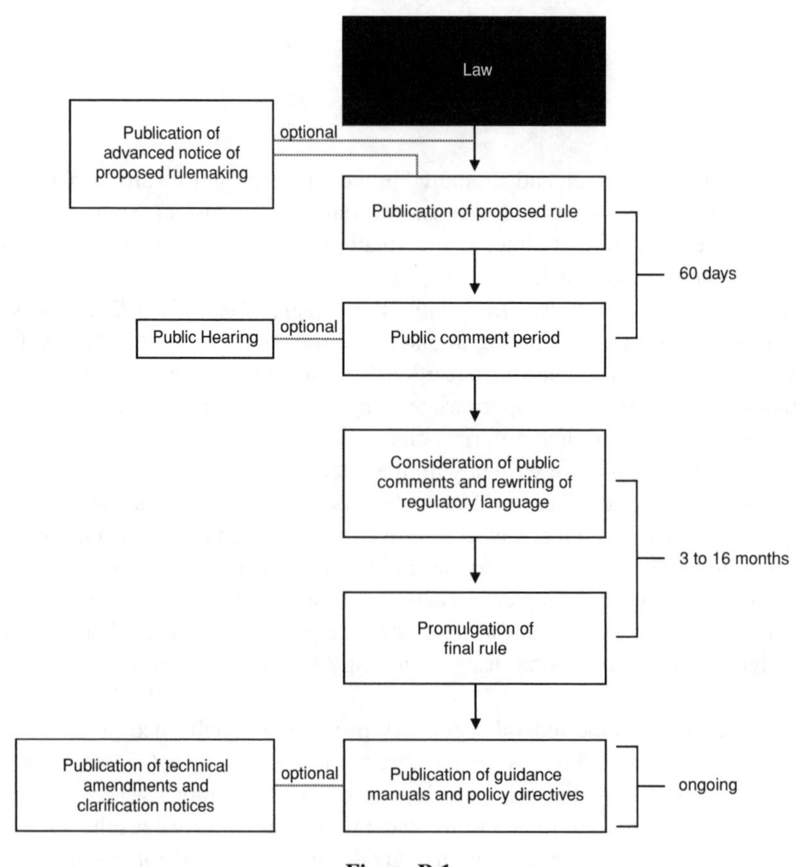

Figure P-1.

to implement the intent of the law. The executive branch (e.g., Environmental Protection Agency) is either directed or authorized by the Congress to develop and implement regulations where explicit authority is provided for in the law. The Resource Conservation and Recovery Act (RCRA), which is a law, has various sections that Congress specifically requires EPA to promulgate implementing regulations. (In this book, the symbol § is used to denote sections in a law.) This process must follow a prescribed pattern that allows for adequate public notice and comment in accordance with the federal Administrative Procedures Act. EPA meets the public notification requirement by publishing notices of its intentions to promulgate regulations in the *Federal Register*, a compendium of notices, announcements, and descriptions of the activities of the federal government, published every business day.

The general rulemaking procedure is as follows. First, a proposed rule is developed by an agency and published in the *Federal Register.* (Occasionally, an agency will issue an advanced notice of proposed rulemaking as a means to solicit comments and input, which is followed by a proposed rule if necessary.) The proposed rule establishes a public comment period, usually 60 days, and possibly a public hearing if requested by at least five people and the issue is deemed sufficiently significant. Immediately after the public comment period closes, the agency reviews the comments and any data it receives. After the consideration of comments, the proposed rule is revised as appropriate. A final rule is subsequently promulgated. A regulation is *promulgated* (made known to the public) by being published as a final rule in the *Federal Register.* An agency, however, is not necessarily obligated to promulgate a final rule unless Congress or the courts have mandated otherwise. The agency may formally withdraw the proposal and close the administrative docket through a *Federal Register* notice. Conversely, an agency may not promulgate a final rule that is significantly different from the proposed rule. Only modest adjustments or deletions of portions of the rule are generally allowed. If there is a need to make a rule significantly more stringent or broader in scope than originally proposed, the agency must publish another notice of proposed rulemaking describing the revised action and soliciting additional public comments. An agency can later change a promulgated final rule by publishing proposed amendments, interim final amendments, technical corrections, technical amendments, and clarification notices.

Volumes of the *Federal Register* are consecutively numbered by year, and each daily issue likewise receives a consecutive number. The following heading of a typical notice serves as an example:

226091/Federal Register/Vol. 64, No. 245/Thursday, December 21, 1999, Final Rule

This heading indicates that this issue of the *Federal Register* was published on Thursday, December 21, 1999; it is the 245th daily issue published and is part of Volume 64 (the 1999 volume); it is from the Final Rules section of this day's issue; and it is page 226091. The generally accepted notation for reference purposes is 64 *FR* 226091. However, because that notation does not identify which daily issue contains that page, the date is often included (December 21, 1999). When a regulation is promulgated, it becomes part of the *Code of Federal Regulations* and carries the full force of the law.

The *Code of Federal Regulations*, often denoted by the acronym CFR, is the compilation of all final federal regulations in effect in the United States. The full text (not including the preamble) of all final regulations promulgated by all federal government agencies is included in the CFR. However, because the CFR is published only once per year, the most recent CFR may not actually include all applicable regulations. (Availability of the CFR is discussed in Appendix E.) The regulations are grouped under "Titles" in the CFR, and each Title is divided into "Chapters." The hazardous waste regulations are found in "Title 40—Protection of Environment" and "Title 49—Transportation." Titles 40 and 49 are generally identified or annotated as 40 CFR and 49 CFR. Each Chapter of a Title is divided into numerous "parts." Each part is further divided into "subparts." The subparts are composed of "sections." For example, the groundwater monitoring requirements for permitted hazardous waste management units are found in 40 CFR part 264, subpart F, section 90, which is annotated as 40 CFR 264.90. Because specific requirements are typically addressed in this book, specific citations are given. For example, the specific requirement for surface impoundments to monitor groundwater, which is contained in 40 CFR part 264, subpart F, is cited as 40 CFR 264.90(a)(2). [The (a)(2) denotes the subsection.)

Beyond the language in the law or act (statutory language) and promulgated regulations, requirements are further clarified through the issuance of guidance documents and policy directives. Guidance documents are issued primarily to elaborate on the implementation of a regulation. They essentially explain how to do something. In contrast, policy directives specify procedures that must be followed pertaining to a regulation.

This book has been designed primarily as a reference source to research specific requirements. However, because of the complex interrelationships between all the federal laws and regulations governing hazardous waste, it is recommended that you first read through the entire book for a basic understanding and a sense of these potential interrelationships.

Should you have any comments concerning the content, organization, or utility of *The Complete Guide to the Hazardous Waste Regulations: Third Edition,* please contact me through the publisher, c/o Rob Garber, John Wiley & Sons, 605 Third Avenue, New York, New York 10158.

PART I

THE RESOURCE CONSERVATION AND RECOVERY ACT

The Solid Waste Disposal Act (SWDA), enacted in 1965, was the first piece of federal legislation that specifically addressed the nation's waste management problem. The act was amended significantly by the Resource Conservation and Recovery Act (RCRA) in 1976 and again by the Hazardous and Solid Waste Amendments of 1984 (HSWA). These three acts collectively are referred to as RCRA (pronounced "wreckra").

The three major parts of RCRA are Subtitles C, D, and I. Subtitle C regulates hazardous waste, Subtitle D regulates solid waste (nonhazardous waste), and Subtitle I regulates underground storage tanks that hold petroleum products and hazardous substances (does not include wastes except for waste oil).

HAZARDOUS WASTE MANAGEMENT

The Resource Conservation and Recovery Act (RCRA), enacted on October 21, 1976, regulates hazardous waste "from the cradle to the grave." The statute requires EPA to establish minimum acceptable requirements for all aspects of hazardous waste for generators and transporters as well as for treatment, storage, and disposal facilities.

The regulatory framework established under Subtitle C was designed to protect human health and the environment from the effects of improper management of hazardous waste. Determining what is a *hazardous waste* is therefore a key question because only those wastes that meet the definition of hazardous waste are subject to Subtitle C. Making this determination can be a very complex exercise. The universe of potential hazardous wastes is large and diverse, consisting of chemical products, generic waste streams, and specialized byproducts. The terms *hazardous waste* and *toxic waste* are often used interchangeably; however, regarding the Subtitle C regulations, there is an important distinction between the two. Hazardous waste denotes a regulated waste; only certain waste streams are designated as hazardous. This designation is not based solely on toxicity; it also includes other physical characteristics that present an environmental or health threat, as well as the quantity generated, damage case history, and environmental fate. Toxic waste is a phrase used primarily by the media and the public. Although every waste stream is toxic to some degree, only some wastes are classified as hazardous specifically because of their toxicity, which is discussed in more detail in Chapter 2. *Toxic* simply means that it has the ability to cause harm. Thus, those wastes that pose a serious threat when mismanaged must be differentiated.

Applicability of RCRA

RCRA was enacted in 1976, but it was not until May 19, 1980, that EPA finalized the first phase of the RCRA hazardous waste regulatory program. The May 19 rule did not become effective until November 19, 1980. By this date, hazardous waste management facilities had to either close or comply with RCRA. Except for a few provisions (e.g., corrective action), waste disposed of before November 19, 1980, is not subject to RCRA but is subject to Superfund. RCRA is written in the present tense and its regulatory scheme is prospective. Therefore, EPA believes that Congress intended the hazardous waste regulatory program under Subtitle C of RCRA to control primarily hazardous waste management activities that take place after the effective date of the Phase I regulations (November 19, 1980). Thus, the Subtitle C regulations did not by their terms apply to inactive (either closed or abandoned) waste management facilities. This statement can be found in the May 19, 1980, *Federal Register,* Volume 45, page 33170 (cited as 45 *FR* 33170). However, hazardous waste placed in a surface impoundment, tank, or drum before November 19, 1980, and that remained in *storage* after that date, is subject to RCRA. (The device or facility that stores the waste is likewise subject to RCRA.) This is considered active storage. Active storage, on or after November 19, 1980, is subject to RCRA requirements. This decision is based on *Environmental Defense Fund v. J. Lamphier,* 714 F.2nd (1983).

Any person that generates a waste is potentially subject to RCRA. However, there are many exemptions and exclusions that affect the applicability of the law and implementing regulations. Determining the intent, scope, and conditions of the exemption or exclusion becomes a key compliance point.

Regulatory Sections

The Subtitle C regulations are contained in Title 40 of the *Code of Federal Regulations*. The regulations are organized into parts. The major, codified regulatory parts of Subtitle C for hazardous waste are as follows:

Part 260	Definitions, petitions for rulemaking changes, and delisting procedures
Part 261	Hazardous waste identification
Part 262	Generator standards
Part 263	Transporter standards
Part 264	Treatment, storage, and disposal facilities: final operating standards
Part 265	Treatment, storage, and disposal facilities: interim status standards
Part 266	Hazardous waste fuel burned for energy recovery, used oil fuel, and process-specific regulatory standards
Part 268	Land disposal restrictions
Part 270	Permits and operation during interim status
Part 271	State programs
Part 273	Universal waste management standards

State Requirements

RCRA encourages states to develop and operate their own hazardous waste programs as an alternative to federal management. Thus, the hazardous waste regulatory program described in this book may be run by EPA or a state agency. However, for a state to have jurisdiction over its hazardous waste program, it must receive EPA approval by showing that its program is at least as stringent as the EPA program; and in many approved states the program is identical. Chapter 10 explains the requirements for a state to obtain authorization.

Petitions

RCRA contains provisions (codified in Part 260) that allow any person to petition the EPA Administrator to modify or revoke any provision in Parts 260 through 273. The general informational requirements for a petition include the following:

- Petitioner's name and address
- A statement of the petitioner's interest in the proposed action
- A description of the proposed action, including (where appropriate) suggested regulatory language
- A statement of the need and justification for the proposed action, including any supporting tests, studies, or other relevant information

Specialized information is required for a petition, depending on the changes being sought. This information is contained in Part 260, Subpart C, of 40 CFR.

The EPA Administrator must make a tentative decision to grant or deny a petition and publish a notice of the tentative decision—in the form of either an advanced notice of proposed rulemaking (ANPR), a proposed rule, or a tentative determination to deny the petition—in the *Federal Register* to allow public comment. An informal public hearing may be held if the Administrator decides that one is warranted or if one is requested.

After evaluating all public comments, the Administrator must make a final decision by publishing in the *Federal Register* a regulatory amendment or a denial of the petition.

RCRA'S Relationship with NEPA

EPA has determined that RCRA permits are not subject to the environmental impact statement (EIS) provisions of §102(2)(C) of the National Environmental Policy Act (NEPA) [40 CFR 124.9(b)(6)]. EPA has asserted that RCRA permit requirements are the functional equivalent of an EIS; thus, an EIS is not required (45 *FR* 33173, May 19, 1980).

1

KEY DEFINITIONS UNDER RCRA

This chapter contains key regulatory definitions of RCRA terms used throughout Part 1 of this book. All of the definitions from the various parts (i.e., 40 CFR Parts 260, 268, 270, 273, and 279) are included, the source of each definition is designated at the end of the entry. Definitions appear here verbatim as they appear in their respective federal statutes or Code of Federal Regulations.

Aboveground tank means a device meeting the definition of "tank" in §260.10 and that is situated in such a way that the entire surface area of the tank is completely above the plane of the adjacent surrounding surface and the entire surface area of the tank (including the tank bottom) is able to be visually inspected. (40 CFR 260.10)

Act or RCRA means the Solid Waste Disposal Act, as amended by the Resource Conservation and Recovery Act of 1976, as amended, 42 U.S.C. section 6901 et seq. (40 CFR 260.10)

Active life of a facility means the period from the initial receipt of hazardous waste at the facility until the Regional Administrator receives certification of final closure. (40 CFR 260.10)

Active portion means that portion of a facility where treatment, storage, or disposal operations are being or have been conducted after the effective date of part 261 of this chapter and which is not a closed portion. (See also "closed portion" and "inactive portion.") (40 CFR 260.10)

Administrator means the Administrator of the United States Environmental Protection Agency, or an authorized representative. (40 CFR 270.2)

Ancillary equipment means any device including, but not limited to, such devices as piping, fittings, flanges, valves, and pumps, that is used to distribute, meter, or control the flow of hazardous waste from its point of generation to a storage or treatment tank(s), between hazardous waste storage and treatment tanks to a point of disposal onsite, or to a point of shipment for disposal off-site. (40 CFR 260.10)

Application means the EPA standard national forms for applying for a permit, including any additions, revisions or modifications to the forms; or forms approved by EPA for use in approved States, including any approved modifications or revisions. Application also includes the information required by the Director under §§270.14 through 270.29 (contents of part B of the RCRA application). (40 CFR 270.2)

Approved program or approved State means a State which has been approved or authorized by EPA under part 271. (40 CFR 270.2)

Aquifer means a geologic formation, group of formations, or part of a formation capable of yielding a significant amount of ground water to wells or springs. (40 CFR 260.10)

Authorized representative means the person responsible for the overall operation of a facility or an operational unit (i.e., part of a facility), e.g., the plant manager, superintendent or person of equivalent responsibility. (40 CFR 260.10)

Battery means a device consisting of one or more electrically connected electrochemical cells which is designed to receive, store, and deliver electric energy. An electrochemical cell is a system consisting of an anode, cathode, and an electrolyte, plus such connections (electrical and mechanical) as may be needed to allow the cell to deliver or receive electrical energy. The term battery also includes an intact, unbroken battery from which the electrolyte has been removed. (40 CFR 260.10)

Boiler means an enclosed device using controlled flame combustion and having the following characteristics:

(1)(i) The unit must have physical provisions for recovering and exporting thermal energy in the form of steam, heated fluids, or heated gases; and

(ii) The unit's combustion chamber and primary energy recovery sections(s) must be of integral design. To be of integral design, the combustion chamber and the primary energy recovery section(s) (such as waterwalls and superheaters) must be physically formed into one manufactured or assembled unit. A unit in which the combustion chamber and the primary energy recovery section(s) are joined only by ducts or connections carrying flue gas is not integrally designed; however, secondary energy recovery equipment (such as economizers or air preheaters) need not be physically formed into the same unit as the combustion chamber and the primary energy recovery section. The following units are not precluded from being boilers solely because they are not of integral design: process heaters (units that transfer energy directly to a process stream), and fluidized bed combustion units- and

(iii) While in operation, the unit must maintain a thermal energy recovery efficiency of at least 60 percent, calculated in terms of the recovered energy compared with the thermal value of the fuel; and

(iv) The unit must export and utilize at least 75 percent of the recovered energy, calculated on an annual basis. In this calculation, no credit shall be given for recovered heat used internally in the same unit. (Examples of internal use are the preheating of fuel or combustion air, and the driving of induced or forced draft fans or feedwater pumps); or

(2) The unit is one which the Regional Administrator has determined, on a case-by-case basis, to be a boiler, after considering the standards in §260.32. (40 CFR 260.10)

Carbon regeneration unit means any enclosed thermal treatment device used to regenerate spent activated carbon. (40 CFR 260.10)

Certification means a statement of professional opinion based upon knowledge and belief. (40 CFR 260.10)

Closed portion means that portion of a facility which an owner or operator has closed in accordance with the approved facility closure plan and all applicable closure requirements. (See also "active portion" and "inactive portion.") (40 CFR 260.10)

Closure means the act of securing a Hazardous Waste Management facility pursuant to the requirements of 40 CFR part 264. (40 CFR 270.2)

Component means any constituent part of a unit or any group of constituent parts of a unit which are assembled to perform a specific function (e.g., a pump seal, pump, kiln liner, kiln thermocouple). (40 CFR 270.2)

Component means either the tank or ancillary equipment of a tank system. (40 CFR 260.10)

Confined aquifer means an aquifer bounded above and below by impermeable beds or by beds of distinctly lower permeability than that of the aquifer itself; an aquifer containing confined ground water. (40 CFR 260.10)

Container means any portable device in which a material is stored, transported, treated, disposed of, or otherwise handled. (40 CFR 260.10)

Containment building means a hazardous waste management unit that is used to store or treat hazardous waste under the provisions of subpart DD of parts 264 or 265 of this chapter. (40 CFR 260.10)

Contingency plan means a document setting out an organized, planned, and coordinated course of action to be followed in case of a fire, explosion, or release of hazardous waste or hazardous waste constituents which could threaten human health or the environment. (40 CFR 260.10)

Corrective action management unit or CAMU means an area within a facility that is designated by the Regional Administrator under part 264 subpart S, for the purpose of implementing corrective action requirements under §264.101 and RCRA section 3008(h). A CAMU shall only be used for the management of remediation wastes pursuant to implementing such corrective action requirements at the facility. (40 CFR 260.10)

Corrosion expert means a person who, by reason of his knowledge of the physical sciences and the principles of engineering and mathematics, acquired by a professional education and related practical experience, is qualified to engage in the practice of corrosion control on buried or submerged metal piping systems and metal tanks. Such a person must be certified as being qualified by the National Association of Corrosion Engineers (NACE) or be a registered professional engineer who has certification or licensing that includes education and experience in corrosion control on buried or submerged metal piping systems and metal tanks. (40 CFR 260.10)

Debris means solid material exceeding a 60 mm particle size that is intended for disposal and that is: A manufactured object; or plant or animal matter; or natural geologic material. However, the following materials are not debris: Any material for which a specific treatment standard is provided in Subpart D, Part 268, namely lead acid batteries, cadmium batteries, and radioactive lead solids; Process residuals such as smelter slag and residues from the treatment of waste, wastewater, sludges, or air emission residues; and Intact containers of hazardous waste that are not ruptured and that retain at least 75% of their original volume. A mixture of debris that has not been treated to the standards provided by §268.45 and other material is subject to regulation as debris if the mixture is comprised primarily of debris, by volume, based on visual inspection. (40 CFR 268.2)

Designated facility means a hazardous waste treatment, storage, or disposal facility which

(1) has received a permit (or interim status) in accordance with the requirements of parts 270 and 124 of this chapter,

(2) has received a permit (or interim status) from a State authorized in accordance with part 271 of this chapter, or

(3) is regulated under §261.6(c)(2) or subpart P of part 266 of this chapter, and (4) that has been designated on the manifest by the generator pursuant to §260.20. If a waste is destined to a facility in an authorized State which has not yet obtained authorization to regulate that particular waste as hazardous, then the designated facility must be a facility allowed by the receiving State to accept such waste. (40 CFR 260.10)

Destination facility means a facility that treats, disposes of, or recycles a particular category of universal waste, except those management activities described in paragraphs (a) and (c) of §§273.13 and 273.33 of this chapter. A facility at which a particular category of universal waste is only accumulated; is not a destination facility for purposes of managing that category of universal waste. (40 CFR 260.10)

Dike means an embankment or ridge of either natural or man-made materials used to prevent the movement of liquids, sludges, solids, or other materials. (40 CFR 260.10)

Director means the Regional Administrator or the State Director, as the context requires, or an authorized representative. When there is no approved State program, and there is an EPA administered program, Director means the Regional Administrator. When there is an approved State program. Director normally means the State Director. In some circumstances, however, EPA retains the authority to take certain actions even when there is an approved State program. In such cases, the term Director means the Regional Administrator and not the State Director. (40 CFR 270.2)

Discharge or hazardous waste discharge means the accidental or intentional spilling, leaking, pumping, pouring, emitting, emptying, or dumping of hazardous waste into or on any land or water. (40 CFR 260.10)

Disposal means the discharge, deposit, injection, dumping, spilling, leaking, or placing of any solid waste or hazardous waste into or on any land or water so that such solid waste or hazardous waste or any constituent thereof may enter the environment or be emitted into the air or discharged into any waters, including ground waters. (40 CFR 260.10)

Disposal facility means a facility or part of a facility at which hazardous waste is intentionally placed into or on any land or water, and at which waste will remain after closure. The term disposal facility does not include a corrective action management unit into which remediation wastes are placed. (40 CFR 260.10)

Do-it-yourselfer used oil collection center means any site or facility that accepts/aggregates and stores used oil collected only from household do-it-your-selfers. (40 CFR 279.1)

Draft permit means a document prepared under §124.6 indicating the Director's tentative decision to issue or deny, modify, revoke and reissue, terminate, or reissue a permit. A notice of intent to terminate a permit, and a notice of intent to deny a permit, as discussed in §124.5, are types of draft permits. A denial of a request for modification, revocation and reissuance, or termination, as discussed in §124.5 is not a "draft permit." A proposed permit is not a draft permit. (40 CFR 270.2)

Drip pad is an engineered structure consisting of a curbed, free-draining base, constructed of non-earthen materials and designed to convey preservative kick-back or drippage from treated wood, precipitation, and surface water run-on to an associated collection system at wood preserving plants. (40 CFR 260.10)

Elementary neutralization unit means a device which:

(1) Is used for neutralizing wastes that are hazardous only because they exhibit the corrosivity characteristic defined in §261.22 of this chapter, or they are listed in subpart D of part 261 of the chapter only for this reason; and
(2) Meets the definition of tank, tank system, container, transport vehicle, or vessel in §260.10 of this chapter. (40 CFR 260.10)

Emergency permit means a RCRA permit issued in accordance with §270.61. (40 CFR 270.2)

Environmental Protection Agency (EPA) means the United States Environmental Protection Agency. (40 CFR 270.2)

EPA hazardous waste number means the number assigned by EPA to each hazardous waste listed in part 261, subpart D, of this chapter and to each characteristic identified in part 261, subpart C, of this chapter. (40 CFR 260.10)

EPA identification number means the number assigned by EPA to each generator, transporter, and treatment, storage, or disposal facility. (40 CFR 260.10)

EPA region means the states and territories found in any one of the following ten regions:

Region I-Maine, Vermont, New Hampshire, Massachusetts, Connecticut, and Rhode Island.

Region II-New York, New Jersey, Commonwealth of Puerto Rico, and the U.S. Virgin Islands.

Region III-Pennsylvania, Delaware, Maryland, West Virginia, Virginia, and the District of Columbia.

Region IV-Kentucky, Tennessee, North Carolina, Mississippi, Alabama, Georgia, South Carolina, and Florida.

Region V-Minnesota, Wisconsin, Illinois, Michigan, Indiana, and Ohio.

Region VI-New Mexico, Oklahoma, Arkansas, Louisiana, and Texas.

Region VII-Nebraska, Kansas, Missouri, and Iowa.

Region VIII-Montana, Wyoming, North Dakota, South Dakota, Utah, and Colorado.

Region IX-California, Nevada, Arizona, Hawaii, Guam, American Samoa, and Commonwealth of the Northern Mariana Islands.

Region X-Washington, Oregon, Idaho, and Alaska.

Equivalent method means any testing or analytical method approved by the Administrator under §§260.20 and 260.21. (40 CFR 260.10)

Existing hazardous waste management (HWM) facility or *existing* facility means a facility which was in operation or for which construction commenced on or before November 19, 1980. A facility has commenced construction if:

(1) The owner or operator has obtained the Federal, State and local approvals or permits necessary to begin physical construction; and either
(2)(i) A continuous onsite, physical construction program has begun; or

(ii) The owner or operator has entered into contractual obligations which cannot be canceled or modified without substantial loss or physical construction of the facility to be completed within a reasonable time. (40 CFR 260.10)

Existing portion means that land surface area of an existing waste management unit, included in the original Part A permit application, on which wastes have been placed prior to the issuance of a permit. (40 CFR 260.10)

Existing tank system or existing component means a tank system or component that is used for the storage or treatment of hazardous waste and that is in operation, or for which installation has commenced on or prior to July 14, 1986. Installation will be considered to have commenced if the owner or operator has obtained all Federal, State, and local approvals or permits necessary to begin physical construction of the site or installation of the tank system and if either (1) a continuous on-site physical construction or installation program has begun, or (2) the owner or operator has entered into contractual obligations-which cannot be canceled or modified without substantial loss or physical construction of the site or installation of the tank system to be completed within a reasonable time. (40 CFR 260.10)

Facility means:

(1) All contiguous land, and structures, other appurtenances, and improvements on the land, used for treating, storing, or disposing of hazardous waste. A facility may consist of several treatment, storage, or disposal operational units (e.g., one or more landfills, surface impoundments, or combinations of them).

(2) For the purpose of implementing corrective action under §264.101, all contiguous property under the control of the owner or operator seeking a permit under subtitle C of RCRA. This definition also applies to facilities implementing corrective action under RCRA Section 3008(h). (40 CFR 260.10)

Facility or activity means any HWM facility or any other facility or activity (including land or appurtenances thereto) that is subject to regulation under the RCRA program. (40 CFR 270.2)

Facility mailing list means the mailing list for a facility maintained by EPA in accordance with 40 CFR 124.10(c)(viii). (40 CFR 270.2)

Federal agency means any department, agency, or other instrumentality of the Federal Government, any independent agency or establishment of the Federal Government including any Government corporation, and the Government Printing Office. (40 CFR 260.10)

Federal, State and local approvals or permits necessary to begin physical construction means permits and approvals required under Federal, State or local hazardous waste control statutes, regulations or ordinances. (40 CFR 270.2)

Final authorization means approval by EPA of a State program which has met the requirements of section 3006(b) of RCRA and the applicable requirements of part 271, subpart A. (40 CFR 270.2)

Final closure means the closure of all hazardous waste management units at the facility in accordance with all applicable closure requirements so that hazardous waste management activities under parts 264 and 265 of this chapter are no longer conducted at the facility unless subject to the provisions in §262.34. (40 CFR 260.10)

Food-chain crops means tobacco, crops grown for human consumption, and crops grown for feed for animals whose products are consumed by humans. (40 CFR 260.10)

Free liquids means liquids which readily separate from the solid portion of a waste under ambient temperature and pressure. (40 CFR 260.10)

Freeboard means the vertical distance between the top of a tank or surface impoundment dike, and the surface of the waste contained therein. (40 CFR 260.10)

Functionally equivalent component means a component which performs the same function or measurement and which meets or exceeds the performance specifications of another component. (40 CFR 270.2)

Generator means any person, by site, whose act or process produces hazardous waste identified or listed in part 261 of this chapter or whose act first causes a hazardous waste to become subject to regulation. (40 CFR 260.10)

Ground water means water below the land surface in a zone of saturation. (40 CFR 260.10)

Halogenated organic compounds or HOCs means those compounds having a carbon-halogen bond which are listed under appendix III to this part. (40 CFR 268.2)

Hazardous constituent or constituents means those constituents listed in appendix VIII to part 261 of this chapter. (40 CFR 268.2)

Hazardous debris means debris that contains a hazardous waste listed in subpart D of part 261 of this chapter, or that exhibits a characteristic of hazardous waste identified in subpart C of part 261 of this chapter. (40 CFR 268.2)

Hazardous waste means a hazardous waste as defined in §261.3 of this chapter. (40 CFR 260.10)

Hazardous waste constituent means a constituent that caused the Administrator to list the hazardous waste in part 261, subpart D, of this chapter, or a constituent listed in table 1 of §261.24 of this chapter. (40 CFR 260.10)

Hazardous waste management facility (HWM facility) means all contiguous land, and structures, other appurtenances, and improvements on the land, used for treating, storing, or disposing of hazardous waste. A facility may consist of several treatment, storage, or disposal operational units (for example, one or more landfills, surface impoundments, or combinations of them). (40 CFR 270.2)

Hazardous waste management unit is a contiguous area of land on or in which hazardous waste is placed, or the largest area in which there is significant likelihood of mixing hazardous waste constituents in the same area. Examples of hazardous waste management units include a surface impoundment, a waste pile, a land treatment area, a landfill cell, an incinerator, a tank and its associated piping and underlying containment system and a container storage area. A container alone does not constitute a unit; the unit includes containers and the land or pad upon which they are placed. (40 CFR 260.10)

Household "do-it-yourselfer" used oil means oil that is derived from households, such as used oil generated by individuals who generate used oil through the maintenance of their personal vehicles. (40 CFR 279.1)

Household "do-it-yourselfer" used oil generator means an individual who generates household "do-it-yourselfer" used oil. (40 CFR 279.1)

In operation refers to a facility which is treating, storing, or disposing of hazardous waste. (40 CFR 260.10)

Inactive portion means that portion of a facility which is not operated after the effective date of part 261 of this chapter. (See also "active portion" and "closed portion.") (40 CFR 260.10)

Incinerator means any enclosed device that:

(1) Uses controlled flame combustion and neither meets the criteria for classification as a boiler, sludge dryer, or carbon regeneration unit, nor is listed as an industrial furnace; or
(2) Meets the definition of incinerator or plasma arc incinerator. (40 CFR 260.10)

Incompatible waste means a hazardous waste which is unsuitable for:

(1) Placement in a particular device or facility because it may cause corrosion or decay of containment materials (e.g., container inner liners or tank walls); or
(2) Commingling with another waste or material under uncontrolled conditions because the commingling might produce heat or pressure, fire or explosion, violent reaction, toxic dusts, mists, fumes, or gases, or flammable fumes or gases. (See part 265, appendix V, of this chapter for examples.) (40 CFR 260.10)

Individual generation site means the contiguous site at or on which one or more hazardous wastes as generated. An individual generation site, such as a large manufacturing plant, may have one or more sources of hazardous waste but is considered a single or individual generation site if the site or property is contiguous. (40 CFR 260.10)

Industrial furnace means any of the following enclosed devices that are integral components of manufacturing processes and that use thermal treatment to accomplish recovery of materials or energy:

(1) Cement kilns
(2) Lime kilns
(3) Aggregate kilns
(4) Phosphate kilns
(5) Coke ovens
(6) Blast furnaces
(7) Smelting, melting and refining furnaces (including pyrometallurgical devices such as cupolas, reverberator furnaces, sintering machine, roasters, and foundry furnaces)
(8) Titanium dioxide chloride process oxidation reactors
(9) Methane reforming furnaces
(10) Pulping liquor recovery furnaces
(11) Combustion devices used in the recovery of sulfur values- from spent sulfuric acid
(12) Halogen acid furnaces (HAFs) for the production of acid from halogenated hazardous waste generated by chemical production facilities where the furnace is located on the site of a chemical production facility, the acid product has a halogen acid content of at least 3%, the acid product is used in a manufacturing process, and, except for hazardous waste burned as fuel, hazardous waste fed to the furnace has a minimum halogen content of 20% as generated.
(13) Such other devices as the Administrator may, after notice and comment, add to this list on the basis of one or more of the following factors:
    (i) The design and use of the device primarily to accomplish recovery of material products;
    (ii) The use of the device to burn or reduce raw materials to make a material product;
    (iii) The use of the device to burn or reduce secondary materials as effective substitutes for raw materials, in processing using raw materials as principal feedstocks;
    (iv) The use of the device to burn or reduce secondary materials as ingredients in an industrial process to make a material product;
    (v) The use of the device in common industrial practice to produce a material product; and (vi) Other factors, as appropriate. (40 CFR 260.10)

Infrared incinerator means any enclosed device that uses electric powered resistance heaters as a source of radiant heat followed by an afterburner using controlled flame combustion and which is not listed as an industrial furnace. (40 CFR 260.10)

Inground tank means a device meeting the definition of "tank" in §260.10 whereby a portion of the tank wall is situated to any degree within the ground, thereby preventing visual inspection of that external surface area of the tank that is in the ground. (40 CFR 260.10)

Injection well means a well into which fluids are injected. (See also "underground injection.") (40 CFR 260.10)

Inner liner means a continuous layer of material placed inside a tank or container which protects the construction materials of the tank or container from the contained waste or reagents used to treat the waste. (40 CFR 260.10)

Installation inspector means a person who, by reason of his knowledge of the physical sciences and the principles of engineering, acquired by a professional education and related practical experience, is qualified to supervise the installation of tank systems. (40 CFR 260.10)

Interim authorization means approval by EPA of a State hazardous waste program which has met the requirements of section 3006(g)(2) of RCRA and applicable requirements of part 271, subpart B. (40 CFR 270.2)

International shipment means the transportation of hazardous waste into or out of the jurisdiction of the United States. (40 CFR 260.10)

Land disposal means placement in or on the land, except in a corrective action management unit, and includes, but is not limited to, placement in a landfill, surface impoundment, waste pile, injection well, land treatment facility, salt dome formation, salt bed formation, underground mine or cave, or placement in a concrete vault, or bunker intended for disposal purposes. (40 CFR 268.2)

Landfill means a disposal facility or part of a facility where hazardous waste is placed in or on land and which is not a pile, a land treatment facility, a surface impoundment, an underground injection well, a salt dome formation, a salt bed formation, an underground mine, a cave, or a corrective action management unit. (40 CFR 260.10)

Landfill cell means a discrete volume of a hazardous waste landfill which uses a liner to provide isolation of wastes from adjacent cells or wastes. Examples of landfill cells are trenches and pits. (40 CFR 260.10)

Land treatment facility means a facility or part of a facility at which hazardous waste is applied onto or incorporated into the soil surface; such facilities are disposal facilities if the waste will remain after closure. (40 CFR 260.10)

Large quantity handler of universal waste means a universal waste handler (as defined in this section) who accumulates 5,000 kilograms or more total of universal waste (batteries, pesticides, or thermostats, calculated collectively) at any time. This designation as a large quantity handler of universal waste is retained through the end of the calendar year in which 5,000 kilograms or more total of universal waste is accumulated. (40 CFR 273.6)

Leachate means any liquid, including any suspended components in the liquid that has percolated through or drained from hazardous waste. (40 CFR 260.10)

Leak-detection system means a system capable of detecting the failure of either the primary or secondary containment structure or the presence of a release of hazardous waste or accumulated liquid in the secondary containment structure. Such a system must employ operational controls (e.g., daily visual inspections for releases into the secondary containment system of aboveground tanks) or consist of an interstitial monitoring device designed to detect continuously and automatically the failure of the primary or secondary containment structure or the presence of a release of hazardous waste into the secondary containment structure. (40 CFR 260.10)

Liner means a continuous layer of natural or man-made materials, beneath or on the sides of a surface impoundment, landfill, or landfill cell, which restricts the downward or lateral escape of hazardous waste, hazardous waste constituents, or leachate. (40 CFR 260.10)

Major facility means any facility or activity classified as such by the Regional Administrator, or, in the case of approved State programs, the Regional Administrator in conjunction with the State Director. (40 CFR 270.2)

Management or hazardous waste management means the systematic control of the collection, source separation, storage, transportation, processing, treatment, recovery, and disposal of hazardous waste. (40 CFR 260.10)

Manifest means the shipping document EPA form 8700-22 and, if necessary, EPA form 8700-22A, originated and signed by the generator in accordance with the instructions included in the appendix to part 262. (40 CFR 260.10)

Manifest document number means the U.S. EPA twelve digit identification number assigned to the generator plus a unique five digit document number assigned to the Manifest by the generator for recording and reporting purposes. (40 CFR 260.10)

Mining overburden returned to the mine site means any material overlying an economic mineral deposit which is removed to gain access to that deposit and is then used for reclamation of a surface mine. (40 CFR 260.10)

Miscellaneous unit means a hazardous waste management unit where hazardous waste is treated, stored, or disposed of and that is not a container, tank, surface impoundment, pile, land treatment unit, landfill, incinerator, boiler, industrial furnace, underground injection well with appropriate technical standards under 40 CFR part 146, containment building, corrective action management unit, or unit eligible for research, development, and demonstration permit under §270.65. (40 CFR 260.10)

Movement means that hazardous waste transported to a facility in an individual vehicle. (40 CFR 260.10)

National Pollutant Discharge Elimination System means the national program for issuing, modifying, revoking and reissuing, terminating, monitoring and enforcing permits, and imposing and enforcing pretreatment requirements, under sections 307, 402, 318, and 405 of the CWA. The term includes an approved program. (40 CFR 270.2)

New HWM facility means a Hazardous Waste Management facility which began operation or for which construction commenced after November 19, 1980. (40 CFR 270.2)

New hazardous waste management facility or new facility means a facility which began operation, or for which construction commenced after October 21, 1976. (See also "existing hazardous waste management facility.") (40 CFR 260.10)

New tank system or new tank component means a tank system or component that will be used for the storage or treatment of hazardous waste and for which installation has commenced after July 14, 1986; except, however, for purposes of §264.193(g)(2) and §265.193(g)(2), a new tank system is one for which construction commenced after July 14, 1986. (See also "existing tank system.") (40 CFR 260.10)

Off-site means any site which is not on-site. (40 CFR 270.2)

On ground tank means a device meeting the definition of "tank" in §260.10 and that is situated in such a way that the bottom of the tank is on the same level as the adjacent surrounding surface so that the external tank bottom cannot be visually inspected. (40 CFR 260.10)

On-site means the same or geographically contiguous property which may be divided by public or private right-of-way, provided the entrance and exit between the properties is at a cross-roads intersection, and access is by crossing as opposed to going along, the right-of-way. Non-contiguous properties owned by the same person but connected by a right-of-way which he controls and to which the public does not have access, is also considered on-site property. (40 CFR 260.10)

Open burning means the combustion of any material without the following characteristics:

(1) Control of combustion air to maintain adequate temperature for efficient combustion,

(2) Containment of the combustion reaction in an enclosed device to provide sufficient residence time and mixing for complete combustion, and

(3) Control of emission of the gaseous combustion products. (See also "incineration" and "thermal treatment.") (40 CFR 260.10)

Operator means the person responsible for the overall operation of a facility. (40 CFR 260.10)

Owner means the person who owns a facility or part of a facility. (40 CFR 260.10)

Owner or operator means the owner or operator of any facility or activity subject to regulation under RCRA. (40 CFR 270.2)

Partial closure means the closure of a hazardous waste management unit in accordance with the applicable closure requirements of parts 264 and 265 of this chapter at a facility that contains other active hazardous waste management units. For example, partial closure may include the closure of a tank (including its associated piping and underlying containment systems), landfill cell, surface impoundment, waste pile, or other hazardous waste management unit, while other units of the same facility continue to operate. (40 CFR 260.10)

Permit means an authorization, license, or equivalent control document issued by EPA or an approved State to implement the requirements of this part and parts 271 and 124. Permit includes permit by rule (§270.60), and emergency permit (§270.61). Permit does not include RCRA interim status (subpart G of this part), or any permit which has not yet been the subject of final agency action, such as a draft permit or a proposed permit. (40 CFR 270.2)

Permit-by-rule means a provision of these regulations stating that a facility or activity is deemed to have a RCRA permit if it meets the requirements of the provision. (40 CFR 270.2)

Person means an individual, trust, firm, joint stock company, Federal Agency, corporation (including a government corporation), partnership, association, State, municipality, commission, political subdivision of a State, or any interstate body. (40 CFR 260.10)

Personnel or facility personnel means all persons who work, at, or oversee the operations of, a hazardous waste facility, and whose actions or failure to act may result in noncompliance with the requirements of part 264 or 265 of this chapter. (40 CFR 260.10)

Pesticide means any substance or mixture of substances intended for preventing, destroying, repelling, or mitigating any pest, or intended for use as a plant regulator, defoliant, or desiccant, other than any article that:

(1) Is a new animal drug under FFDCA section 201(w), or

(2) Is an animal drug that has been determined by regulation of the Secretary of Health and Human Services not to be a new animal drug, or

(3) Is an animal feed under FFDCA section 201 (x) that bears or contains any substances described by paragraph (1) or (2) of this definition. (40 CFR 260.10)

Petroleum refining facility means an establishment primarily engaged in producing gasoline, kerosene, distillate fuel oils, residual fuel oils, and lubricants, through fractionation, straight distillation of crude oil, redistillation of unfinished petroleum derivatives, cracking or other processes (i.e., facilities classified as SIC 2911). (40 CFR 279.1)

Physical construction means excavation, movement of earth, erection forms or structures, or similar activity to prepare an HWM facility to accept hazardous waste. (40 CFR 270.2)

Pile means any non-containerized accumulation of solid, nonflowing hazardous waste that is used for treatment or storage and that is not a containment building. (40 CFR 260.10)

Plasma arc incinerator means any enclosed device using a high intensity electrical discharge or arc as a source of heat followed by an afterburner using controlled flame combustion and which is not listed as an industrial furnace. (40 CFR 260.10)

Point source means any discernible, confined, and discrete conveyance, including, but not limited to any pipe, ditch, channel, tunnel, conduit, well, discrete fissure, container, rolling stock, concentrated animal feeding operation, or vessel or other floating craft, from which pollutants are or may be discharged. This term does not include return flows from irrigated agriculture. (40 CFR 260.10)

Polychlorinated biphenyls or PCBs are halogenated organic compounds defined in accordance with 40 CFR 761.3. (40 CFR 268.2)

Processing means chemical or physical operations designed to produce from used oil, or to make used oil more amenable for production of, fuel oils, lubricants, or other used oil-derived product. Processing includes, but is not limited to: blending used oil with virgin petroleum products, blending used oils to meet the fuel specification, filtration, simple distillation, chemical or physical separation and re-refining. (40 CFR 279.1)

Publicly owned treatment works or POTW means any device or system used in the treatment (including recycling and reclamation) of municipal sewage or industrial wastes of a liquid nature which is owned by a "State" or "municipality" (as defined by section 502(4) of the CWA). This definition includes sewers, pipes, or other conveyances only if they convey wastewater to a POTW providing treatment. (40 CFR 260.10)

Qualified ground-water scientist means a scientist or engineer who has received a baccalaureate or post-graduate degree in the natural sciences or engineering, and has sufficient training and experience in ground-water hydrology and related fields as may be demonstrated by state registration, professional certifications, or completion of accredited university courses that enable that individual to make sound professional judgements regarding ground-water monitoring and contaminant fate and transport. (40 CFR 260.10)

Re-refining distillation bottoms means the heavy fraction produced by vacuum distillation of filtered and dehydrated used oil. The composition of still bottoms varies with column operation and feedstock. (40 CFR 279.1)

Regional Administrator means the Regional Administrator for the EPA Region in which the facility is located, or his designee. (40 CFR 260.10)

Remediation waste means all solid and hazardous wastes, and all media (including groundwater, surface water, soils, and sediments) and debris, which contain listed hazardous wastes or which themselves exhibit a hazardous waste characteristic, that are managed for the purpose of implementing corrective action requirements under §264.101 and RCRA section 3008(h). For a given facility, remediation wastes may originate only from within the facility boundary, but may include waste managed in implementing RCRA sections 3004(v) or 3008(h) for releases beyond the facility boundary. (40 CFR 260.10)

Replacement unit means a landfill, surface impoundment, or waste pile unit (1) from which all or substantially all of the waste is removed, and (2) that is subsequently reused to treat, store, or dispose of hazardous waste. "Replacement unit" does not apply to a unit from which waste is removed during closure, if the subsequent reuse solely involves the disposal of waste from that unit and other closing units or corrective action areas at the facility, in accordance with an approved closure plan or EPA or State approved corrective action. (40 CFR 260.10)

Representative sample means a sample of a universe or whole (e.g., waste pile, lagoon, ground water) which can be expected to exhibit the average properties of the universe or whole. (40 CFR 260.10)

Run-off means any rainwater, leachate, or other liquid that drains over land from any part of a facility. (40 CFR 260.10)

Run-on means any rainwater, leachate, or other liquid that drains over land onto any part of a facility. (40 CFR 260.10)

Saturated zone or zone of saturation means that part of the earth's crust in which all voids are filled with water. (40 CFR 260.10)

Schedule of compliance means a schedule of remedial measures included in a permit, including an enforceable sequence of interim requirements (for example, actions, operations, or milestone events) leading to compliance with the Act and regulations. (40 CFR 270.2)

Site means the land or water area where any facility or activity is physically located or conducted, including adjacent land used in connection with the facility or activity. (40 CFR 270.2)

Sludge means any solid, semi-solid, or liquid waste generated from a municipal, commercial, or industrial wastewater treatment plant, water supply treatment plant, or air pollution control facility exclusive of the treated effluent from a wastewater treatment plant. (40 CFR 260.10)

Sludge dryer means any enclosed thermal treatment device that is used to dehydrate sludge and that has a maximum total thermal input, excluding the heating value of the sludge itself, of 2,500 Btu/lb of sludge treated on a wet-weight basis. (40 CFR 260.10)

Small quantity generator means a generator who generates less than 1000 kg of hazardous waste in a calendar month. (40 CFR 260.10)

Small quantity handler of universal waste means a universal waste handler (as defined in this section) who does not accumulate more than 5,000 kilograms total of universal waste (batteries, pesticides, or thermostats, calculated collectively) at any time. (40 CFR 273.6)

Solid waste means a solid waste as defined in §261.2 of this chapter. (40 CFR 260.10)

Sorbent means a material that is used to soak up free liquids by either adsorption or absorption, or both. Sorb means to either adsorb or absorb, or both. (40 CFR 260.10)

State means any of the several States, the District of Columbia, the Commonwealth of Puerto Rico, the Virgin Islands, Guam, American Samoa, and the Commonwealth of the Northern Mariana Islands. (40 CFR 260.10)

State Director means the chief administrative officer of any State agency operating an approved program, or the delegated representative of the State Director. If responsibility is divided among two or more State agencies, State Director means the chief administrative officer of the State agency authorized to perform the particular procedure or function to which reference is made. (40 CFR 270.2)

State/EPA Agreement means an agreement between the Regional Administrator and the State which coordinates EPA and State activities, responsibilities and programs. (40 CFR 270.2)

Storage means the holding of hazardous waste for a temporary period, at the end of which the hazardous waste is treated, disposed of, or stored elsewhere. (40 CFR 260.10)

Sump means any pit or reservoir that meets the definition of tank and those troughs/trenches connected to it that serve to collect hazardous waste for transport to hazardous waste storage, treatment, or disposal facilities; except that as used in the landfill, surface impoundment, and waste pile rules, "sump" means any lined pit or reservoir that serves to collect liquids drained from a leachate collection and removal system or leak detection system for subsequent removal from the system. (40 CFR 260.10)

Surface impoundment or impoundment means a facility or part of a facility which is a natural topographic depression, man-made excavation, or diked area formed primarily of earthen materials (although it may be lined with man-made materials), which is designed to hold an accumulation of liquid wastes or wastes containing free liquids, and which is not an injection well. Examples of surface impoundments are holding, storage, settling, and aeration pits, ponds, and lagoons. (40 CFR 260.10)

Tank means a stationary device, designed to contain an accumulation of hazardous waste which is constructed primarily of non-earthen materials (e.g., wood, concrete, steel, plastic) which provide structural support. (40 CFR 260.10)

Tank system means a hazardous waste storage or treatment tank and its associated ancillary equipment and containment system. (40 CFR 260.10)

Thermal treatment means the treatment of hazardous waste in a device which uses elevated temperatures as the primary means to change the chemical, physical, or biological character or composition of the hazardous waste. Examples of thermal treatment processes are incineration, molten salt, pyrolysis, calcination, wet air oxidation, and microwave discharge. (See also "incinerator" and "open burning.") (40 CFR 260.10)

Thermostat means a temperature control device that contains metallic mercury in an ampule attached to a bimetal sensing element, and mercury containing ampules that have been removed from these temperature control devices in compliance with the requirements of 40 CFR 273.13(c)(2) or 273.a3(c)(2). (40 CFR 260.10)

Totally enclosed treatment facility means a facility for the treatment of hazardous waste which is directly connected to an industrial production process and which is constructed and operated in a manner which prevents the release of any hazardous waste or any constituent thereof into the environment during treatment. An example is a pipe in which waste acid is neutralized. (40 CFR 260.10)

Transfer facility means any transportation related facility including loading docks, parking areas, storage areas and other similar areas where shipments of hazardous waste are held during the normal course of transportation. (40 CFR 260.10)

Transport vehicle means a motor vehicle or rail car used for the transportation of cargo by any mode. Each cargo-carrying body (trailer, railroad freight car, etc.) is a separate transport vehicle. (40 CFR 260.10)

Transportation means the movement of hazardous waste by air, rail, highway, or water. (40 CFR 260.10)

Transporter means a person engaged in the offsite transportation of hazardous waste by air, rail, highway, or water. (40 CFR 260.10)

Treatability study means a study in which a hazardous waste is subjected to a treatment process to determine:

(1) Whether the waste is amenable to the treatment process,
(2) what pretreatment (if any) is required,
(3) the optimal process conditions needed to achieve the desired treatment,
(4) the efficiency of a treatment process for a specific waste or wastes, or
(5) the characteristics and volumes of residuals from a particular treatment process. Also included in this definition for the purpose of the §261.4 (e) and (f) exemptions are liner compatibility, corrosion, and other material compatibility studies and toxicological and health effects studies. A "treatability study" is not a means to commercially treat or dispose of hazardous waste. (40 CFR 260.10)

Treatment means any method, technique, or process, including neutralization, designed to change the physical, chemical, or biological character or composition of any hazardous waste so as to neutralize such waste, or so as to recover energy or material resources from the waste, or so as to render such waste non-hazardous, or less hazardous; safer to transport, store, or dispose of; or amenable for recovery, amenable for storage, or reduced in volume. (40 CFR 260.10)

Treatment zone means a soil area of the unsaturated zone of a land treatment unit within which hazardous constituents are degraded, transformed, or immobilized. (40 CFR 260.10)

Underground injection means the subsurface emplacement of fluids through a bored, drilled or driven well; or through a dug well, where the depth of the dug well is greater than the largest surface dimension. (See also "injection well.") (40 CFR 260.10)

Underground source of drinking water (USDW) means an aquifer or its portion: (a)

(1) Which supplies any public water system; or

(2) Which contains a sufficient quantity of ground water to supply a public water system; and

  (i) Currently supplies drinking water for human consumption; or

  (ii) Contains fewer than 10,000 mg/l total dissolved solids; and which is not an exempted aquifer. (40 CFR 270.2)

Underground tank means a device meeting the definition of "tank" in §260.10 whose entire surface area is totally below the surface of and covered by the ground. (40 CFR 260.10)

Underlying-hazardous constituent means any constituent listed in §268.48, Table UTS-Universal Treatment Standards, except vanadium and zinc, which can reasonably be expected to be present at the point of generation of the hazardous waste, at a concentration above the constituent-specific UTS treatment standards. (40 CFR 268.2)

Unfit-for use tank system means a tank system that has been determined through an integrity assessment or other inspection to be no longer capable of storing or treating hazardous waste without posing a threat of release of hazardous waste to the environment. (40 CFR 260.10)

Universal waste means any of the following hazardous wastes that are subject to the universal waste requirements of 40 CFR part 273: (a) Batteries as described in 40 CFR 273.2; (b) Pesticides as described in 40 CFR 273.3; and (c) Thermostats as described in 40 CFR 273.4. (40 CFR 273.6)

Universal waste handler

(1) Means:

  (i) A generator (as defined in this section) of universal waste; or

  (ii) The owner or operator of a facility, including all contiguous property, that receives universal waste from other universal waste handlers, accumulates universal waste, and sends universal waste to another universal waste handler, to a destination facility, or to a foreign destination.

(2) Does not mean:

(i) A person who treats [except under the provisions of 40 CFR 273.13 (a) or (c), or 273.33(a) or (c)], disposes of, or recycles universal waste; or

(ii) A person engaged in the off-site transportation of universal waste by air, rail, highway, or water, including a universal waste transfer facility. (40 CFR 260.10)

Universal waste transfer facility means any transportation-related facility including loading docks, parking areas, storage areas and other similar areas where shipments of universal waste are held during the normal course of transportation for ten days or less. (40 CFR 273.6)

Universal waste transporter means a person engaged in the off-site transportation of universal waste by air, rail, highway, or water. (40 CFR 260.10)

United States means the 50 States, the District of Columbia, the Commonwealth of Puerto Rico, the U.S. Virgin Islands, Guam, American Samoa, and the Commonwealth of the Northern Mariana Islands. (40 CFR 260.10)

Unsaturated zone or zone of aeration means the zone between the land surface and the water table. (40 CFR 260.10)

Uppermost aquifer means the geologic formation nearest the natural ground surface that is an aquifer, as well as lower aquifers that are hydraulically interconnected with this aquifer within the facility's property boundary. (40 CFR 260.10)

Used oil means any oil that has been refined from crude oil, or any synthetic oil, that has been used and as a result of such use in contaminated by physical or chemical impurities. (40 CFR 260.10)

Used oil aggregation point means any site or facility that accepts, aggregates, and/or stores used oil collected only from other used oil generation sites owned or operated by the owner or operator of the aggregation point, from which used oil is transported to the aggregation point in shipments of no more than 55 gallons. Used oil aggregation points may also accept used oil from household do-it-yourselfers. (40 CFR 279.1)

Used oil burner means a facility where used oil not meeting the specification requirements in §279.11 is burned for energy recovery in devices identified in §279.61(a). (40 CFR 279.1)

Used oil collection center means any site or facility that is registered/licensed/permitted/recognized by a state/county/municipal government to manage used oil and accepts/aggregates and stores used oil collected from used oil generators regulated under subpart C of this part who bring used oil to the collection center in shipments of no more than 55 gallons under the provisions of §279.24. Used oil collection centers may also accept used oil from household do-it-yourselfers. (40 CFR 279.1)

Used oil fuel marketer means any person who conducts either of the following activities:

(1) Directs a shipment of off-specification used oil from their facility a used oil burner; or

(2) First claims that used oil to be burned for energy recovery the used oil fuel specifications set forth in §279.11 of this part. (40 CFR 279.1)

Used oil generator means any person, by site, whose act or process produces used oil or whose act first causes used oil to become subject to regulation. (40 CFR 279.1)

Used oil processor/re-refiner means a facility that processes used oil. (40 CFR 279.1)

Used oil transfer facility means any transportation related facility including loading docks, parking areas, storage areas and other areas where shipments of used oil are held for more than 24 hours and not longer than 35 days during the normal course of transportation or prior to an activity performed pursuant to 279.20(b)(2). Transfer facilities that store used oil for more than 35 days are subject to regulation under subpart P of this part. (40 CFR 279.1)

Used oil transporter means any person who transports used oil, any person who collects used oil from more than one generator and transports the collected oil, and owners and operators of used oil transfer facilities. Used oil transporters may consolidate or aggregate loads of used oil for purposes of transportation but, with the following exception, may not process used oil. Transporters may conduct incidental processing operations that occur in the normal course of used oil transportation (e.g., settling and water separation), but that are not designed to produce (or make more amenable for production of) used oil derived products or used oil fuel. (40 CFR 279.1)

Vessel includes every description of watercraft, used or capable of being used as a means of transportation on the water. (40 CFR 260.10)

Wastewaters are wastes that contain less than 1% by weight total organic carbon (TOC) and less than 1% by weight total suspended solids (TSS), with the following exceptions:

(1) F001, F002, F003, F004, F005, wastewaters are solvent-water mixtures that contain less than 1% by weight TOC or less than 1% by weight total F001, F002, F003, F004, F005 solvent constituents listed in §268.41, Table CCWE.

(2) K011, K013, K014 wastewaters contain less than 5% by weight TOC and less than 1% by weight TSS, as generated.

(3) K103 and K104 wastewaters contain less than 4% by weight TOC and less than 1%by weight TSS. (40 CFR 268.2)

Wastewater treatment unit means a device which:

(1) Is part of a wastewater treatment facility that is subject to regulation under either section 402 or 307(b) of the Clean Water Act; and

(2) Receives and treats or stores an influent wastewater that is a hazardous waste as defined in §261.3 of this chapter, or that generates and accumulates a wastewater treatment sludge that is a hazardous waste as defined in §261.3 of this chapter, or treats or stores a wastewater treatment sludge which is a hazardous waste as defined in §261.3 of this Chapter; and

(3) Meets the definition of tank or tank system in §260.10 of this chapter. (40 CFR 260.10)

Water (bulk shipment) means the bulk transportation of hazardous waste which is loaded or carried on board a vessel without containers or labels. (40 CFR 260.10)

Well means any shaft or pit dug or bored into the earth, generally of a cylindrical form and often walled with bricks or tubing to prevent the earth from caving in. (40 CFR 260.10)

Zone of engineering control means an area under the control of the owner/operator that, upon detection of a hazardous waste release, can be readily cleaned up prior to the release of hazardous waste or hazardous constituents to ground water or surface water. (40 CFR 260.10)

2

HAZARDOUS WASTE IDENTIFICATION/CLASSIFICATION

The waste identification and classification process is crucial to determining whether a waste is subject to RCRA requirements. Only those wastes that meet the RCRA definition of "hazardous" are subject to Subtitle C of RCRA, the focus of this book. Unfortunately, determining whether a waste is a RCRA hazardous waste, the most important process under Subtitle C of RCRA, is also one of the most complex.

In general, a waste is classified as hazardous if it has been specifically listed by EPA as hazardous or if it exhibits one of the EPA-defined characteristics of hazardous waste (i.e., corrosive, ignitable, reactive, toxic). But for a waste to meet either one of these criteria, it must meet the definition of a *solid* waste, which is defined as "any *discarded material* that is not specifically excluded." *Discarded* means that the material is either

- Abandoned (i.e., disposed, burned, or incinerated; or accumulated, stored, or treated prior to disposal, burning, or incineration)
- Recycled (i.e., recycled, reclaimed, burned for energy recovery, or accumulated speculatively)
- Considered inherently wastelike (i.e., applied to or placed on the land or used in products that are applied or placed on the land)

In addition to solid wastes that are hazardous because they either are specifically listed or exhibit a hazardous characteristic, there are special categories of hazardous waste (e.g., universal wastes, low-level radioactive mixed waste, hazardous waste mixtures) subject to various levels of control.

Although the general approach may initially appear to be straightforward, the definitions, conditions, and special requirements of, as well as exclusions from, each of the above categories can seem overwhelming. It is important to understand the general identification/classification flow and recognize that it is a very complex but conquerable exercise.

THE IDENTIFICATION/CLASSIFICATION PROCESS

This book uses a step-by-step identification process; however, because so many factors can affect the classification of a waste, it may not necessarily appear to be a logical, sequential process. The main reason is that EPA has made numerous modifications to the basic framework for identifying waste as solid and hazardous since promulgating that framework, as well as changes to the regulations governing the recycling of hazardous wastes. In addition, certain hazardous wastes (e.g., lead-acid batteries, used oil, universal wastes) are eligible to be managed in accordance with less stringent, special regulations. Figure 2-1 depicts a simplified, basic waste identification procedure, which is explained in the following series:

- *Is the waste a "solid" waste?* If the material is "discarded," i.e., treated, disposed, recycled, burned, or abandoned, it is a "RCRA" solid waste. *If yes, go to the next question in this list. If no, consult state authorities to determine if they have additional requirements.*
- *Is the solid waste specifically excluded from RCRA?* Is the solid waste specifically excluded from the definition of solid waste in 40 CFR 261.2 or 261.1(4)(a)? Or, is the solid waste specifically excluded from the definition of hazardous waste in 40 CFR 261.4(b)? *If yes, consult state authorities to determine if they have additional requirements. If no, go to the next question in this list.*
- *Is the solid waste a "hazardous" waste?* Is the nonexcluded solid waste specifically listed as a hazardous waste in 40 CFR Part 261, Subpart D, or does it exhibit any of the four characteristics of hazardous waste (40 CFR Part 261 Subpart C)? *If yes, the waste is subject to full regulation corresponding to the amount generated per calendar month* (see Chapter 3), *and, if it will be recycled, go to the next question in this list. If no, consult state authorities to determine if they have additional requirements.*
- *Is the hazardous waste recycled?* Certain hazardous wastes are excluded from regulation or are subject to reduced regulation depending on what the waste is and how it is recycled. *If yes, determine what type of secondary material it is and how it will be recycled. If no, the management requirements will depend on how much is generated per calendar month* (see Chapter 3).

This identification procedure and the supporting criteria are explained in detail in the following sections.

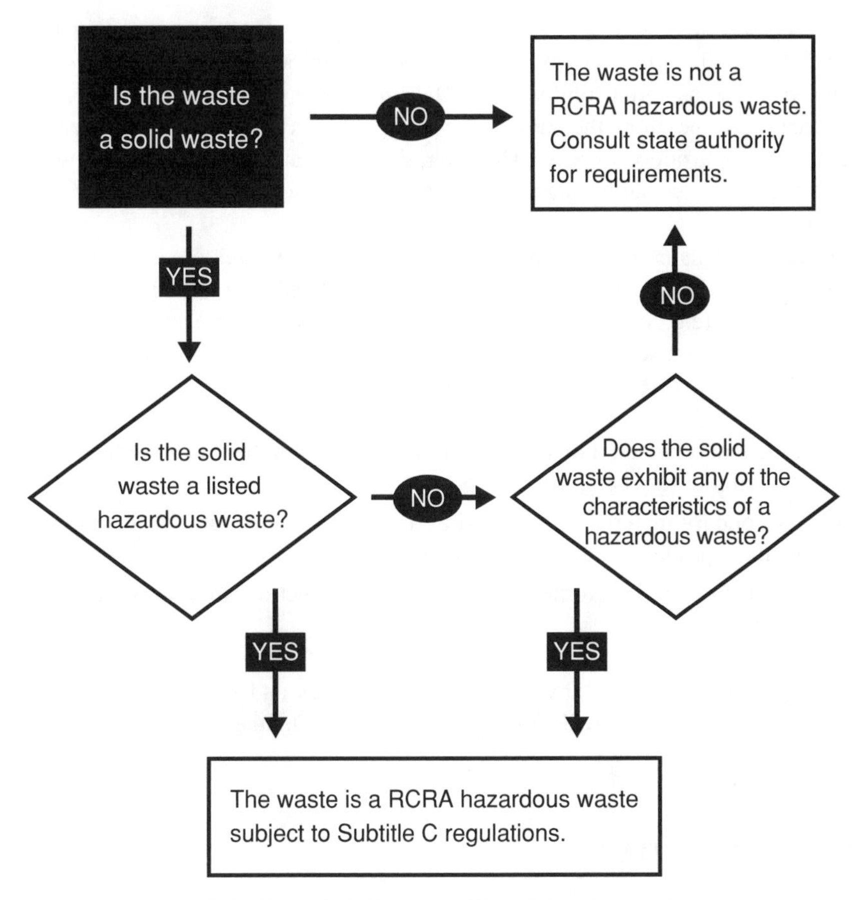

Figure 2-1. The RCRA Hazardous Waste Identification Process

SOLID WASTE CLASSIFICATION

For a material to be classified as a hazardous waste, it must first meet the definition of solid waste under 40 CFR 261.2. Hazardous waste is a subset of solid waste. Thus, if a material does not meet the definition of solid waste, it cannot be classified as a hazardous waste. However, the term *solid waste* does not refer to a material's physical state per se; it is a regulatory term only. Thus, industrial wastewater may be a solid waste.

Note: Being subject to the requirements of Subtitle C hinges on the *regulatory* definitions of solid and hazardous waste. These regulatory definitions are derived from the *statutory* definitions found in RCRA itself. For the most part, only the regulatory definitions, which are more narrow in focus, need to be followed. However, there are instances, such as enforcement, when EPA applies the much broader statutory definitions of solid and hazardous waste.

As depicted in Figure 2-2, a solid waste is any discarded material that is not excluded by 40 CFR 261.4(a). In addition, hazardous wastes that have been "delisted" are considered solid wastes. (*Delisting* is the process by which a generator successfully petitions the EPA to reclassify a facility-specific, listed hazardous waste as a nonhazardous waste.) However, delisted materials are still solid wastes unless they are recyclable materials granted a variance under 40 CFR 260.30 or 260.31.

A *discarded* material is any material that is disposed, stored, or treated before its disposal; that is burned as a fuel, treated, recycled, abandoned, or considered inherently wastelike (e.g., certain dioxin wastes); or that is stored or accumulated before recycling. There are some exceptions to the definition of solid waste regarding specific recycling activities [40 CFR 261.2(e)], including the following:

- A secondary material that is used directly as an ingredient in a production process without first being reclaimed.
- A secondary material that is used as an effective substitute for a commercial chemical product without first being reclaimed.
- A material that is returned to the original production process from which it was generated.

Any person engaged in any of the above activities and who claims his or her waste is excluded must provide documentation supporting the claim if requested [40 CFR 261.2(f)].

EXCLUDED WASTES

A variety of wastes have been excluded from the definitions of solid and hazardous waste by Congress and EPA, and by the fact that they are already regulated under another law. This section discusses the solid and hazardous waste exclusions separately. See Table 2-1 for a summary listing of excluded wastes.

Note: These exclusions from the definition of solid and hazardous wastes does not necessarily mean that they are universally excluded. Most of the exclusions are conditional or limited. Also, individual states may not honor the exclusions because states may be more stringent as discussed in Chapter 10. It is also important to note that wastes excluded from RCRA may still be subject to Superfund if the waste meets the definition of hazardous substance as discussed in Part II.

Wastes Excluded from the Definition of Solid Waste

The following materials are excluded from the statutory or regulatory definition of solid waste:

1. Domestic sewage [40 CFR 261.4(a)(1)(i)] and any mixture of domestic sewage and any other waste that passes through a sewer system to a pub-

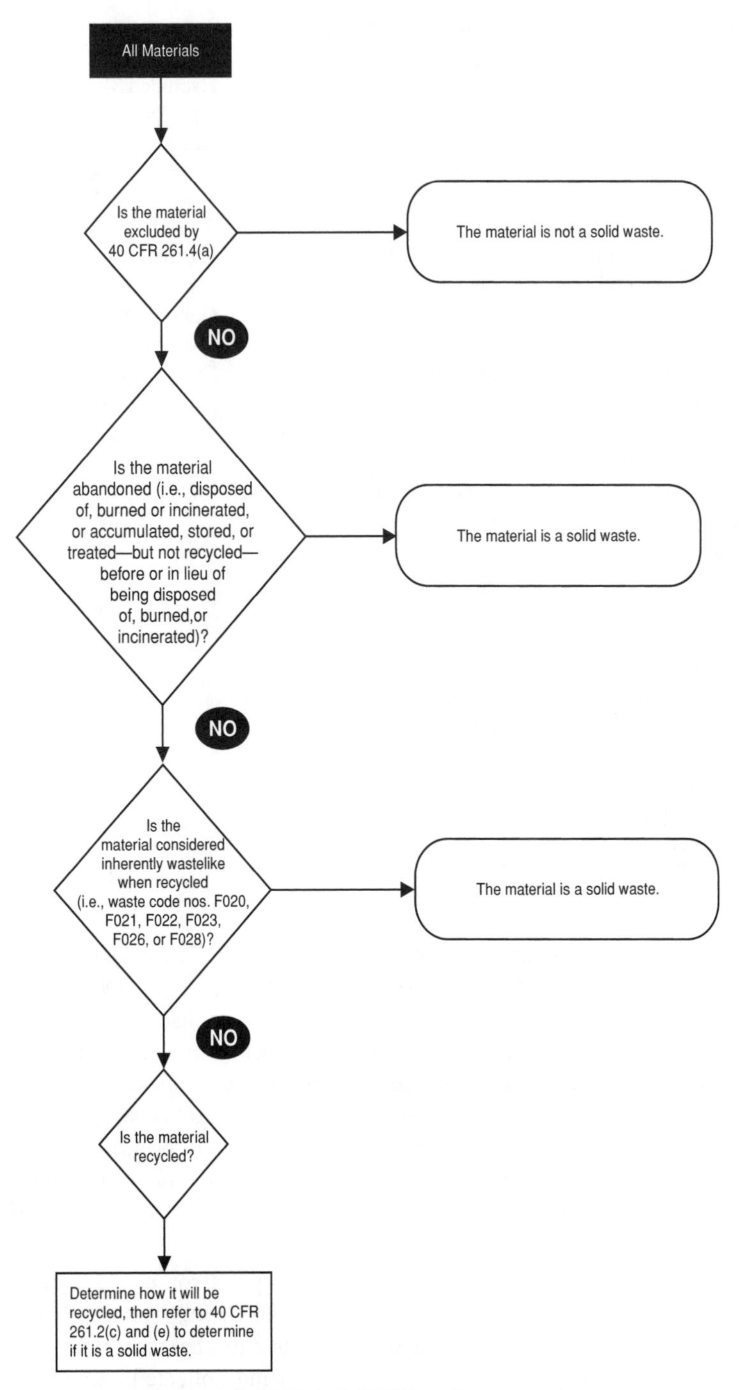

Figure 2-2. RCRA Solid Waste Determination

TABLE 2-1. Exclusions from Subtitle C of RCRA

Excluded from Solid Waste Definition	Excluded from Hazardous Waste Definition	Excluded Materials Requiring Special Management
Domestic sewage	Household wastes	Product storage wastes
Mixtures of domestic sewage and wastes going to POTW	Agricultural wastes used as fertilizers	Waste identification samples
Industrial point source dischargesunder §402 CWA	Mining overburden returned to site	Treatability samples
Irrigation return flows	Discarded wood treated with arsenic	Residues remaining in empty containers
Source, special nuclear, or by-product material under AEA	Tannery and leather scrap wastes	Conditionally exempt small-quantity-generator wastes
In situ mining waste	Specific ore and mineral beneficiating wastes	Farm wastes (pesticides)
Reclaimed pulping liquors from Kraft paper process	Fossil fuel combustion wastes	*PCB-mixed waste
Spent sulfuric acid used to produce new acid	Oil and gas exploration, development, and production wastes	*Low-level radioactive mixed waste
Secondary materials retunred to the original process under certain conditions	Cement kiln dust	
Spent, reused wood preservatives	Petroleum-contaminated media and debris from underground tank cleanup	
Certain coke byproducts	Spent chlorofluorocarbon refrigerant	
Splash condenser dross residue	Used oil filters	
Recovered refinery oil wastes	Still bottoms from the re-refining of used oil	

*These wastes are not explicitly excluded from RCRA, but require special management.

licly owned treatment works (POTW) [40 CFR 261.4(a)(1)(ii)]. A person may legally dispose of hazardous waste into a POTW system under RCRA, except that storage or treatment before discharging into a POTW is not excluded from RCRA and is subject to applicable storage regulations. However, the discharge of hazardous waste is subject to applicable pretreatment standards under the Clean Water Act (CWA). (POTWs have local jurisdiction. Thus, they can legally control or prohibit discharges that interfere with the operation of their system or that may cause them to violate their National Pollutant Discharge Elimination System [NPDES] permit.)

2. Industrial wastewater discharges that are point source discharges subject to §402 of the Clean Water Act (CWA) [40 CFR 261.4(a)(2)]. This exclusion applies only to the actual point source discharge. It does not exclude industrial wastewaters while they are being collected, stored, or treated before discharge, nor does it exclude sludges that are generated by industrial wastewater treatment. However, if the unit in which the industrial

wastewater is treated is defined as a wastewater treatment unit (40 CFR 260.10), the unit is excluded, but not the contents.

3. Irrigation return flows [40 CFR 261.4(a)(3)]. This exclusion was intended for excess water captured in basins collected from irrigation runoff.
4. Materials defined as source, special nuclear, or by-product material by the Atomic Energy Act [40 CFR 261.4(a)(4)].
5. Materials generated as a result of in situ mining techniques that are not removed from the ground during extraction [40 CFR 261.4(a)(5)]. If materials are removed, they are subject to RCRA regulation.
6. Pulping liquors used in the production of paper in the Kraft paper process [40 CFR 261.4(a)(6)].
7. Spent sulfuric acid used to produce virgin sulfuric acid [40 CFR 261.4(a)(7)].
8. Secondary materials that are reclaimed and returned to the original process or processes in which they were generated, provided that only tank storage is used, and the material is not burned, not used to produce a fuel, and is not accumulated for more than 12 months before reclamation (i.e., "closed-loop recycling") [40 CFR 261.4(a)(8)].
9. Spent wood preservatives that are reused for their original purpose [40 CFR 261.4(a)(9)].
10. Listed and characteristic coke by-product waste that exhibits the toxicity characteristic when recycled by being returned to the coke oven as a feedstock to produce coke, returned to the tar recovery process as a feedstock to produce tar, or mixed with coal tar before coal tar refining or sale as a product [40 CFR 261.4(a)(10)]. (The terms *listed* and *characteristic* refer to two specific classifications of hazardous waste, which is explained in detail later in this chapter under "Hazardous Waste Classification.")

    Listed coke by-product wastes eligible for this exclusion are EPA waste codes K087, K141, K142, K143, K144, K145, K147, and K148. (See Appendix A for a description of these wastes.) For a coke by-product waste to be eligible for this exclusion, it may not be placed on the ground from the time it is generated to the time it is recycled.
11. Splash condenser dross residue from the treatment of K061 (emission control dust from the primary production of steel in electric arc furnaces), provided it is shipped in drums if sent off-site, and not disposed of on the land [40 CFR 261.(a)(11)].
12. Recovered oil (e.g., slop oil and emulsions, oil from refinery process units, oil skimmed from ballast water tanks, and oil/water separator skimmings from plant wastewaters) that is returned to the petroleum refinery along with normal process streams, provided the oil is not managed on the land or speculatively accumulated [40 CFR 261.4(a)(12)].

Wastes Excluded from the Definition of Hazardous Waste

The following materials, which are classified as solid wastes, are excluded from the statutory or regulatory definition of hazardous wastes (but this does not necessarily mean that they are universally exempt from RCRA because the exclusions are conditional):

1. All household wastes and resource recovery facilities that burn only household waste. (Hotel, motel, septic sewage, and campground waste all are considered household waste [40 CFR 261.4(b)(1)].
2. Agricultural wastes, including manure and crops returned to the soil as fertilizers [40 CFR 261.4(b)(2)].
3. Mining overburden that is overlying a mineral deposit returned to the mine site from mining operations [40 CFR 261.4(b)(3)].
4. Fossil fuel combustion waste, which includes fly ash waste, bottom ash, boiler slag, and flue gas emission control waste generated primarily from the combustion of coal or other fossil fuels (the "utility waste" exemption) [40 CFR 261.4(b)(4)].
5. Drilling fluids, produced waters, and other wastes associated with the exploration, development, or production of crude oil, natural gas, or geothermal energy [40 CFR 261.4(b)(5)].
6. Tannery and leather scrap wastes and wastewater treatment sludges from the production of titanium oxide pigment using chromium-bearing wastes that contain primarily trivalent chromium instead of hexavalent chromium [40 CFR 261.4(b)(6)].
7. Specified solid wastes from the extraction and beneficiation of ores and minerals (Bevill Wastes) [40 CFR 261.4(b)(7)]. The definition of *beneficiation* of ores and minerals is restricted to crushing; grinding; washing; dissolution; crystallization; filtration; sorting; sizing; drying; sintering; pelletizing; briquetting; calcining to remove water and/or carbon dioxide; roasting; autoclaving; and/or chlorination in preparation for leaching (except where the roasting—and/or autoclaving and/or chlorination—leaching sequence produces a final or intermediate product that does not undergo further beneficiation or processing); gravity separation; magnetic separation; electrostatic separation; flotation; ion exchange; solvent extraction; electrowinning; precipitation; amalgamation; and heap, dump, vat, and in situ leaching.

   The specific wastes generated from the processing of ores and minerals that are excluded under this provision are the following:

   - Slag from primary copper, lead, and zinc smelting
   - Red and brown muds from bauxite refining
   - Phosphogypsum and process wastewater from phosphoric acid production

- Slag and furnace off-gas solids from elemental phosphorus production
- Roast/leach ore residue from primary chromite production
- Gasifier ash from coal gasification
- Slag tailings from primary copper smelting
- Calcium sulfate wastewater treatment plant sludge from primary copper smelting/refining
- Fluorogypsum and process wastewater from hydrofluoric acid production
- Air pollution control dust/sludge and slag from iron blast furnaces
- Process wastewater from primary lead production
- Air pollution control dust/sludge from lightweight aggregate production
- Process wastewater from primary magnesium processing by the anhydrous process
- Basic oxygen furnace and open-hearth furnace slag from carbon steel production
- Basic oxygen furnace and open-hearth furnace air pollution control dust/sludge from carbon steel production
- Sulfate processing waste acids and solids from titanium dioxide production
- Chloride processing waste solids from titanium tetrachloride production

8. Cement kiln dust [40 CFR 261.4(b)(8)].
9. Discarded wood that fails only the toxicity characteristic test (a test to determine if a waste exhibits a hazardous characteristic) for arsenic as a result of being treated with arsenical compounds [40 CFR 261.4(b)(9)].
10. Petroleum-contaminated media and debris that fail only the toxicity characteristic test and are subject to the corrective action requirements under 40 CFR Part 280, which covers underground storage tanks [40 CFR 261.4(b)(10)].
11. Spent chlorofluorocarbon (CFC) refrigerant that exhibits a characteristic of hazardous waste and is being recycled [40 CFR 261.4(b)(11)].
12. Used oil filters (nonterne-plated) provided the filters are gravity hot-drained by one of the following methods [40 CFR 261.4(b)(12)]:
    - puncturing the filter antidrain back valve or the filter dome end and hot draining,
    - hot draining and crushing,
    - dismantling followed by hot draining, or
    - any equivalent method of hot draining that will remove the oil.

    Terne-plated filters are not excluded because they often exhibit the toxicity characteristic as a result of their lead content. Furthermore, the drained used oil is subject to the used-oil management standards in 40 CFR Part 279.

13. Distillation bottoms from the re-refining of used oil that are used for feedstock for the production of asphalt paving and roofing materials [40 CFR 261.4(b)(13)].
14. Any waste generated in a product or raw material storage tank, product transport vehicle or vessel, or manufacturing process unit is not subject to Subtitle C until it exits from that unit, or the waste remains in the unit 90 days after the unit ceases operation [40 CFR 261.4(c)]. This exclusion does not apply to surface impoundments.
15. Waste identification samples are conditionally excluded under 40 CFR 261.4(d). For a sample to be excluded, it must be in the process of being analyzed for the sole purpose of hazardous waste identification, in which case the sample is excluded from RCRA during its storage and transportation. However, there may be Hazardous Materials Transportation Act (HMTA) requirements. Once the sample has been analyzed, it must be sent immediately back to the sample collector. If the laboratory keeps the sample or does not send it back to the original sample collector, the sample is no longer excluded and is subject to applicable regulations. If the sample is sent back to the collector, the collector becomes the generator. If the laboratory keeps the sample, the laboratory will be considered the generator.
16. Hazardous waste samples used in small-scale treatability studies are conditionally excluded from Subtitle C regulations [40 CFR 261.4(e)]. Generators of the waste samples and owners/operators of laboratories or treatment facilities conducting treatability studies will be excluded from the Subtitle C hazardous waste regulations, including the permitting requirements, provided that certain conditions are met.

    A *treatability study* is a study in which a hazardous waste is subjected to treatment process to determine

    - Whether the waste is amenable to the treatment process.
    - What pretreatment (if any) is required.
    - The optimal process conditions needed to achieve the desired treatment.
    - The efficiency of a treatment process for a specific waste or wastes.
    - The characteristics and volumes of residuals from a particular treatment process.

    Also included in the definition of a treatability study for the purpose of the 40 CFR 261.4(e) and (f) exemptions are liner compatibility, corrosion, and liner material compatibility studies (e.g., leachate collection systems, geotextile materials, pumps, and personal protective equipment) and toxicological and health effects studies. A treatability study is not a means to commercially treat or dispose of hazardous waste. In addition, this definition does not apply if the treatability study could result in a significant uncontrolled release of hazardous constituents to the environment. The definition of a treatability study does not include open burning or any type of treatment involving placement of hazardous waste on the land, such as in situ stabilization (53 *FR* 27293, July 19, 1988).

Under the treatability-study exclusion, waste samples are excluded from Subtitle C when the sample is being

- Collected and prepared for transportation by the generator or sample collector.
- Accumulated or stored by the generator or sample collector before it is transported to a laboratory or testing facility.
- Transported to the laboratory or testing facility at which a treatability study will be conducted.

The preceding exclusions are applicable to samples of hazardous waste collected and shipped for the purpose of conducting treatability studies, provided the following:

- The generator sends, or the sample collector uses in the treatability study, no more than 1,000 kg of any nonacute hazardous waste; 1 kg of acute hazardous waste; or 250 kg of soil, water, or debris contaminated with acute hazardous waste for each process being evaluated for each generated waste stream (i.e., not limited to the waste code, but to the waste stream).
- The sample is packaged so that it will not leak or vaporize from its packaging during shipment.

The transportation of the sample must comply with HMTA or U.S. Postal Service (USPS) shipping requirements. However, if HMTA, USPS, or other shipping requirements do not apply, the following information must accompany the sample:

- The name, mailing address, and telephone number of the originator of the sample.
- The name, address, and telephone number of the facility that will perform the treatability study.
- The quantity of the sample.
- The date of shipment.
- A description of the sample, including its EPA hazardous waste number.

A facility or laboratory conducting treatability testing is excluded from the hazardous waste regulatory requirements of RCRA, provided that certain conditions are met. EPA has determined (53 *FR* 27297, July 19, 1988) that mobile treatment units (MTUs) conducting treatability studies may qualify for this exemption. However, each MTU or group of MTUs operating at the same location is subject to the treatment rate, storage, and time limitations and the notification, record-keeping, and reporting requirements that are applicable to stationary laboratories or testing facilities conducting treatability studies. That is, a group of MTUs operating at one location will be treated as one MTU facility for purposes of 40 CFR

261.4(e) and (f). Furthermore, these requirements apply to each location where an MTU will conduct treatability studies.

To comply with the exclusion, the MTU must do the following:

- Notify EPA at least 45 days before conducting any treatability studies.
- Obtain an EPA identification number.
- Test no more than 250 kg of hazardous waste per day.
- Maintain records that document compliance with the treatment rate limits, storage time, and quantity limits.
- Maintain on-site all treatability contracts and shipping papers for at least three years.
- The information must include the following:
  - —The name, address, and EPA identification number of the generator or sample collector of each waste sample.
  - —The date the shipment was received.
  - —The quantity of waste received and in storage.
  - —The date the treatment study was initiated and the amount of waste introduced to treatment each day.
  - —The date the treatability study was concluded.
  - —The date any unused sample or residues generated from the treatability study were returned to the generator or sample collector, or sent off-site to a designated facility (including the designated facility's name and EPA identification number).
- Submit an annual report to EPA, by March 15, estimating the number of studies and the amount of waste expected to be used in treatability studies during the current year. The report covering the previous year must include information on the following:
  - —The name, address, and EPA identification number of the facility conducting the treatability studies.
  - —The types (by process) of treatability studies conducted.
  - —The names and addresses of persons for whom studies have been conducted (including their EPA identification numbers).
  - —The total quantity of waste in storage each day.
  - —The quantity and types of waste subjected to treatability studies.
  - —When each treatability study was concluded.
  - —The final disposition of residues and unused samples from each treatability study.
- Notify EPA by letter when a facility is no longer planning to conduct treatability studies at the site.

17. Any hazardous waste remaining in a container that is considered empty according to these definitions (as depicted in Figure 2-3):
    - *Nonacutely Hazardous Waste Containers.* A nonacutely hazardous waste container is considered empty if it is thoroughly emptied using common industry practices and contains less than one inch of residue on the bottom, or less than 3 percent by weight for containers less than

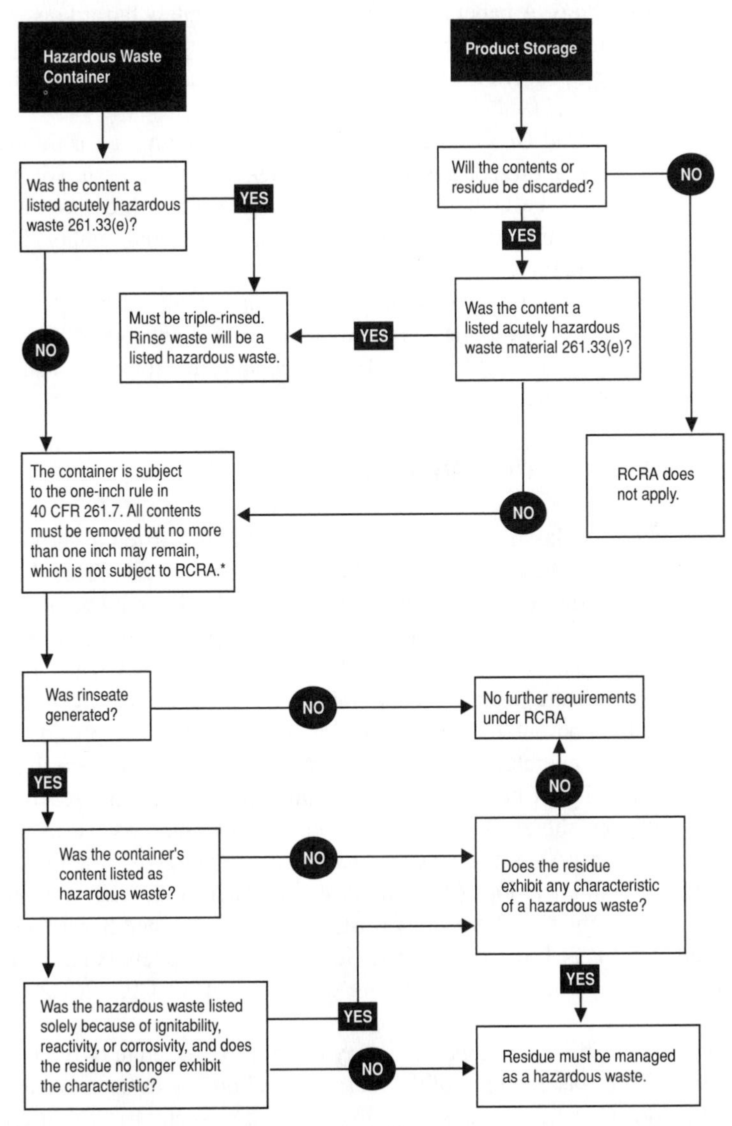

Figure 2-3. Container Residue Classification. (*However, CERCLA does not recognize this RCRA exclusion, and thus, the generator is still liable for the contents in perpetuity.)

110 gallons, or less than 0.3 percent by weight for containers greater than 110 gallons [40 CFR 261.7(b)(1)].

- *Acutely Hazardous Waste Containers.* A container or liner holding an acutely hazardous waste (P-code and certain F-code wastes) must be triple-rinsed with an appropriate solvent, rinsed using another method shown to be equivalent, or have the liner removed to be considered empty [40 CFR 261.7(b)(3)].

- *Paper Bags.* A paper bag that contained an acutely hazardous waste (P-code and certain F-code wastes) is considered empty by repeatedly beating an inverted bag to ensure thorough emptiness (OSWER Directive 9441.15).
- *Compressed Gas Containers.* A container holding a hazardous waste that is a compressed gas must be emptied until the pressure of the material inside the container is equal to atmospheric pressure. When the pressures are equal, the container is considered empty [40 CFR 261.7(b)(2)].
- *Tanks.* There is no definition for an empty tank. Although the empty container definition is commonly applied, the tank owner should check with the state or appropriate EPA regional office for a conclusive determination.

HAZARDOUS WASTE CLASSIFICATION

A hazardous waste is classified as hazardous if it either is a listed hazardous waste or if it exhibits one of the four characteristics of hazardous waste. This determination follows a sequential protocol as described in the following paragraph.

To determine if a waste is hazardous (assuming it is a solid waste and is not excluded), first it must be determined if the waste is a RCRA-listed hazardous waste. (*Listed wastes* are wastes predetermined to be hazardous by EPA. These wastes are contained on one of five lists, which are presented in Appendix A.) If a waste is not listed, a generator then must determine if the solid waste exhibits any of the four *characteristics* of hazardous waste: ignitability, corrosivity, reactivity, and toxicity. In addition, there are other hazardous wastes, such as waste mixtures, wastes derived from processing wastes, and wastes contained in nonwastes, that meet the definition of hazardous (these are discussed later in this chapter). Table 2-2 presents all the categories of hazardous wastes. The RCRA listed hazardous wastes are named in 40 CFR 261.30 through 261.33, as well as in Appendix A of this book. The characteristics of hazardous waste are contained in 40 CFR 261.20 through 261.24.

As discussed in Chapter 3, generators may use their knowledge of the hazardous characteristic of the waste in light of the materials and processes used to classify the waste as hazardous, or may test the waste to classify it as such (40 CFR 262.11).

Note: If a generator relies on knowledge to determine if a waste is hazardous, and erroneously classifies a waste as nonhazardous, the generator is liable for this error regardless of whether a good-faith effort was made.

Should the generator test the waste, approved test methods must be used. These approved methods are contained in the EPA manual *Test Methods for Evaluating Solid Waste, Physical/Chemical Methods,* more commonly known as SW-846; which provides information on sampling and analyzing procedures for complying with RCRA. However, the general use of SW-846 for testing purposes is not necessarily required. One must refer to the regulatory text to determine if and what spe-

TABLE 2-2. The Categories of Hazardous Waste

RCRA Hazardous Wastes		
Listed Hazardous Wastes	Characteristic Hazardous Wastes*	Other Hazardous Wastes
• Nonspecific sources (F codes) • Specific sources (K codes) • Commercial chemical products—Acutely hazardous (P codes) • Commercial chemical products—Nonacutely hazardous (U codes)	• Ignitable • Corrosive • Reactive • Toxic	• Mixtures (hazardous and nonhazardous) • Derived-from wastes (treatment residues) • Materials containing listed hazardous wastes

*All are D codes.

cific test method is required; in some cases specific test methods from SW-846 are mandatory. This manual is available from the U.S. Government Printing Office. (See Appendix E for further information.)

Listed Hazardous Wastes

EPA has listed specific hazardous wastes based on the criteria set forth in 40 CFR 261.11.[1] If a waste meets these criteria, it is presumed to be hazardous regardless of the concentration of the hazardous constituents in the waste, and then becomes a listed hazardous waste. However, generators may demonstrate that a listed waste is not hazardous by petitioning EPA to delist a waste at a particular generator's site based on specified criteria. A more detailed discussion of delisting appears later in this chapter.

Note: The hazardous waste lists can be confusing. It is important to understand that a waste must meet the listing definition for the waste to be classified as that particular waste code.

[1] The Appendix VIII constituents, found in Appendix VIII of 40 CFR Part 261 (and Appendix B of this book), are known as the *hazardous constituents.* Their presence in solid wastes are used as a basis by EPA for making hazardous waste listing determinations if they are at concentrations of concern. The presence of Appendix VIII constituents in an unregulated waste stream provides justification for EPA to list a waste stream as hazardous. If a generator tests an unregulated waste and finds that it contains Appendix VIII constituents, this finding does not render the waste hazardous. A waste may be hazardous *only* if it is specifically listed or exhibits an EPA-defined hazardous characteristic. If a waste does not meet any of these two criteria, it is not hazardous regardless of the presence of Appendix VIII constituents. The Appendix VIII hazardous constituents are also used in the RCRA corrective action program to determine if a past release may have occurred. (See Chapter 9 for further information.)

Thus, the mere presence of a chemical that is both in the waste and that appears on the list of hazardous wastes does not automatically mean that that generator's waste meets the listing.

EPA has listed wastes based on toxicity, reactivity, corrosivity, and ignitability. In the case of hazardous wastes listed because they meet the criteria of toxicity, EPA's principal focus is on the identity and concentration of the waste's constituents and the nature of the toxicity presented by the constituents. If a waste contains significant concentrations of hazardous waste constituents, EPA is likely to list the waste as hazardous unless it is evident that the waste constituents are incapable of migrating in significant concentrations even if improperly managed, or that the waste constituents are not mobile or persistent should they migrate.

A detailed justification for listing each hazardous waste is contained in *EPA's Listing Background Documents.* Each listing background document is organized in the following sequence: (1) the EPA Administrator's basis for listing the waste or waste stream; (2) a brief description of the industries generating the listed waste stream; (3) a description of the manufacturing process or other activity that generates the waste; (4) identification of waste composition, constituent concentrations, and annual quantity generated; (5) a summary of the adverse health effects of each of the waste constituents of concern; and (6) a summary of damage case histories involving the waste. The listing background documents, prepared for each listed waste, are located at EPA's RCRA regulatory dockets and are available for public review.

The listed hazardous wastes, contained in Appendix A of this book and 40 CFR 261.31, 261.32, and 261.33, are separated into the following categories:

- Wastes from nonspecific sources (F codes)[2]
- Wastes from specific sources (K codes)
- Commercial chemical products (U and P codes)

Some of the listed wastes are classified as acutely hazardous waste, designated with an H. These wastes are subject to more stringent weight limits regarding generator categories, as discussed in Chapter 3, and more stringent requirements concerning the determination of empty containers, as discussed earlier in this chapter.

Wastes from Nonspecific Sources. The first category of listed hazardous wastes includes commonly generated wastes from generic industrial processes. This category of wastes includes the following:

- Solvent wastes
- Electroplating wastes
- Metal heat treating wastes
- Dioxin containing wastes

[2] The various letter codes refer to the EPA hazardous waste numbers. Each hazardous waste—listed and characteristic—is assigned a four-digit code, which identifies the waste.

- Chlorinated aliphatic production wastes
- Wood preserving wastes
- Petroleum refinery wastewater treatment waste
- Hazardous waste landfill leachate

Solvent Wastes. Solvent wastes are designated as wastes F001 through F005 (see Appendix A). (They are also known as "F-listed" wastes.) For a waste to be classified as a *solvent* waste, the purpose of the material must have been to mobilize or solubilize a constituent. Thus, if a material was used solely as a reactant or a feedstock, it is not classified as a solvent waste (OSWER Directive 9444.08).

The mere presence of any of the components listed in F001 through F005 in a waste does not necessarily render the waste a listed hazardous waste. A material will meet the listing criteria if it is a solvent with a sole active ingredient that is specifically listed. A solvent mixture can also meet the F listing if it is a commercial blend with any of the F-listed solvent wastes (F001 through F005), except for F003 solvent wastes, alone or in any combination equal to 10 percent or more of the blend before use (i.e., a *solvent mixture*). Such a blend, when spent, is classified as an F-listed, solvent mixture hazardous waste. (*Spent* refers to any material that has been used and as a result of contamination can no longer serve its intended purpose without processing.) It is important to note that if a mixture had 10 percent of solvent before use, and the spent solvent mixture contains less than 10 percent solvent, it is still a hazardous waste. This is because the solvent mixture rule's 10 percent threshold is applicable before the solvent is used, not when it is spent.

Electroplating Wastes. The wastes F006, F007, F008, F009, and F019 are the electroplating wastes. EPA excludes electroless plating, chemical conversion coating, or printed circuit board manufacturing in its definition of electroplating (51 *FR* 43352, December 2, 1986). However, these processes may fall under another listing category, and the wastes generated from these processes may exhibit a characteristic of hazardous waste. Further clarification of the individual electroplating wastes are included in *EPA's Listing Background Documents.*

Metal Heat Treating Wastes. EPA has listed several wastes generated from metal heat treating processes. Metal heat treatment involves case hardening by carbonizing, which adds carbon to the surface of steel. Liquid carbonizing, which uses cyanides as the source of carbon, is accomplished by submerging the metal in a molten salt bath containing sodium cyanide. Sodium cyanide also is used in the case hardening of steel through the liquid nitriding or carbonizing processes. The associated wastes include quenching bath residues from oil baths (F010), salt bath pot cleaning (F011), and quenching wastewater treatment sludges (F012). These wastes are generated from metal heat treating operations only when cyanides (complex or free) are used in the process.

Dioxin-Containing Wastes. The wastes F020, F021, F022, and F023 are known as the "dioxin-containing wastes" and are classified as acutely hazardous. These wastes are hazardous whether dioxin is present or not. Dioxin itself is not considered a hazardous waste but may be an inadvertently generated component of the above-listed wastes.

Chlorinated Aliphatic Production Wastes. The waste codes F024 and F025 are specific hazardous wastes generated from the manufacture of chlorinated aliphatic hydrocarbons.

Wood Preserving Wastes. The waste codes F032, F034, and F035 pertain to wood preserving operations. Typically, after wood is treated with pentachlorophenol, creosote, or arsenic-based preservatives, it is stored on drip pads to allow excess to drip off. The preservative drippage is the hazardous waste; the waste code depends on the type of preservative used.

Petroleum Refinery Wastewater Treatment Waste. Waste codes F037 and F038 apply to specific waste streams generated by petroleum refining operations. F037 applies to the sludges and float created by gravitational treatment of petroleum refinery wastewater, whereas F038 applies to sludges and float created during the chemical or physical treatment of refinery wastewater.

Hazardous Waste Landfill Leachate. The waste code F039 applies to multisource leachate, that is, the leachate collected at the bottom of a hazardous waste landfill.

Wastes from Specific Sources. The second category of listed hazardous wastes includes those generated from specific industrial processes. These wastes, under the designation of K codes, are listed according to the specific industrial process that generates the waste, such as *Untreated process wastewater from the production of toxaphene* (K098), whereas most hazardous wastes are listed by a chemical name, such as benzene. The categories of hazardous waste under the K-code designations include waste from the following processes (see Appendix A):[3]

- Wood preservation wastes
- Inorganic pigment manufacturing wastes
- Organic chemical manufacturing wastes
- Inorganic chemical manufacturing wastes
- Pesticide manufacturing wastes
- Explosives manufacturing wastes
- Petroleum refining wastes
- Iron and steel production wastes
- Primary copper production wastes

[3] It is important to note that these manufacturing categories contain specific, designated waste streams. Thus, not all wastes generated from these processes are hazardous.

- Primary lead production wastes
- Secondary lead processing wastes
- Primary zinc production wastes
- Primary aluminum production wastes
- Ferroalloy production wastes
- Veterinary pharmaceutical manufacturing wastes
- Ink formulation wastes
- Coking wastes

Commercial Chemical Products. The third category of listed hazardous wastes encompasses commercial chemical products, designated by either a U or P code. All of the P-code wastes are considered acutely hazardous (H) and are subject to more stringent requirements concerning empty containers and weight limits for determining generator category. (The P-code and U-code lists are contained in Appendix A of this book.) For a waste to be categorized as a commercial chemical product waste, it must be in an *unused* form. The definition of commercial chemical products includes technical grades, pure forms, off-specification products, or sole active ingredient products. If a material is used or spent, it is *not* defined as a commercial chemical product and, thus, it is not listed as a U- or P-code waste. However, a spent waste may meet one of the other listings or exhibit a hazardous characteristic, which would classify it as hazardous. For example, unused technical-grade toluene that is to be discarded is identified as a listed hazardous waste, U220. However, if that same toluene were used as a solvent, the spent material would be classified as a different listed hazardous waste, F005. If a material contains more than one active ingredient that is a listed U-code substance, it does *not* meet a U or P listing and can be deemed hazardous only by another listing or because it exhibits a hazardous characteristic.

Note: In terms of hazardous waste lists, the commercial chemical product lists generate the most confusion by far. Although these U- and P-code lists are very specific as to what constitutes a waste, many people are confused by the listing definition. Even if a particular waste does not meet the commercial chemical product listing, it can possibly be hazardous due to a different listing (F or K) or if it exhibits a characteristic of hazardous waste.

A commercial chemical product is not considered a hazardous waste unless it is intended to be discarded or is spilled, in which case the spill-cleanup residue also satisfies the appropriate U- or P-code listing. Thus, a commercial chemical product may be stored indefinitely without RCRA constraints if the intent is to use it or have it recycled, provided the material does not spill.

Hazardous Waste Characteristics

Solid wastes that are not specifically listed as hazardous wastes can still be regulated as such if the waste exhibits a characteristic of hazardous waste. Section 3001 of RCRA required EPA to develop criteria for identifying characteristics of hazardous waste. Using standard available testing protocols, EPA has established ignitability, corrosivity, reactivity, and toxicity as the characteristics of hazardous waste. The generator is responsible for determining whether a waste exhibits a hazardous characteristic.

Ignitability. A waste is an ignitable hazardous waste, designated as D001, if it meets any of the following four criteria (40 CFR 261.21):

- It is a *liquid* that has a flashpoint of less than 60°C (or 140°F) as determined by either a Pensky-Martens or a Setaflash closed-cup test.

  A *liquid* is defined as the material, called the liquid phase, obtained from a waste in section 7.1.1 (Separation Procedure) of the Extraction Procedure Test, Method 1311 (a test method contained in SW-846, *Test Methods for Evaluating Solid Waste, Physical/Chemical Methods*). The liquid extract is the *liquid* subject to the ignitability determination.

  There is an exclusion in 40 CFR 261.21(a)(1) from the ignitability characteristic for aqueous solutions that fail the flash-point test and contain less than 24 percent alcohol. Originally, this exclusion was intended for alcoholic beverages such as wine (45 *FR* 33108, May 19, 1980). However, the regulatory language is ambiguous regarding the extent of this exclusion. OSWER Directive 9443.02 states that "while the Agency's intent was that this exemption apply to potable beverages only, because the term alcohol was used instead of ethanol, all aqueous wastes which are ignitable only because they contain alcohols (here using the term alcohol to mean any chemical containing the hydroxyl-[-OH] functional group) are excluded from regulation." The directive also defines the term "aqueous solution" by stating: "With respect to what constitutes an aqueous solution, such a solution is one in which water is the primary component. This means that water constitutes at least 50 percent by weight of the sample."
- It is a *nonliquid* that can spontaneously combust through friction, absorption, or loss of moisture. (At this time, there is no standardized test for this determination.)
- It is an *ignitable* (i.e., flammable) *compressed gas* as defined by the Department of Transportation (DOT) in 49 CFR 173.300.

  An ignitable compressed gas must first meet the definition of a compressed gas, which is "any material or mixture having in the container an absolute pressure exceeding 40 psi at 70°F or, regardless of the pressure at 70°F, having an absolute pressure exceeding 104 psi at 130°F; or any liquid flammable mixture having a vapor pressure exceeding 40 psi absolute at 100°F." A material meeting the definition of a compressed gas is considered ignitable if it is either a mixture of 13 percent or less (by volume) with air and forms a flammable mixture, or if the flammable range with air is wider than 12 percent regardless of the lower limit; if the flame projects more than 18 inches beyond the ignition source with the valve opened fully, or the flame flashes back and burns at the valve with any degree of valve opening; or if there is any significant propagation of flame away from the ignition source.

- It is an *oxidizer* as defined by DOT in 49 CFR 173.151.

  An *oxidizer* is a substance that yields oxygen readily when involved in a fire, thereby accelerating and intensifying the combustion of organic material. There is no prescribed test for determining this classification. DOT provides examples of oxidizers, which include chlorate, permanganate, inorganic peroxide, and nitrates.

Corrosivity. A waste is a corrosive hazardous waste, designated as D002, if it meets either of the following two criteria (40 CFR 262.22):

- It is *aqueous* with a pH less than or equal to 2.0, or a pH of 12.5 or more.
- It is a *liquid* and corrodes steel at a rate of 6.35 mm or more per year as determined by the National Association of Corrosion Engineers (NACE).

Wastes in a solid phase, as determined by the Separation Procedure (§7.1.1) of the Extraction Procedure Test, Method 1311, are not considered corrosive wastes (45 *FR* 33108, May 19, 1980).

Reactivity. A waste is a reactive hazardous waste, designated as D003, if it has the capability to explode or undergo violent chemical change in a variety of situations. This characteristic is used to identify wastes that, because of their extreme instability and tendency to react violently or explode, pose a threat to human health and the environment at all stages of the waste-handling process. If a waste meets any of the following eight criteria, it is classified as a reactive hazardous waste (40 CFR 262.23):

- Instability and readiness to undergo violent change.
- Violent reactions when mixed with water.
- Formation of potentially explosive mixtures when mixed with water.
- Generation of toxic fumes in quantities sufficient to present a danger to human health or the environment when mixed with water.
- Cyanide- or sulfide-bearing material that generates toxic fumes when exposed to acidic conditions.

  Note: If, using the appropriate test methods in SW-846 for sulfide and cyanide gas generation, the total available amount of HCN is generated at 250 mg/kg of waste or more, or if sulfide is generated at 500 mg/kg of waste, it is classified as reactive.

- Ease of detonation or explosive reaction when exposed to pressure or heat.
- Ease of detonation or explosive decomposition or reaction at standard temperature and pressure.
- Definition as a forbidden explosive or a Class A or Class B explosive by DOT in 49 CFR part 173.

Toxicity. EPA has determined that the leaching of toxic constituents of land disposed wastes into groundwater poses a significant danger. The toxicity characteristic (TC) is designed to identify wastes that are likely to leach hazardous constituents into groundwater because of improper management conditions. EPA established a standardized testing procedure that extracts constituents from a solid waste in a manner that simulates the leaching action that can occur in a landfill. EPA made the assumption that industrial waste would be codisposed of with nonindustrial waste in an actively decomposing municipal landfill situated over an aquifer. However, for the purpose of determining whether a waste is hazardous when using the toxicity test, it is irrelevant whether generators actually codispose of their hazardous waste in a municipal landfill or in any type of landfill.

The extraction procedure of the TC is also known as the toxicity characteristic leaching procedure test (TCLP).[4] The TCLP requires the following basic steps:

1. If the waste is liquid (i.e., contains less then 0.5 percent solids), the waste itself is considered the extract after it is filtered (simulated leachate).
2. If the waste contains greater than 0.5 percent solid material, the solid phase is separated from the liquid phase, if any. If required, the particle size of the solid phase is reduced until it passes through a 9.5 mm sieve.
3. For analysis other than for volatiles, the solid phase is then placed in a rotary agitation device with an acidic solution and rotated at 30 rpm for 18 hours. The pH of the solution is approximately 5, unless the solid is more basic, in which case a solution with a pH of approximately 3 is used. After extraction (rotation), solids are filtered from the liquid extract and discarded.
4. For volatiles analysis, a solution of pH 5 and a Zero Headspace Extraction Vessel (ZHE), which does not allow headspace to develop, are used for liquid/solid separation, agitation, and filtration.
5. Liquid extracted from the solid/acid mixture is combined with any original liquid separated from the solid material and is analyzed for the presence of the contaminants listed in Table 2-3.

If any of the contaminants in the extract meet or exceed any of the maximum concentration levels listed in Table 2-3, the waste is classified as a toxicity characteristic hazardous waste.

Special Categories of Hazardous Waste

The following sections outline the regulatory status of special categories of solid and hazardous waste. These categories are found in different sections of 40 CFR but have been grouped here for easier reference. They include the following:

[4] The TCLP replaced the Extraction Procedure Toxicity Test on March 29, 1990 (55 *FR* 11798).

TABLE 2-3. Toxicity Characteristic Levels

EPA HW No.	Contaminant	Limit (mg/l)
D004	Arsenic	5.000
D005	Barium	100.0
D018	Benzene	0.5
D006	Cadmium	1.0
D019	Carbon tetrachloride	0.5
D020	Chlordane	0.03
D021	Chlorobenzene	100.0
D022	Chloroform	6.0
D007	Chromium	5.0
D023	*o*-Cresol	200.0*
D024	*m*-Cresol	200.0*
D025	*p*-Cresol	200.0*
D026	Cresol	200.0*
D016	2,4-D	10.0
D027	1,4-Dichlorobenzene	7.5
D028	1,2-Dichloroethane	0.5
D029	1,1-Dichloroethylene	0.7
D030	2,4-Dinitrotoluene	0.13
D012	Endrin	0.02
D031	Heptachlor	0.008
D032	Hexachlorobenzene	0.13
D033	Hexachlorobutadiene	0.5
D034	Hexachloroethane	3.0
D008	Lead	5.0
D013	Lindane	0.4
D009	Mercury	0.2
D014	Methoxychlor	10.0
D035	Methyl ethyl ketone	200.0
D036	Nitrobenzene	2.0
D037	Pentachlorophenol	100.0
D038	Pyridine	5.0
D010	Selenium	1.0
D011	Silver	5.0
D039	Tetrachloroethylene	0.7
D015	Toxaphene	0.5
D040	Trichloroethylene	0.5
D041	2,4,5-Trichlorophenol	400.0
D042	2,4,6-Trichlorophenol	2.0
D017	2,4,5-TP Silvex	1.0
D043	Vinyl chloride	0.2

*If the concentration for *o*-, *m*-, and *p*-cresol cannot be differentiated, the total cresol concentration is used.

- Hazardous waste mixtures
- Wastes derived from the management of hazardous wastes
- Hazardous wastes contained in nonwastes
- Low-level radioactive mixed wastes
- Delisted hazardous wastes

Hazardous Waste Mixtures. A mixture of a listed hazardous waste and solid (nonhazardous) waste is considered a hazardous waste [40 CFR 261.3(a)(2)(iii)]. (There is no *de minimis* amount that qualifies for an exclusion from the mixture rule except for some *de minimis* mixtures in wastewater treatment systems meeting certain conditions.) However, if the hazardous waste contained in the mixture is hazardous solely because it exhibits a hazardous characteristic, and the resultant mixture no longer retains that characteristic, it is not considered a hazardous waste. An example is F003 (a listed waste), which is listed solely because of the ignitability characteristic. Hence, a mixture of F003 and a nonignitable, nonhazardous waste will become nonhazardous provided that the mixture no longer exhibits the ignitability characteristic (46 *FR* 56588, November 17, 1981).[5]

Wastewater treatment systems subject to either a National Pollutant Discharge Elimination System (NPDES) permit or pretreatment standards have specific exclusions from the hazardous waste mixture rule for the effluent under 40 CFR 261.3(a)(2)(iv). These exclusions include wastewater mixed with specified spent solvents, *de minimis* losses, and laboratory wastes. All of these exclusions are subject to certain conditions.

Waste Derived from the Management of Hazardous Waste. Any solid waste generated from the treatment (including reclamation), storage, or disposal of a listed hazardous waste, including any sludge (pollution control residue), spill residue, ash, leachate, or emission control dust, is a hazardous waste (and retains the waste code of the original listed waste) unless it is delisted. Residual solid waste generated as a result of treating, storing, or disposing of a characteristic waste is deemed hazardous if the residual continues to exhibit a characteristic [40 CFR 261.3(c)(2)].

There are specific exclusions from the derived-from rule for certain wastes (e.g., pickle liquor sludge, biological treatment sludges, and residues from high-temperature metals recovery), which are found in 40 CFR 261.3(c)(2)(ii).

Hazardous Waste Contained in Nonwaste. The hazardous waste mixture rule discussed previously applies to those situations in which a hazardous waste is mixed with a solid waste. However, in some instances a hazardous waste is mixed with a material that is not considered a solid waste. In this case, the hazardous waste is considered to be *contained in* the other material, and the mixture is regulated as a hazardous waste. For example, if a surface impoundment leaks a listed hazardous waste

[5] It should be noted that if the waste will be land disposed, it must still meet the universal treatment standards in accordance with the land disposal restrictions, which are discussed in Chapter 8.

into the groundwater, the resulting contaminated groundwater containing the listed waste is regulated as a hazardous waste, although the groundwater is not by itself a solid waste, because the groundwater is not "discarded." However, the groundwater, once contaminated, contains a listed hazardous waste. The hazardous waste component itself remains hazardous, and must be managed as hazardous waste even though the groundwater itself, if it did not contain any of the hazardous waste, would not be considered hazardous. Hence, the groundwater contaminated with hazardous waste must be handled as if the groundwater itself were hazardous because the hazardous waste leachate contained in the groundwater is subject to regulation under Subtitle C of RCRA (OSWER Directive 9481.00-6).

Low-Level Radioactive Mixed Wastes. A *low-level radioactive waste (LLRW)* is radioactive material that (a) is not a high-level radioactive waste, spent nuclear fuel, or by-product material (i.e., uranium or thorium mill tailings) as defined in §11(e)(2) of the Atomic Energy Act (AEA), and (b) is classified by the Nuclear Regulatory Commission (NRC) as a low-level radioactive material.

If an LLRW contains a listed RCRA hazardous waste, or if the LLRW exhibits a characteristic of hazardous waste, the material is classified as a mixed low-level waste and must be managed and disposed of in compliance with EPA regulations, 40 CFR parts 124 and 260 through 280, *and* with NRC regulations, 10 CFR parts 20, 30, 40, 61, and 70. The management and disposal of mixed LLRW must also comply with state requirements in states that are EPA-authorized for the hazardous components of the waste, and with the agreement between the NRC and state radiation control programs for the low-level radioactive portion of the waste (OSWER Directive 9440-1).

Delisted Hazardous Waste. Delisting is a formal rulemaking request to EPA, by petition, to reclassify a listed hazardous waste as a nonhazardous waste at a specific generation site.[6] The requirements to follow for submitting a delisting petition are set forth in 40 CFR 260.20 and 260.22.

In general, a waste is listed by EPA if it contains hazardous constituents or if it exhibits a characteristic of hazardous waste (that is, if it is either ignitable, corrosive, reactive, or toxic).[7] In 40 CFR part 261, Appendix VII, the hazardous constituents for which the waste is listed are named. A petitioner must demonstrate that the hazardous constituents for which the waste was listed are not present in the waste and that the waste does not exhibit a characteristic of hazardous waste. A facility must also demonstrate that the waste will not be hazardous for other reasons, including other hazardous constituents.

[6] According to EPA, the majority of wastes that have been excluded through delisting are the metal-bearing wastes (e.g., F016, F019, and K061). Historically, only 15 to 20 percent of submitted petitions have been granted (EPA/530-F-93-005).

[7] The list of designated hazardous substances is contained in 40 CFR part 261, Appendix VIII.

EPA has outlined the specific information to be submitted with the delisting petition in *Petitions to Delist Hazardous Waste: A Guidance Manual* (EPA/530-R-93-007), which shoud include the following:

- A detailed description of the manufacturing and treatment processes generating the petitioned waste and the volume of waste generated.
- A discussion of why the waste is listed as hazardous and a description of how the waste is managed.
- A discussion of why samples collected in support of the demonstration are thought to represent the full range of variability of the petitioned waste.
- Results from the analyses of a minimum of four representative samples of the petitioned waste for the following:
  — the total oil and grease content of the waste;
  — applicable hazardous waste characteristics (ignitability, corrosivity, toxicity, or reactivity);
  — total and leachable concentrations of all hazardous constituents likely to be present in the petitioned waste; and
  — chain-of-custody controls and quality control data for all analytical data.
- A statement signed by an authorized representative of the facility certifying that all information is accurate and complete.
- Groundwater monitoring information if the petitioned waste has been disposed of in a land-based hazardous waste management unit.

After receipt of a delisting petition, EPA conducts a series of review steps. Because the delisting process is a rulemaking procedure subject to the Administrative Procedures Act, it typically takes EPA two years to review and finalize a delisting petition. The major review steps are as follows:

- Completeness check and a request for additional information if needed.
- Technical evaluation of the waste analysis and process data.
- Publication of a proposed rule providing public notice of EPA's decision in the *Federal Register.*
- Review of public comments in response to the proposed rule and promulgation of a final decision in the *Federal Register.*

There are three different types of exclusions that can be finalized as a result of the delisting process, based on site conditions and the waste:

Standard Exclusion—requires no testing of waste as a condition of the exclusion and is granted when a petitioner demonstrates that the waste meets the delisting criteria and the variability of the waste is not a concern.

Conditional Exclusion—granted when the waste being generated is expected to be highly variable in composition. Thus, as a condition of granting the exclusion, EPA may require the facility to periodically test the waste to ensure that the excluded waste continues to meet the delisting criteria.

Upfront Exclusion—a special form of conditional exclusion that is granted for a waste not yet generated. These exclusions typically require extensive verification testing once the full-scale generating process is operational.

RECYCLING HAZARDOUS WASTES

It is intuitively understood that the legitimate reuse, recycling, and reclamation of hazardous waste is far more beneficial than treatment and/or disposal. However, recycling activities also have environmental and public health impacts. In fact, many of the early Superfund sites were former recycling facilities. Thus, EPA has been faced with a major challenge: promoting the reuse, recycling, and reclamation of hazardous waste while ensuring that it is done in a manner protective of human health and the environment. This balancing act has been a constant struggle for EPA since the inception of the RCRA program.

In addition, the legal constraints of RCRA are such that subsequent recycling regulations have become quite complex. Under Subtitle C of RCRA, EPA has the authority to regulate hazardous wastes; hazardous wastes, however, are defined in the statute as a subset of *solid* wastes. Thus, EPA has developed the hazardous waste recycling regulations around a series of regulations that either include wastes and secondary materials in the definition of solid waste, or exempt those materials from that definition.

Under 40 CFR 261.2(a), a *solid waste* is "any discarded material that is not excluded . . ." A *discarded material* is "any material that is abandoned, . . . recycled . . . , or considered inherently waste-like . . ."

For the regulated community, determining if a material is a solid waste is fairly straightforward. However, determining if a waste to be recycled is still a solid waste is a totally different matter. In making an initial classification of a waste, a person should ignore the fact that the waste will be recycled. Given this approach, the following protocol should be used:

- Is it a solid waste?
- Is it excluded from the definition of hazardous waste?
- Is it a listed hazardous waste?
- Is it a characteristic hazardous waste?

If the waste is to be recycled, the regulations do not expressly require the following protocol, but inherently force one to answer the following questions:

- Would it be a solid waste if the recycling question were ignored?
- Is it excluded from the definition of hazardous waste?
- Would it be a listed hazardous waste?
- Would it be a characteristic hazardous waste?
- Would it be a solid waste when recycled based on the manner of recycling?

Notice the extra question at the end of the recycling list, required because materials to be recycled are in fact classified as solid waste depending on the type of hazardous waste (i.e., listed or characteristic). To answer the final question, one must know if the material would be a listed or characteristic waste. Admittedly, this procedure is confusing and cumbersome; however, by following a simple procedure, confusion can be lessened or avoided:

> If you intend to recycle a waste, first determine if it would be a listed or characteristic hazardous waste. Then, classify it accordingly under the definition of solid waste. Whether the waste is or is not a solid waste will depend on the type of secondary material the waste is and the recycling activity to be used.

Recycling Process

Under RCRA, actual recycling processes, except those that burn wastes as fuel or entail use constituting disposal, are generally unregulated (subject to state restrictions). For example, a generator may distill solvents on-site without the distillation unit being regulated (OSWER Directive 9441.24).[8] However, generation, transportation, and storage before recycling are regulated unless the specific waste is excluded from the definition of a solid waste. Thus, a facility that distills solvents from off-site sources must have interim status or a permit for the storage of the waste solvents. A generator may recycle and/or store hazardous wastes before recycling them without interim status or a permit, provided that the hazardous waste is generated on-site and the accumulation is done in accordance with 40 CFR 262.34 (i.e., accumulation is for less than 90/180 days).

In most cases, the waste generated from the treatment and storage of hazardous waste remains a hazardous waste. However, there is an exclusion for products derived from hazardous waste (unless they are burned for energy recovery or used in a manner constituting disposal). For example, if a person recycles spent solvent in a distillation unit, the material distilled will no longer be regulated if it is used as a solvent. Only the still bottoms, the residues remaining from distilling dirty or contaminated solvent, remain regulated because they were derived from hazardous waste and are themselves waste [40 CFR 261.3(c)(2)(i)].

Excluded Recyclable Wastes

Certain recycled materials are excluded from regulation under 40 CFR parts 262 through 266 [40 CFR 261.6(a)(3)]. These materials are as follows:

- Used oil that is recycled in some way other than burning for energy recovery. (This is regulated under 40 CFR part 273.)

[8] Large-quantity generators and hazardous waste treatment facilities using steam strippers, distillation units, fractionation units, thin-film evaporation units, solvent extraction, or air strippers are required to install specific air emission control requirements [40 CFR 262.34(a)(1)(ii) and 264/265.1032].

- Industrial ethyl alcohol.
- Scrap metal.
- Fuels produced from the refining of oil-bearing wastes from normal processes at petroleum refineries.
- Oil reclaimed from hazardous waste generated as a result of normal petroleum refining operations and reinserted into petroleum refining processes.

In addition, 40 CFR 261.2(e) excludes certain recyclable materials. These materials include the following:

- Materials used or reused as ingredients to make a product, provided they are not reclaimed before use.
- Wastes used or reused as effective substitutes for commercial products without prior reclamation.
- Wastes returned to the original process from which they were generated without first being reclaimed.

Classification of Recyclable Materials

When a material is recycled, its regulatory classification (i.e., whether it is a solid waste and potentially a regulated hazardous waste) depends on two factors: the type of secondary material and the specific recycling activity. Table 2-4 identifies

Table 2-4. Classification of Secondary Materials When Recycled

	**Recycling Activity*			
Secondary Material	Use Constituting Disposal	Reclamation	Speculative Accumulation	Burned as Fuel
Spent materials, listed and exhibiting a characteristic	Yes*	Yes	Yes	Yes
Sludges, listed	Yes	Yes	Yes	Yes
Sludges, exhibiting a characteristic	Yes	No†	Yes	Yes
By-products, exhibiting a characteristic	Yes	No	Yes	Yes
By-products, listed	Yes	Yes	Yes	Yes
Commercial chemical products	Yes	No	No	Yes

**Yes* means the material is a solid waste.

†*No* means the material is not a solid waste.

whether the material being recycled in a specific process is a solid waste, and thus a hazardous waste subject to regulation.

Secondary Materials A *secondary material* is one that potentially can be a solid and hazardous waste when recycled (50 *FR* 616, January 4, 1985). Under 40 CFR 261.2, there are five types of secondary materials, as summarized in Table 2-5:

- Spent materials
- Sludges
- By-products
- Commercial chemical products
- Scrap metal

Table 2-6 provides examples of secondary materials by EPA waste codes.

Spent Materials. Spent materials are used materials that, as a result of such use, have become contaminated by physical or chemical impurities to the point they can no longer serve the purpose for which they were produced without regeneration [40 CFR 261.1(c)(1)]. The following materials are examples: wastewater, solvents, catalysts, acids, pickle liquor, foundry sands, lead-acid batteries, and activated carbon. (Spent activated carbon can also be classified as a sludge if it was used within or generated from a pollution-control device.)

Sludges. Sludges are residues from pollution control technology (40 CFR 260.10). Examples include bag house dusts, flue dusts, wastewater treatment sludges, and filter cakes.

By-products. By-products include residual materials resulting from industrial, commercial, or agricultural operations that are not primary products, are not produced separately, are not fit for a desired end use without substantial further processing, and are not spent materials, sludges, commercial chemical products, or scrap metals [40 CFR 261.1(c)(3)]. Examples of by-products include mining slags, drosses, and distillation column bottoms.

Commercial Chemical Products. Commercial chemical products include commercial chemical products and intermediates, off-specification variants, spill residues, and container residues that are listed in 40 CFR 261.33 or that exhibit a hazardous waste characteristic.

Scrap Metal. For the purposes of RCRA recycling requirements, excluded scrap metal specifically was redefined on May 12, 1997 (62 *FR* 25988). Excluded scrap metal includes processed scrap metal, unprocessed home scrap metal, and unprocessed prompt scrap metal. *Processed scrap metal* is scrap metal that has been manually or physically altered to separate it into distinct materials for economical

Table 2-5. Examples of Wastes by Type of Secondary Material

Waste Type	Examples
By-products	Distillation column bottoms
	Mining slags
	Drosses
Commercial Chemical Products	Commercial chemical products or manufacturing intermediates listed in 40 CFR 261.33(e) or (f)
	Off-specification variants of the above substances
	Residue or contaminated debris from the cleanup of a spill of the above substances
	Containers or inner liners from containers used to hold the above substances (unless rendered empty as defined by 40 CFR 261.7)
Scrap Metal	Processed scrap metal (baled, shredded, sheared, chopped, crushed, flattened, cut, melted, and absorbed metal; and agglomerated fines, drosses, and slags)
	Home scrap metal (turnings, cuttings, punchings, and borings)
	Prompt scrap metal (turnings, cuttings, punchings, and borings)
Sludges	Bag house dusts
	Flue dusts
	Wastewater treatment sludges
Spent Materials	Spent acids
	Spent activated carbon*
	Spent catalysts
	Spent foundry sands
	Spent pickle liquor
	Spent solvents
	Spent lead-acid batteries
	Wastewater

*Spent activated carbon is a sludge if it results from pollution control technology.

recovery or better handling. Processed scrap metal includes metal that has been baled, shredded, sheared, chopped, crushed, flattened, cut melted, and sorted, as well as fines, drosses, and slags that have been agglomerated.

Note: Shredded circuit boards being sent for recycling are not considered processed scrap metal and thus retain their exclusion under 40 CFR 261.4(a)(13).

Table 2-6. Examples of EPA Waste Codes for Secondary Materials

Spent Materials	Commercial Chemical Products	Sludges	By-products	
*F001	All U and P wastes	F006	F008	K034
*F002		F012	F010	K036
*F003		F019	F024	K039
*F004				K042
*F005		K001	K008	K043
F007		K002	K009	K049
F009		K003	K010	K050
F011		K004	K011	K052
		K005	K013	K060
K021		K006	K014	K071
K028		K007	K015	K073
K033		K032	K016	K083
K038		K035	K017	K085
K045		K037	K018	K087
K047		K040	K019	K093
K062		K041	K020	K094
K086		K044	K022	K095
K098		K046	K023	K096
K099		K048	K024	K097
K104		K051	K025	K101
K111		K061	K026	K102
K117		K069	K027	K103
K118		K084	K029	K105
		K100	K030	K112
		K106	K031	K114
			K113	K116
			K115	K136

*The still bottoms from the recovery of these spent solvents are by-products.

Home scrap metal is scrap metal generated by steel mills, foundries, and refineries and includes turnings, cuttings, punchings, and borings. *Prompt scrap metal* (also called new scrap metal) is scrap metal generated by the metal workings/fabrication industries and includes turnings, cuttings, punchings, and borings.

Recycling Activities. The definition of solid waste under 40 CFR 261.2 identifies four types of recycling activities for which EPA asserts jurisdiction under Subtitle C of RCRA:

- Speculative accumulation
- Use constituting disposal
- Reclamation
- Burning of wastes or waste-derived fuels (discussed in a separate subsection later in this chapter)

Speculative Accumulation. Any hazardous secondary material not otherwise defined as a solid waste is a solid waste if it is accumulated speculatively before recycling, unless it can be shown that (1) the material is potentially recyclable and has a feasible means of being recycled, and (2) at least 75 percent of the accumulated material is recycled in one calendar year. The exceptions to this rule are hazardous commercial chemical products (listed or characteristic), which are not considered wastes when stored before recycling, and other statutorily specified materials. The 75 percent requirement may be calculated on the basis of either volume or weight and applies to waste accumulated during a calendar year beginning on the first day of January [40 CFR 261.1(c)(8)].

Use Constituting Disposal. A waste that is managed in such a way that it is considered to be used in a manner constituting disposal is a hazardous waste. Either of the following definitions [40 CFR 261.2(c)(1)] is *use constituting disposal*:

- Applying materials to the land or placing them on the land in a manner constituting disposal, or
- Applying materials contained in a product to the land or placing them on the land in a manner constituting disposal.

Examples of such use include use as a fill, cover material, fertilizer, soil conditioner, or dust suppressor, or use in asphalt or building foundation materials.

As already noted, hazardous secondary materials are considered solid wastes when applied to the land in these ways, except for listed commercial chemical products whose ordinary use involves application to the land. Therefore, hazardous waste generator, transporter, and storage requirements apply before use, and applicable land disposal requirements under 40 CFR parts 264, 265, 266, and 268 apply to the activity itself.

Products that include listed hazardous wastes as ingredients are classified as solid wastes when placed directly on the land for beneficial use, unless and until the products are formally delisted. Products that include characteristic hazardous wastes as ingredients are classified as solid wastes only if the products themselves exhibit hazardous waste characteristics.

Reclamation. Reclamation is defined as the regeneration of waste materials or the recovery of material with value from wastes [40 CFR 261.1 (c)(4)]. Reclamation includes such activities as dewatering, ion exchange, distillation, and smelting. However, simple collection, such as collection of solvent vapors, is not considered reclamation. Use of materials as feedstocks or ingredients, such as the use of material as a reactant in the production of a new product, also is not considered reclamation.

Requirements for Recycling Hazardous Waste

EPA has established reduced regulatory requirements for the recycling of certain nonexcluded hazardous wastes in parts 266, 273, and 279. This section describes the requirements for recycling specific types of hazardous waste, which include the following:

- Precious metal recovery
- Recycling of lead-acid batteries
- Burning and blending of used oil
- Burning of hazardous waste in boilers and industrial furnaces
- Universal waste recycling

Precious Metal Recovery. Hazardous wastes that contain precious metals are subject to reduced requirements when they are recycled (40 CFR 266.70). The *precious metals* include gold, silver, platinum, palladium, iridium, osmium, rhodium, and ruthenium. These reduced requirements are as follows:

- Generators and transporters must have an EPA identification number and must use a uniform hazardous waste manifest when transporting the materials.
- Treatment and storage facilities must have an EPA identification number, comply with manifest requirements, and keep records to demonstrate that 75 percent of all received wastes are being recycled per calendar year to satisfy the speculative accumulation provision.

The reclaimer is not required to comply with the technical standards of 40 CFR parts 264 or 265 for precious metal recovery operations. A reclaimer can be considered a designated facility (see Chapter 1 for definition) if a Part A permit application is filed with an attached statement explaining that only precious metal reclamation will be conducted.

Lead-Acid Batteries. Persons who generate, transport, or collect spent lead-acid batteries, and who do not also recover the lead from the batteries, are not subject to regulation under 40 CFR parts 262 through 265 (40 CFR 266.80). Reclaimers of lead from these batteries must notify EPA, and, if they store lead-acid batteries before their reclamation, must comply with the storage requirements of 40 CFR parts 264 and 265, subparts A through L, with the exceptions of the waste characterization (40 CFR 264.14 and 265.23) and manifest-related requirements (40 CFR 264/265.71 and 72).

It is important to note that cracking batteries is considered reclamation and thus is regulated. Generators that crack a battery to recover the lead plates will be considered recyclers subject to regulation.

Burning and Blending of Used Oil. Another form of recycling is the burning of used oil as a fuel for legitimate energy recovery, which is regulated under 40 CFR part 279. *Legitimate energy recovery* means that the waste being burned has a heating value of 5,000 to 8,000 Btu/lb in an industrial furnace or boiler (see Chapter 1 for definitions). *Used oil* means oil that is refined from crude oil, is used, and as a result of that use is contaminated by physical or chemical means (40 CFR 279.1). Currently, used oil is not considered a hazardous waste unless it exhibits a hazardous waste characteristic and is discarded. If, however, the intent is to dispose of used oil, the generator must comply with the requirements under RCRA if the used oil exhibits a characteristic of hazardous waste. The following regulations govern only the burning of used oil. Other used oil recycling activities (not including disposal) are not currently regulated as a hazardous waste. However, used oil mixed with any hazardous waste (other than an "ignitable-only" hazardous waste when the resulting mixture is not defined as ignitable) is subject to full regulation as a hazardous waste [40 CFR 279.10(b)].

Any used oil that contains more than 1,000 ppm of total halogens (fluorine, chlorine, bromine, iodine, and astatine) is presumed to have been mixed with a hazardous waste and is regulated as a hazardous waste [40 CFR 279.10(b)(ii)]. However, there are provisions to rebut this presumption. If it can be demonstrated that mixing has not occurred (i.e., by labels, other documentation, or analysis), the oil may be handled under the used oil rules. For example, commercial cutting oils may contain chlorinated substances as additives. Thus, used cutting oils containing chlorine at a concentration higher than 1,000 ppm may not have been mixed with a hazardous waste.

Used oils that do not meet the used oil specification criteria are regulated under the less stringent 40 CFR part 279 requirements. *Specification oil* is any used oil that does not exceed any of the specification levels in Table 2-7 (40 CFR 279.11). However, all used oil burned for energy recovery is presumed to be off-specification oil unless demonstrated otherwise. This demonstration may be accomplished by a laboratory analysis. Once the analysis is completed and documented, used oil is no

Table 2-7. Specification Levels for Used Oil Fuels

Specification	Maximum level
Arsenic concentration	5 ppm
Cadmium concentration	2 ppm
Chromium concentration	10 ppm
Lead concentration	100 ppm
Total halogen concentration	4,000 ppm
Flash point	100°F (minimum)

longer subject to the federal used-oil recycling regulations (40 CFR part 279) if the used oil is burned for energy recovery (40 CFR 279.11).

It may be demonstrated that used oil with more than 1,000 ppm of total halogens has not been mixed with hazardous wastes. If this is demonstrated, and the total halogen level is below 4,000 ppm, the used oil is considered specification oil. If the total halogen level is above 4,000 ppm, the used oil is considered waste fuel (40 CFR 279.11). This demonstration may be accomplished by a laboratory analysis. A total chlorine test may be used instead of a total halogen test to satisfy this requirement.

Off-specification oil. is any oil that exceeds any specification level identified in Table 2-7. Persons that blend (process) off-specification oil, including generators, to meet the specifications must comply with 40 CFR part 279, subpart F. Although a permit or interim status is not required for this particular activity, the facility must comply with requirements similar to hazardous waste management facilities, including the following:

- EPA identification number
- Emergency preparedness and prevention
- Emergency response equipment
- Emergency communications
- Contingency plan
- Emergency procedures
- Waste analysis plan
- Container management
- Tank management
- Shipment tracking

Generators who "direct" shipments of off-specification oil from their facility (or first claim that used oil to be burned for energy recovery meets specification) must comply with the requirements for used-oil marketers in 40 CFR part 279, subpart H [40 CFR 279.20(b)(4)].

Generators who burn off-specification oil on-site, except for in on-site space heaters, must comply with the requirements for burners of used oil in 40 CFR part 279, subpart H [40 CFR 279.20(b)(3)].

Generators may burn used oil in used-oil space heaters provided that the heater burns only used oil generated at the site or used oil received from do-it-yourself used-oil generators. Also, the heater must be designed to have a maximum capacity of 0.5 million Btu or less per hour, and the combustion gases must be vented to the ambient air (40 CFR 279.23).

Burning Hazardous Waste in Boilers and Industrial Furnaces. On February 21, 1991 (56 *FR* 7134), EPA promulgated the boilers and industrial furnaces (BIF) rule. The BIF rule dramatically changed the requirements for burning hazardous waste in boilers and industrial furnaces, as it subjected BIFs to almost all the same standards as hazardous waste management facilities. As a result of this final rule, 40 CFR part 266, subpart D, was removed, and the regulations governing burning hazardous waste in BIFs were codified in 40 CFR part 266, subpart H.

Those regulations apply to all nonexempted hazardous waste burned or processed in BIFs regardless of the purpose of burning or processing (e.g., recycling or destruction). The wastes and devices exempt from BIF requirements are as follows:

- Used oil that is burned for energy recovery under 40 CFR part 279
- Gas recovered from hazardous or solid waste landfills when the gas is burned for energy recovery
- Hazardous wastes exempt from regulation under 40 CFR 261.4 and 261.6(a)(3)(iii)–(v)
- Hazardous waste from conditionally exempt small-quantity generators
- Coke ovens that burn only K087 (decanter tank tar sludge from coking operations)

The management of hazardous waste fuels before burning them in a BIF is subject to all applicable RCRA regulations (40 CFR 266.101). Generators of hazardous waste fuels are required to comply with the 40 CFR part 262 regulations, and transporters of hazardous waste fuels are subject to 40 CFR part 263. In addition, any storage before burning is subject to the hazardous waste storage regulations in 40 CFR parts 264, 265, and 270, except under limited circumstances. This management requirement includes any storage activities conducted by the burner as well as any intermediaries.

Before 1991, BIFs were virtually exempt from regulation, but they are now regulated similar to TSDFs. Both permitted and interim status BIFs are now required to comply with strict air emissions standards, which are divided into four pollutant categories: organics, particulate matter, metals, and hydrogen chloride and chlorine. Permitted BIFs are also subject to all the general TSDF standards, including general facility, preparedness and prevention, contingency plan, manifest system, closure and financial assurance, corrective action, and air emission standards [40 CFR 266.102(a)(2)].

BIF owners or operators must perform a waste analysis to identify the type and quantity of the hazardous constituents that may reasonably be expected to exist in the waste. The analysis must include all hazardous constituents found in Appendix VIII of 40 CFR part 261. In addition, periodic sampling and analysis must be undertaken while a BIF is operating to ensure that the constituent levels for hazardous constituents in the hazardous waste are within the limits of the facility's permit [40 CFR 266.102(b)(2)].

Once a BIF is permitted, it may burn only those types of hazardous waste specified in its permit. In addition, owners/operators must manage the unit in accordance with all the operating conditions described in the permit. Compliance with the interim status standards is analogous to compliance with a Part B permit. Throughout interim status, the BIF is required to comply with the operating limits established during compliance testing. Interim status BIFs must be operated much in the same way as those facilities with permits. Because interim status facilities have not yet conducted trial burns to ensure compliance with the standards, EPA has placed some restrictions on their use and what types of hazardous waste these facilities may burn.

Facilities that transfer hazardous waste directly from a transport vehicle (e.g., tank truck) to the BIF without first storing the waste must comply with special requirements (40 CFR 266.111). Generally, the direct transfer operations must be managed in a manner similar to that required by the regulations for hazardous waste tanks and containers. In addition, the direct transfer equipment must have a secondary containment system, the owner/operator must visually inspect the operation at least once every hour, and the facility must keep records of these inspections.

Universal Waste. Because of the unique conditions associated with certain hazardous wastes, EPA established a new program for these wastes, which are known as *universal wastes.* Universal wastes are so named because they are hazardous wastes that are

- frequently generated in a wide variety of settings other than the industrial settings
- generated by a vast community, and
- present in significant volumes in non hazardous-waste-management systems.

Universal wastes currently include certain recalled and collected pesticides, spent batteries managed in accordance with 40 CFR part 266, subpart G, and mercury containing thermostats. (Some states have also designated spent fluorescent lights as universal waste.)

There are four types of regulated entities in the universal waste system:

- Small-quantity handlers of universal waste
- Large-quantity handlers of universal waste
- Universal waste transporters
- Destination facilities

There are specific requirements for each of the above entities, depending on the type of universal waste generated or managed. The following sections, however, describe the general requirements for each regulated entity handling universal waste.

Small-Quantity Handlers of Universal Waste. Small-quantity handlers of universal waste (SQHUWS) are those handlers that accumulate no more than 5,000 kg total (all universal waste categories combined) of universal waste at their location at any time. An SQHUW may store universal waste up to one year from the date it was generated or received from another handler. The waste may be stored longer than one year if the sole purpose for such practice is to facilitate proper recovery, treatment, or disposal. However, the burden of proof for such practice rests with the SQHUW [40 CFR 273.16(b)].

The general requirements for an SQHUW include the following (40 CFR part 273, subpart B):

- An SQHUW is not allowed to dispose of or dilute universal waste.
- Employees must be informed on proper handling and emergency response procedures.
- All releases of universal wastes must be responded to immediately.
- Universal waste may be sent off-site only to a universal waste handler, destination facility, or foreign source.
- Self-transported universal waste must comply with universal transporter requirements.
- Shipments of universal waste must comply with the Hazardous Materials Transportation Act (see Chapter 4).
- The waste-specific packaging and labeling requirements must be followed.
- Before universal waste is shipped to an authorized handler, assurance must be received that the shipment will be accepted.
- If a shipment is rejected, the originator must receive the waste.
- If a shipment is received and it is discovered that the shipment contains non-universal waste, the EPA Regional Office must be notified immediately.
- All exports are subject to the export requirements in 40 CFR part 262.

Large-Quantity Handlers of Universal Waste. Large-quantity handlers of universal waste (LQHUWs) are those handlers that accumulate more than 5,000 kg (all universal waste categories combined) of universal waste at their location at any time. The LQHUW designation is retained through the remainder of the calendar year in which the 5,000 kg level is met (40 CFR 273.6). An LQHUW may store universal waste up to one year from the date it was generated or received from another handler. The waste may be stored longer than one year if the sole purpose for such practice is to facilitate proper recovery, treatment, or disposal. However, the burden of proof for such practice rests with the handler [40 CFR 273.35(b)].

The other general requirements for an LQHUW include the following (40 CFR part 273, subpart C):

- An LQHUW is not allowed to dispose of or dilute universal waste.
- EPA must be notified of the handler's activity, and an EPA identification number must be obtained.
- Employees must be informed on proper handling and emergency response procedures.
- All releases of universal wastes must be responded to immediately.

- Universal waste may be sent off-site only to a universal waste handler, destination facility, or foreign source.
- Self-transported universal waste must comply with universal transporter requirements.
- Shipments of universal waste must comply with the Hazardous Materials Transportation Act (see Chapter 4).
- The waste-specific packaging and labeling requirements must be followed.
- Before shipping universal waste to an authorized handler, assurance must be received that the shipment will be accepted.
- If a shipment is rejected, the originator must receive the waste.
- If a shipment is received and it is discovered that the shipment contains non-universal hazardous waste, the EPA Regional Office must be notified immediately.
- A record must be kept of each shipment of universal waste received, including the name and address of the originator, quantity by waste type, and date received.
- Records (e.g., logs, invoices, manifests, bills of lading) must be kept for each off-site shipment, including the name and address of the receiving facility, quantity by waste type, and date sent.
- All exports are subject to the export requirements in 40 CFR part 262.

Universal Waste Transporters. Universal waste transporters must comply with the following requirements (40 CFR part 273, subpart D):

- A transporter may not dispose of or dilute universal waste.
- All requirements under the Hazardous Materials Transportation Act must be followed.
- Universal waste may be stored for only 10 days at a transfer facility; otherwise, the transporter becomes a universal waste handler.
- All releases of universal wastes must be responded to immediately.
- Universal waste may be sent off-site only to a universal waste handler, destination facility, or foreign source.
- All exports are subject to the export requirements in 40 CFR part 262.

Destination Facilities. Under the universal waste regulations, a *destination facility* (not to be confused with designated facilities under 40 CFR 262.20) is a facility that treats, disposes of, or recycles a particular category of universal waste, except those management activities described in 40 CFR 273.13(a) and (c) and 40 CFR 273.33(a) and (c). A facility at which a particular category of universal waste is only *accumulated* is not a destination facility for purposes of managing that category of universal waste (40 CFR 273.6).

The owner or operator of a destination facility is subject to the same requirements as a hazardous waste management facility, including the duties of notifying EPA of his or her waste management activities, obtaining an EPA identification number, and obtaining a permit. Destination facilities that do not store universal waste before recycling are subject only to notification and manifesting requirements [40 CFR 273.60(c)(2)].

Destination facilities also must comply with the following requirements (40 CFR part 273, subpart E):

- Universal waste may be sent off-site only to a universal waste handler, destination facility, or foreign source.
- A destination facility may reject a shipment if the shipper is notified and arrangements are made to handle the shipment properly.
- If a shipment is received and it is discovered that the shipment contains non-universal hazardous waste, the EPA Regional Office must be notified immediately.
- A record (e.g., logs, invoices, manifests, bills of lading) must be kept of each shipment of universal waste received that includes the name and address of the originator, the quantity by waste type, and the date received.

3

GENERATORS

RCRA, the Hazardous Materials Transportation Act (HMTA), and Superfund all place the greatest accountability for the environmentally sound management of hazardous wastes on the generator. This chapter outlines the RCRA requirements for each generator category. (The HMTA requirements for generators acting as shippers are described in part in this chapter and more fully in Chapter 4. The liability provisions of Superfund are described in Chapter 13.)

An important difference under RCRA between generators and other entities that manage hazardous waste (excluding transporters) is that, provided a generator complies with the specified conditions for on-site management, that party is not required to obtain a federal RCRA permit. However, this conditional exclusion is limited to managing wastes *generated at the site.* A generator may *not* accept shipments of hazardous waste generated from off-site sources regardless of where the waste was generated (e.g., even if the waste was generated at another facility under the same ownership).

Generators of hazardous waste must comply with the regulations set out in 40 CFR part 262, which include, the following:

- Accurate identification of all waste.
- Proper storage and/or authorized on-site accumulation.
- Adequate preparedness and prevention and emergency procedures.
- Accurate marking and labeling of waste accumulation units.
- Use of the Uniform Hazardous Waste Manifest.
- Compliance with land disposal restrictions.

- Delivery of the waste to a permitted treatment, storage, or disposal facility.
- Proper record keeping and reporting.

The requirements for hazardous waste generators are based on how much hazardous waste is generated in kilograms (kg) at a facility *per calendar month* (mo).[1] The three categories of hazardous waste generators, based on the amount of such waste generated (as shown in Table 3-1), are as follows:

Conditionally Exempt Small-Quantity Generators generate 100 kg/mo or less of hazardous waste or less than 1 kg/mo of acutely hazardous waste.[2]

Small-Quantity Generators generate between 100 and 1,000 kg/mo.

Large-Quantity Generators generate 1,000 kg/mo of hazardous waste or 1 kg/mo or more of acutely hazardous waste.

The unique set of requirements for each generator category is shown in Table 3-2. As expected, the stringency of regulation increases with the amount of hazardous waste generated.

It is important to note that regardless of the regulatory status of a generator's hazardous waste under RCRA, the liability provisions of Superfund (discussed in Chapter 13) do not exclude a generator's hazardous waste based on the amount or management of the waste. Thus, it is in the generator's best interest to ensure that the hazardous waste is managed in the most secure manner possible. If a generator's hazardous waste is ever involved in a release, the generator is potentially liable for all or part of the cleanup costs. (See Chapter 13 for further discussion.)

DEFINITION OF A GENERATOR

A *generator* is any person, by site, whose act or process produces hazardous waste or whose act first causes a hazardous waste to become subject to regulation (40 CFR 260.10). Although the term *site* is not explicitly defined, EPA typically views a property under the control of an owner or operator to be a site.

There are instances in which two or more parties may fit the definition of a generator. An example: An owner of a raw-material product storage tank hires a contractor to clean out the tank residue; the contractor generates the waste, but the tank owner owns the facility, tank, and residue. EPA would like for the parties to select a person to accept the duties of the generator, although in this example EPA would normally define the owner of the tank as the generator (45 *FR* 72026, October 30, 1980). However, EPA may apply the definition of generator to each of these parties because it is the act or process of each of these parties that produces the hazardous waste. EPA has stated (OSWER Directive 9451.01) that it will hold the persons who

[1] One kilogram is equal to 2.2 pounds.

[2] Acutely hazardous waste includes all hazardous wastes with a hazard code of H and all wastes with a hazardous waste number beginning with P.

TABLE 3-1. Hazardous Waste Generator Categories (per calendar month)

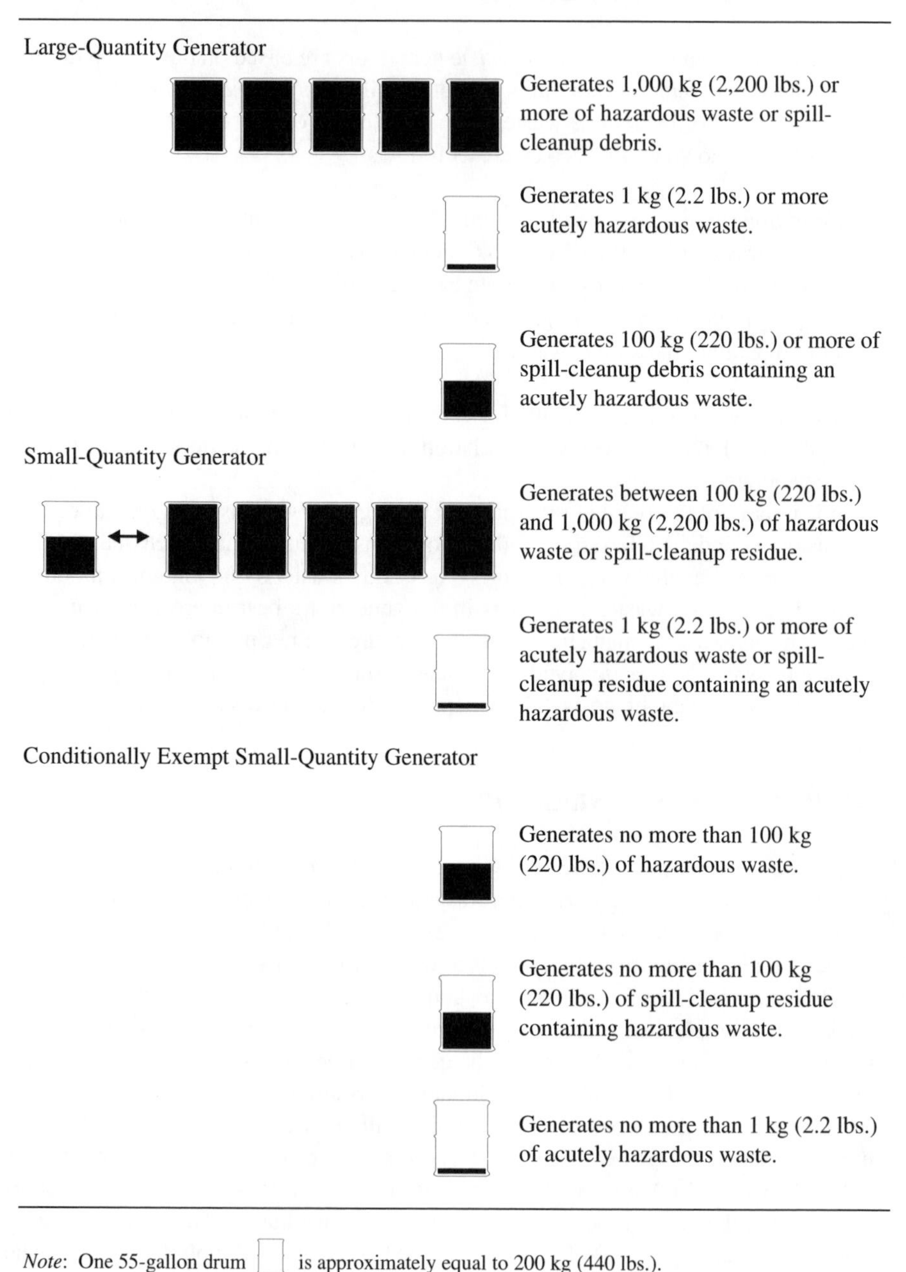

Large-Quantity Generator

Generates 1,000 kg (2,200 lbs.) or more of hazardous waste or spill-cleanup debris.

Generates 1 kg (2.2 lbs.) or more acutely hazardous waste.

Generates 100 kg (220 lbs.) or more of spill-cleanup debris containing an acutely hazardous waste.

Small-Quantity Generator

Generates between 100 kg (220 lbs.) and 1,000 kg (2,200 lbs.) of hazardous waste or spill-cleanup residue.

Generates 1 kg (2.2 lbs.) or more of acutely hazardous waste or spill-cleanup residue containing an acutely hazardous waste.

Conditionally Exempt Small-Quantity Generator

Generates no more than 100 kg (220 lbs.) of hazardous waste.

Generates no more than 100 kg (220 lbs.) of spill-cleanup residue containing hazardous waste.

Generates no more than 1 kg (2.2 lbs.) of acutely hazardous waste.

Note: One 55-gallon drum is approximately equal to 200 kg (440 lbs.).

TABLE 3-2. Requirements for Hazardous Waste Generators[*]

Requirement	Conditionally Exempt Small-Quantity Generators	Small-Quantity Generators	Large-Quantity Generators
Generation Limits per Month	<100 kg nonacutely HW[+] <1 kg acutely HW	100–1,000 kg nonacutely HW <1 kg acutely HW	None
Management of Waste	State-approved or RCRA-permitted facility	RCRA-permitted facility	RCRA-permitted facility
On-site Storage Limits	May accumulate up to 1,000 kg	May accumulate up to 6,000 kg for up to 180 days (270 days if waste will be transported >200 miles)	May accumulate any amount up to 90 days
Storage Require-ments	None specified	Basic requirements with the technical standards in 40 CFR Part 265 for tanks or containers	With some minor exceptions, full compliance with 40 CFR Part 265 for containers, tanks, containment buildings, and drip pads
EPA ID Number	Not required	Required	Required
Manifest	Not required	Required	Required
Manifest Exception Report	Not required	Required after 60 days	Required after 45 days
Biennial Report	Not required	Not required	Required
Personnel Training	Not required	Basic training required	Full training required
Emergency Procedures	Not required	Basic procedures and a designated emergency coordinator	Full procedures required
Contingency Plan	Not required	Basic plan required	Full plan required

Notes: *Individual states may have more stringent requirements. Table 10-2 summarizes the RCRA requirements for all states.

+HW = hazardous waste

generated the waste jointly and severally liable even though the persons may not be the owners or operators of the facility. Thus, all persons who may fit the regulatory definition of a generator are potentially liable as the generator, even though they may not have accepted the duties of the generator.

Requirements for All Generators

A generator of any amount of waste must determine if the waste is hazardous (40 CFR 261.11) by following the protocol described previously in Chapter 2 and in Figure 2-1. Briefly, the generator must first determine if the waste is a listed hazardous waste, and, if not, if it exhibits a characteristic of hazardous waste. This determination may be made either by analyzing the waste using approved methods or by applying knowledge of the characteristics of the waste in light of the materials and processes used [40 CFR 262.11(c)]. However, EPA does not recognize a defense of a good-faith mistake in identifying one's waste (45 *FR* 12727, February 26, 1980). If the waste is determined to be hazardous, the generator must refer to 40 CFR parts 264, 265, and 268 to determine if the waste's management is restricted in any way. These management restrictions focus on land disposal restrictions, which are discussed later in this chapter [40 CFR 262.11(d)].

Counting Hazardous Waste

To determine the appropriate generator category and the associated requirements, the quantity of hazardous wastes generated must be counted on a calendar-month basis. A generator must tally the weight of all *countable* hazardous wastes generated in a calendar month.

Hazardous wastes are countable if

- They are generated and accumulated on-site for any period of time before their subsequent management.
- They are recyclable wastes subject to 40 CFR part 266.
- They are placed directly into an on-site Subtitle C regulated treatment, storage, or disposal unit.
- They are generated from a product storage tank or manufacturing process unit [40 CFR 261.5(c)].

Hazardous wastes do *not* have to be counted if

- They are specifically excluded from regulation. (These wastes include spent lead-acid batteries, used oil, and commercial chemical products sent off-site for reclamation.)
- They are nonacutely hazardous waste remaining in an empty container.
- They are managed in an elementary neutralization unit, a totally enclosed treatment unit or a wastewater treatment unit, as these units are defined in 40 CFR 261.10.

- They are discharged directly to a publicly owned treatment works (POTW) without being accumulated or treated before discharge.
- They are produced from the on-site treatment of previously counted hazardous waste. (For example, a facility generates 995 pounds of methyl ethyl ketone (MEK), then distills the spent MEK in the facility's on-site distillation unit. The distillation unit, after the distillation process, generates 15 pounds of waste MEK still bottoms. However, because the facility already counted the spent MEK before it was treated, the still bottoms do not have to be counted, as, technically, this is still the same waste.)
- They are universal waste (e.g., batteries, certain pesticides, mercury-containing lamps) [40 CFR 261.5(c)].

Changing Generator Categories

Because a generator is subject to regulations that correspond to the amount of hazardous waste generated *per calendar month,* the generator's category may change monthly. An example: A generator normally produces less than 100 kg/mo; however, once a year, the generator conducts a facility-wide product storage and manufacturing process unit cleanout. This annual operation generates more than 1,000 kg during that month. The generator is considered a large-quantity generator for that month and must manage the waste in accordance with the requirements for large-quantity generators. After that month, if we assume that less than 100 kg of hazardous waste is generated, the generator may revert back to the conditionally exempt small-quantity generator category.

CONDITIONALLY EXEMPT SMALL-QUANTITY GENERATORS

A conditionally exempt small-quantity generator (CESQG) is a generator that, in a calendar month, generates no more than 100 kg of nonacutely hazardous waste or no more than 1 kg of an acutely hazardous waste [40 CFR 261.5(a)]. (Acutely hazardous wastes include the P-code wastes and the following wastes with a hazard code H: F020, F021, F022, F023, F026, and F027.)

According to 40 CFR 261.5, a CESQG is excluded from 40 CFR Parts 262 through 270 if the waste is identified to determine if it is hazardous and the generator does not accumulate at any one time hazardous wastes in quantities of 1,000 kg or more. In addition, a CESQG may treat or dispose of the hazardous waste either on-site or off-site. In either case, the units must

- be permitted or have interim status under 40 CFR part 270;
- be authorized to manage hazardous waste by a state with an authorized program under 40 CFR part 271;
- be permitted, licensed, or registered by a state to manage municipal solid waste and, if managed in a municipal solid waste landfill, be subject to 40 CFR part 258;

- be permitted, licensed, or registered by a state to manage nonmunicipal non-hazardous waste subject to 40 CFR 257.5 through 257.30;
- beneficially use, reuse, or legitimately recycle or reclaim the hazardous waste; and
- be a universal waste handler (for universal waste regulated under Part 273) or a destination facility regulated under 40 CFR part 273.

If a CESQG mixes CESQG-excluded hazardous waste with a nonhazardous waste, the resultant mixture will still retain the exclusion [40 CFR 261.5(h)]. However, if a CESQG mixes excluded waste with nonexcluded hazardous waste, and the resultant mixture is 100 kg or greater, the exclusion is no longer retained and the generator is subject to full regulation as a large-quantity generator [40 CFR 261.5(i)].

The CESQG's waste itself, not just the generator, is excluded from regulation. This means that if a transporter collects waste from multiple CESQGS, the waste itself will still be excluded. However, such waste loses its exclusion when it is mixed with non-CESQG hazardous waste. Additionally, this exclusion does not apply when a CESQG mixes a hazardous waste with any other waste that is or will be burned in an industrial furnace or boiler for energy recovery [40 CFR 261.5(b)].

If a conditionally exempt small-quantity generator accumulates 1,000 kg or more of hazardous waste on-site, the generator loses the CESQG exclusion, and all of the accumulated waste is subject to the regulations for small-quantity generators in 40 CFR 262.34(d) [40 CFR 261.5(g)(2)]. There are no federal technical standards specified for accumulation units used by CESQGs.

SMALL- AND LARGE-QUANTITY GENERATORS

Small-quantity generators (SQGs) and large-quantity generators (LQGs) are subject to similar provisions except that LQGs are more stringently regulated under each provision as shown previously in Table 3-2. The remainder of this chapter describes the requirements common to both SQGs and LQGs. At the end of each section, the additional requirements for LQGs are described. The applicable provisions for both classes of generators are as follows:

- EPA identification number
- On-site accumulation
- Personnel training
- Preparedness and prevention
- Contingency plan
- Emergency procedures
- Pretransport requirements
- Manifest system

- Land disposal restrictions
- Waste minimization
- Biennial reporting
- Record keeping

EPA Identification Number

Small- and large-quantity generators may not treat, store, or dispose of hazardous waste, or offer it for transportation, without an EPA identification number (40 CFR 262.12). These 12-digit identification numbers are obtained from the state or EPA Regional office by submitting EPA Form 8700-12, *Notification of Hazardous Waste Activity.* EPA assigns an identification number to each generation site. Thus, if the facility relocates, the generator must obtain a new identification number for the new site. If the ownership or operational control changes for a facility, a new identification number is not required, although it is recommended that the facility request a new number.

Note: There are no additional requirements for large-quantity generators for the EPA identification number requirement.

On-site Accumulation

Both small- and large-quantity generators are allowed to accumulate waste on-site without a permit if the following provisions are met. This authorization for on-site accumulation also includes treatment if it occurs in a tank, container, or containment building and if the generator maintains compliance with 40 CFR 262.34 (March 24, 1986, 51 *FR* 10168, and August 18, 1992, 57 *FR* 37194). (Some states, such as California and Rhode Island, are more stringent and do not allow on-site treatment by generators without a permit.) All treatment residue resulting from this practice is subject to full regulation under RCRA.

The remainder of this section addresses accumulation time limits and authorized accumulation units.

Accumulation Time Limits. The date on which accumulation began must be clearly marked on individual containers and tanks (51 *FR* 10160, March 24, 1986). The accumulation period starts the minute the first drop hits the tank or drum, except for satellite accumulation containers discussed in the next section (47 *FR* 1250, January 11, 1982). Each container and tank must also have the words HAZARDOUS WASTE clearly marked on the unit. (See Chapter 4, "Shipping and Transportation," to ensure that the markings used satisfy both RCRA and HMTA and to ensure that efforts are not duplicated.) By the end of the time period, the hazardous waste must be sent to a designated facility authorized to accept and manage hazardous waste.

Small-quantity generators may accumulate hazardous waste on-site without a permit or interim status for up to 180 days (or 270 days if the waste will be transported over 200 miles), provided that the generator does not store 6,000 kg or more of hazardous waste at any time. Should the weight limit be exceeded, the generator would need a RCRA storage permit, which is discussed in Chapters 5 and 7.

Large-quantity generators may accumulate hazardous waste on-site for up to 90 days without a permit or interim status regardless of the amount.

Small- and large-quantity generators may receive up to a 30-day extension to the 90-, 180-, or 270-day accumulation time limit if uncontrollable and unforeseen circumstances will cause them to accumulate waste longer than the allowed time period [40 CFR 262.34(b)]. This time extension may be obtained from the state or EPA Regional Office.

Satellite Accumulation. Small- and large-quantity generators may accumulate up to 55 gallons of hazardous waste (or 1 quart of acutely hazardous waste) at "satellite" areas, which are accumulation areas subject to reduced regulatory requirements (subject to state restrictions). A *satellite accumulation area* is a place where wastes are generated in an industrial process or laboratory and are initially accumulated prior to removal to a central area. A satellite accumulation area must be under the control of the operator of the process generating the waste (49 *FR* 49569, December 20, 1984).

Note: There is no limit as to the number of satellite accumulation areas a facility may have. However, establishing multiple satellite accumulation areas adjacent to each other to circumvent the 90-day accumulation requirements is specifically prohibited. In addition, a collection of satellite accumulation areas must be at the same facility (see 40 CFR 260.10 for definition of facility). Thus, hazardous waste may not be accumulated at a satellite area and transported across town to a central accumulation facility (even if under the same ownership) unless the accumulation facility has a RCRA permit and the shipment is manifested.

To qualify for this provision, the following requirements must be met [40 CFR 262.34(c)]:

- Containers must be in good condition and not leak.
- The accumulated waste must be compatible with the container.
- The accumulation container must always be closed except when it is necessary to add or remove waste.
- The accumulation container must be marked with the words HAZARDOUS WASTE or with other words that identify the contents.
- Any hazardous waste generated in excess of the 55-gallon/1-quart limit must be moved to the central hazardous waste accumulation area (i.e., a permitted storage area or a conforming 90-day accumulation area) within 3 days.
- Containers holding hazardous waste generated in excess of the 55-gallon/1-quart limit must be marked with the date the excess amount began accumulating, which in turn initiates the 90/180-accumulation time limit.

Thus, provided the 55-gallon/1-quart limit is never exceeded (e.g., periodic transfers of accumulated waste to the central accumulation area occur), the last two bullets do not need to be addressed, which further reduces the regulatory requirements.

Accumulation Units. Small- and large-quantity generators may accumulate hazardous waste in containers, tanks, drip pads, and containment buildings subject to specific conditions.

Containers. Small-quantity generators accumulating hazardous waste in containers must comply with Subpart I of 40 CFR Part 265, except for 40 CFR 265.176 (which requires ignitable and reactive wastes to be placed at least 50 feet from the facility's property boundary) and 40 CFR 265.178 (which deals with air emission standards). (Large-quantity generators must comply with all of Subpart I, including 40 CFR 265.176 and 178.) The general requirements for containers under Subpart I include the following:

- Containers must be in good condition. If a container leaks or is not in good condition, its contents must be transferred to a sound container.
- Containers must be compatible with the stored contents. Containers holding hazardous waste that is incompatible with other waste or other materials must be protected or physically separated.
- Containers must always be closed unless their contents are being transferred. In addition, they must always be handled in such a way as to prevent rupture or leaks.
- Container storage areas must be inspected at least weekly for signs of deterioration, corrosion, or leaks.

Large-quantity generators must also comply with the air emission standards contained in 40 CFR Part 265, subparts AA, BB, or CC. The requirements that must be followed depend on the organic concentration of the waste. Hazardous waste containing at least 10 ppm organics requires compliance with Subpart AA, waste with at least 10 percent organics by weight requires compliance with Subpart BB, and waste with an average volatile organic concentration of 500 ppm or greater requires compliance with Subpart CC. Containers with a capacity of 0.1 m^3 (approximately 26 gallons) or less are excluded from the air emission control requirements.

A large-quantity generator may place hazardous waste with a high organic concentration in a container

- that is equipped with a vapor leak-tight cover; and
- that is designed with a capacity less than or equal to 0.46 m^3 (approximately 119 gallons) and equipped with a cover in compliance with HMTA requirements under 49 CFR Part 178; or
- that is attached to or forms part of any truck, trailer, or railcar, and must have demonstrated organic-vapor tightness within the preceding 12 months.

In addition, each container must be maintained in a closed, sealed position at all times except when necessary to add or remove waste, inspect or repair equipment located inside the container, or vent gases or vapors to a control device.

Tanks. There are different tank requirements for small- and large-quantity generators.

Small-Quantity Generator Requirements. Small-quantity generators accumulating hazardous waste in a tank must comply with the following requirements as outlined in 40 CFR 262.34(d)(3):

- Treatment must not generate any extreme heat, explosions, fire, flammable fumes, mists, dusts, or gases; damage the tank's structural integrity; or threaten human health or the environment.
- Hazardous wastes or treatment reagents that may cause rupture, corrosion, or structural failure must not be placed in the tank.
- At least 2 feet of freeboard (the distance between the top of the tank and the surface of the contents) must be maintained in an uncovered tank unless sufficient overfill containment capacity (i.e., capacity equals or exceeds volume of top two feet of tank) is supplied.
- Continuously fed tanks must have a waste-feed cutoff or bypass system.
- Ignitable, reactive, or incompatible wastes must not be placed into a tank unless these wastes are rendered nonignitable, nonreactive, or compatible.
- At least once each operating day, the waste-feed cutoff and bypass systems, monitoring equipment data, and waste level must be inspected.
- At least weekly, the construction materials and the surrounding area of the tank system must be inspected for visible signs of erosion or leakage.
- Incompatible wastes may not be placed in the same tank.
- The applicable air emission requirements under 40 CFR part 265, subparts AA, BB, or CC are followed.
- At closure, all hazardous wastes must be removed from the tank, containment system, and discharge control systems.

Large-Quantity Generator Requirements. Large-quantity generators accumulating hazardous waste in tanks must comply with most of the provisions of Subpart J of 40 CFR part 265 (see Chapter 6), including the following:

- A one-time assessment of the tank system, including results of an integrity test.
- Installation standards for new tank systems.
- Design standards, including an assessment of corrosion potential.
- Secondary containment phase-in provisions.
- Periodic leak testing if the tank system does not have secondary containment.
- Air emission controls.
- Closure of the tank system.
- Response requirements regarding leaks, including reporting to the EPA Regional Administrator the extent of any release and requirements for repairing or replacing of leaking tanks.

However, owners or operators of 90-day accumulation tanks are not required to prepare closure or post-closure plans or contingent closure or post-closure plans, maintain financial responsibility, or conduct waste analysis and trial tests.

Pertaining to air emission controls, large-quantity generators storing hazardous waste in tanks must place their waste into one of the following units:

- tanks equipped with a cover (fixed roof) vented to a control device;
- tanks equipped with a fixed roof and an internal floating roof;
- tanks equipped with an external floating roof; or
- pressure tanks designed to operate as a closed system with no detectable emissions at all times.

Large-quantity generators using tanks for recycling or treatment, which are defined as strippers, distillation units, fractionation units, thin-film evaporation units, solvent extraction units, or air strippers, must reduce or destroy organics from all such devices at a facility to an aggregate maximum of 3.0 lbs./hour or 3.1 tons/year, or by 95 percent by weight [40 CFR 262.34(a)(1)(ii)]. See 40 CFR 265.1032 for further details.

Drip Pads. Drip pads are hazardous waste management units unique to the wood preserving industry. A hazardous waste drip pad is a nonearthen structure consisting of a curbed, free-draining base designed to convey excess preservative drippage, precipitation, and surface water runon from treated wood operations to an associated collection system. Both small- and large-quantity generators are subject to the same requirements for drip pads. Generators are allowed to temporarily accumulate hazardous waste on drip pads provided that

- The unit conforms to the technical standards in 40 CFR part 265, subpart W.
- Written procedures are developed to ensure that wastes are removed from the pad and collection system at least once every 90 days.
- Records are kept documenting that those procedures are followed.

Drip pads used for temporary accumulation of wastes by a generator are exempt from all requirements in 40 CFR part 265, subparts G and H, except for those in 40 CFR 265.111 and 265.114, which relate to the closure performance standard and disposal or decontamination of all equipment, structures, and soils.

The technical requirements for drip pads include the following criteria:

- Drip pads must be designed and constructed of nonearthen materials with enough structural strength to prevent failure of the unit under the weight of the waste, preserved wood products, personnel, and any moving equipment.
- A raised curb or berm must be constructed around the perimeter of the pad.
- The pad's surface must be sloped toward a collection unit, such as a sump.

- Unless the pad is protected from precipitation, a stormwater runon and runoff control system must be used.
- The pad's surface must be maintained so that it remains free of cracks, gaps, corrosion, or other deterioration.
- The pad must be sealed, coated, or covered with an impermeable material.
- Drip pads must also be cleaned frequently to allow for weekly inspections of the entire drip pad surface without interference from accumulated wastes and residues.
- Drippage and precipitation must be emptied into a collection system as often as necessary to prevent waste from overflowing the curb around the perimeter of the unit.
- Drip pads must be inspected to ensure the unit is protective of human health and the environment and thus fit for continued use.
- Drip pads must be inspected weekly and after storms to ensure proper operation and to detect any deterioration or leaks. If a drip pad shows any deterioration, the affected portion of the unit must be removed from service.

Containment Buildings. Generators may accumulate hazardous waste in a containment building for up to 90 days without a federal permit. A *containment building* is a completely enclosed structure (i.e., four walls, a roof, and a floor) that houses an accumulation of noncontainerized waste primarily bulky waste (e.g., slag, spent potliners, contaminated debris). The maximum generator accumulation time period in containment buildings is 90 days. Small-quantity generators using containment buildings must also comply with the large-quantity generator requirements for personnel training, development of a full contingency plan, and biennial reporting. Further, SQGs may accumulate hazardous waste in containment building for only 90 days, instead of the standard 180/270 time limit. Containment buildings must comply with the following:

- The technical standards in 40 CFR part 265, subpart DD.
- Certification from a professional engineer that the building conforms to the design standards specified in 40 CFR 265.1101.
- A written description of the procedures used to ensure that wastes remain in the unit for no more than 90 days.
- Documentation that these procedures are followed.

Generators accumulating hazardous waste in containment buildings are exempt from most of the closure and financial assurance requirements in 40 CFR part 265, subparts G and H. However, after the useful life of the building has expired, the building must be closed in compliance with 40 CFR 265.111 and 265.114 (i.e., closure performance standard and disposal or decontamination of equipment, structures, and soils).

The technical standards for containment buildings include the following:

- The containment building must be completely enclosed with four walls, a floor, and a roof, which must be constructed of man-made materials possessing sufficient structural strength to withstand movement of wastes, personnel, and heavy equipment within the unit.
- Dust control devices must be used as necessary to prevent fugitive dust from escaping through building exits.
- All surfaces in the containment building that come into contact with waste during treatment or storage must be chemically compatible with that waste.
- Incompatible wastes that could cause unit failure may not be placed in containment buildings.
- If the containment building is used to manage hazardous wastes containing free liquids, the unit must be equipped with a liquid collection system, a leak detection system, and a secondary barrier.
- The floor of the unit must be free of significant cracks, corrosion, or deterioration.
- Containment buildings must be inspected at least once a week, and all results must be recorded.

If a release is discovered during an inspection, the generator must remove the affected portion of the unit from service and take all appropriate steps for repair and release containment. EPA must be notified of the discovery and of the proposed schedule for repair. Upon completion of all necessary repairs and cleanup, a qualified, registered professional engineer must verify that the plan submitted to EPA was followed.

Personnel Training

Small-quantity generators need only ensure that all employees are thoroughly familiar with proper waste handling and emergency procedures relevant to their responsibilities during normal facility operations and emergencies [40 CFR 262.34(d)(5)(iii)]. However, if the small-quantity generator has an emergency response team, that generator must also comply with the OSHA hazardous waste operations training requirements of 29 CFR 1910.120(p)(8). (See also the training requirements under HMTA, described in Chapter 4; the training requirements under both laws may be combined.)

Large-quantity generators must establish a formal training program for appropriate facility personnel to reduce the potential for errors that might threaten worker and public health and the environment by ensuring that facility personnel acquire expertise in their assigned areas, and also in emergency response procedures. The training program must also include training to ensure facility compliance with all applicable regulations. The program must contain an initial training program and annual updates (40 CFR 265.16). The specific content and format of the program are unspecified in the regulations, but the following subjects must be addressed:

- Procedures for using, inspecting, repairing, and replacing facility emergency and monitoring equipment.
- Key parameters for automatic waste-feed cutoff systems.
- Communications or alarm systems.
- Response to fires or explosions.
- Responses to groundwater contamination incidents.
- Shutdown of operations.

EPA accepts the use of on-the-job training as a substitute for, or supplement to, formal classroom instruction (45 *FR* 33182, May 19, 1980). However, the job title, employee name, training program content, schedule of training, and techniques used for training must be described in the training records, which must be maintained at the generator's facility [40 CFR 265.16(d)]. (See also the training requirements under HMTA, described in Chapter 4; the training can be combined.) In addition, if the generator has an emergency response team, the generator must also comply with the OSHA hazardous waste operations training requirements of 29 CFR 1910.120(p)(8).

Preparedness and Prevention

The preparedness and prevention requirements for small- and large-quantity generators are the same. These requirements state that a facility must be operated and maintained in a manner that will minimize the possibility of any fire, explosion, or unplanned sudden or nonsudden release [40 CFR 262.34(d)(4) and 265.31].

Each facility must have certain equipment to respond to emergencies. The equipment includes an alarm system, a communication device to contact emergency personnel (e.g., telephone, radio), portable fire extinguishers, fire control equipment, and an adequate firefighting water supply system in the form of hydrants, hoses, or an auto-sprinkler system. This equipment must be routinely tested and maintained in proper working order (40 CFR 265.32). Whenever hazardous waste is being poured, mixed, spread, or otherwise handled, all personnel involved in the operation must have immediate access to an internal alarm or emergency communication device. If there is ever just one employee at the facility while it is in operation, that employee must have immediate access to a device (e.g., radio, telephone) capable of summoning outside emergency assistance (40 CFR 265.34).

There must be adequate aisle space to allow the unobstructed movement, deployment, and evacuation of emergency equipment and personnel to any area of the facility (40 CFR 265.35). Although the regulations do not specify the aisle space, it should be determined upon consultation with appropriate emergency organizations.

The facility must make prior arrangements with local emergency organizations (e.g., hospitals, local emergency planning committee, hazmat team, police, and fire departments) and personnel for an emergency response. The arrangements should include notification of the types of waste handled, a detailed map of the facility, a list of facility contacts, and specific agreements with various state and local emergency response organizations. Refusal of any state or local authorities to enter into such arrangements must be documented and maintained in the facility's files (40 CFR 265.37).

Contingency Plan

Only large-quantity generators are required to have a contingency plan. (The requirements of 40 CFR 262.34(a)(4) reference the generator to comply with the requirements of subpart D of 40 CFR part 265.) The contingency plan must be designed to minimize hazards in the case of a sudden or nonsudden release, fire, explosion, or similar emergency. As shown in Table 3-3, the plan must include a description of actions that will be undertaken by facility personnel, a detailed list of emergency equipment with locations, and evacuation procedures.

On June 5, 1996 (61 *FR* 28642), the National Response Team announced the Integrated Contingency Plan (ICP). The intent of the ICP is to provide a mechanism for consolidating multiple plans (e.g., RCRA, SPCC, OSHA's HAZWOPER) under various federal statutes into one functional emergency response plan. Thus, instead of preparing multiple emergency response plans, one "super" emergency response plan, following the ICP guidance, may be prepared instead. Table 3-4 contains the basic elements of the ICP as suggested by the National Response Team.

The provisions of the plan must be carried out immediately whenever there has been a fire, explosion, or release of hazardous waste or constituents. A copy of the contingency plan must be maintained at the facility and submitted to all local police, fire, hospital, and emergency response teams. The plan must also include a list of all available emergency equipment at the facility, including the equipment's location and physical description and a brief outline of its capabilities (40 CFR 265.52 and 53).

Emergency Procedures

There are different emergency procedure requirements for small- and large-quantity generators. Such procedures establish a set of minimum response actions to a facility emergency.

Small-Quantity Generator Requirements. At all times there must be at least one employee either at the facility or on call (i.e., able to respond to an emergency by reaching the facility within a short period) who is responsible for coordinating all emergency response measures. This employee, deemed the emergency coordinator [40 CFR 262.34(d)(5)(i)], must respond to any emergencies that may arise and, if appropriate, institute the following measures [40 CFR 262.34(d)(5)(iv)]:

- Contact the fire department and/or attempt to extinguish the fire.
- For any spill, contain the flow and commence cleanup wherever possible.

Table 3-3. Basic Elements of the Hazardous Waste Contingency Plan

Component	Requirements
General Information	Company name, address, telephone number, owner, and EPA identification number. Brief description of operations involving hazardous waste.
Primary Emergency Coordinator	Name, address, home and office telephone numbers.
Secondary Emergency Coordinators	Names, addresses, home and office telephone numbers.
Arrangements with Emergency Response Organizations	List of organization for which arrangements have been made. Name, address, and telephone number of point of contacts.
Emergency Equipment	List and describe available emergency equipment, its capabilities, and its exact location.
Emergency Response Personnel	Job descriptions, duties, and responsibilities of each employee assigned as emergency response personnel.
Evacuation Plan	Conditions necessitating evacuation. Signals to initiate evacuation. Evacuation paths and setback distances. Notification of special populations (e.g., schools)
Medical Services	Availability and capabilities of on-site and local medical services.
Emergency Response Personnel	General procedures to follow in responding to various emergencies (e.g., spills, releases, fires, explosions, vehicle accidents).

- For any fire, explosion, or release that meets a Superfund or EPCRA reportable quantity (see Chapter 12) or for a release that threatens human health or the environment, notify the National Response Center (1-800-424-8802).

According to 40 CFR 262.34(d)(5)(ii), the following emergency information must be posted next to facility telephones:

- The name and telephone number of the designated emergency coordinator.
- The telephone number of the fire department and appropriate emergency response organizations.
- The location of fire extinguishers, spill-control equipment, and fire alarms.

Large-Quantity Generator Requirements. In accordance with 40 CFR 262.34(a)(4), there must be at least one employee on-site, or close by and on call, who is the

TABLE 3-4. Suggested Elements of an Integrated Contingency Plan

ICP Element	RCRA Citation(s) (40 CFR)	Applicability of Other Statutes
Section I—Plan Introduction Elements		
Purpose and Scope of Plan Coverage	264/265.51, 264/ 265.52(a)	D,E
Table of Contents	NSR	A,B,C
Current Revision Date	NSR	A,B
General Facility Information	NSR	A,C
• Facility name	NSR	A,B
• Owner/operator/agent	NSR	A,B,C
• Physical address and directions	NSR	A,B,C
• Mailing address	NSR	A,B,C
• Key contacts for plan development	NSR	D,E
• Facility phone number	NSR	A,B
• Facility fax number	NSR	B
Section II—Care Plan Elements		
Discovery	NSR	A,B,C,D,E,F
Initial Response	NSR	A,B,C,D,E,F
• Procedures for internal and external notifications	264/265.52(d), 264/265.55, 264/265.56(a)(1) 264/265.56(d)(1)	A,B,C,D,E,F
• Establishment of a response management structure	264/265.37, 264/ 265.52(c)	A,B,C,E
• Preliminary assessment	264/265.56(b) and (c)	A,B,C,D,E
• Establishment of objectives and priorities for response	264/265.52(e)	A,B,C,D,E
• Implementation of tactical plan	264/265.52(e)	A,B,C,E
• Mobilization of resources	264/265.52(e)	A,B,C,D,E,F
Sustained Actions	NSR	A,B,E,F
Termination and Follow-up Actions	264/265.56(I)	
Section III—Annexes		
Facility and Locality Information	NSR	A,B,C
• Facility maps	NSR	A,C
• Facility drawings	NSR	A,B,C
Facility description/layout	NSR	A,B,C,E

continued

TABLE 3-4. Continued

ICP Element	RCRA Citation(s) (40 CFR)	Applicability of Other Statutes
Notification	264/265.52(d), 264/ 265.56(a)(1), 264/ 265.56(d)(1)	A,C,D,E,F
• Internal	NSR	A,B,C,D,E
• Community	NSR	A,B,D,E
• Federal and state agency	NSR	A,B,C,E
Response Management Structure	NSR	A,B,C,E
• General	264/265.52(c)	B,E
• Command	264/265.55, 264/ 265.52(f), 264/ 265.56(a)(1)	A,B,C,D,E
• Operations	264/265.56(b), (c), (d), (e), (f), and (h)(1)	A,B,C,D,E
• Planning	264/265.56(g), (h)(1)	A,B,C,D,E,F
• Logistics	NSR	A,B,C,D,E,F
• Finance/procurement/ administration	264/265.52(e)	A,B,C,E
Incident Documentation	NSR	D,E
• Postaccident investigation	264/265.56(j)	D,E,F
• Incident history	NSR	A,D,F
Training and Exercises/Drills	NSR	A,B,C,D,E,F
Response Critique, Plan Review, and Modification Process	264/265.54	A,B,C,D,E,F
Prevention	NSR	E

Explanation of Codes

NSR = Not specifically required

A = EPA's Oil Pollution Prevention Regulations (SPCC) (40 CFR part 112)

B = USCG's Facility Response Plans (33 CFR part 154)

C = DOT/HMTA Facility Response Plan (49 CFR part 194)

D = OSHA's Emergency Action Plan (29 CFR 1910.38(a) and .119)

E = OSHA's HAZWOPER (29 CFR 1910.120)

F = EPA's CAA Risk Management Plan (40 CFR part 68)

Source: 61 *FR* 26850, June 5, 1996

designated emergency coordinator.[3] The contingency plan must list current names, addresses, and phone numbers of all persons qualified as facility emergency coordinators. Whenever there is an imminent or actual emergency situation, the emergency

[3] This section requires an LQG to follow the requirements in Subpart D of 40 CFR part 265.

coordinator must immediately activate the facility alarm system, notify all facility personnel, and, if needed, notify appropriate state or local agencies (e.g., Local Emergency Planning Committee, police and fire departments, hazmat team).

In the event of a release, fire, or explosion, the emergency coordinator must identify the source, character, and amount of materials involved. If any hazardous waste containing a Superfund hazardous substance in an amount that meets or exceeds a reportable quantity is released into the environment, the owner or operator must contact the National Response Center (1-800-424-8802) immediately (40 CFR 265.56). (See Chapter 12 for further information on reportable quantities.)

Immediately after the incident, the emergency coordinator must provide for the treatment, storage, or disposal of any material contaminated as a result of the emergency [40 CFR 265.56(g)]. The owner or operator must report to EPA within 15 days any incident that requires the implementation of the contingency plan (40 CFR 265.560).

Pretransport Requirements

The pretransport requirements, in essence, compel generators to comply with the regulations under the Hazardous Materials Transportation Act (HMTA) for transporting hazardous materials (40 CFR part 262, subpart C). These provisions, which are described in Chapter 4, require proper

- classification,[4]
- packaging,
- labeling,
- marking,
- placarding.

Note: The pretransport requirements are the same for small- and large-quantity generators.

Manifest System

A generator that either transports hazardous waste off-site or offers it for off-site transportation must use the Uniform Hazardous Waste Manifest (EPA Form 8700-22) [40 CFR 262.20(a)].

Note: The manifest requirements are the same for small- and large-quantity generators except for manifest exception reporting and the contractual relationship provision.

The Uniform Hazardous Waste Manifest is the result of a joint program between DOT and EPA. This manifest gives states only limited rights to require information beyond the federal requirements. The standard federal manifest has white numbered boxes and shaded lettered boxes. The shaded lettered boxes are for optional infor-

[4] The classification system under HMTA is different from RCRA. However, all information required to identify a waste as hazardous under RCRA will assist in the transportation classification process.

mation that a state may require; the white numbered boxes must be filled out by all users of a manifest. A transporter must comply only with the requirements in the state of origin and the consignment (destination) state. The transporter is not bound to comply with the state manifest requirements of the states through which the transporter travels.

The manifest, a one-page form with several carbon copies for the participants in the shipment, must identify the type and quantity of waste, the generator, the transporter, and the facility to which the waste is being shipped. The manifest must accompany the waste wherever it travels, as shown in Figure 3-1.

Each individual involved in a shipment must sign the manifest and keep one copy. When the waste reaches its final destination, the owner or operator of the designated facility signs the manifest and returns a copy to the generator to confirm arrival. Each person involved in the movement, storage, or receipt of hazardous waste requiring a manifest must retain a copy of that manifest for at least three years

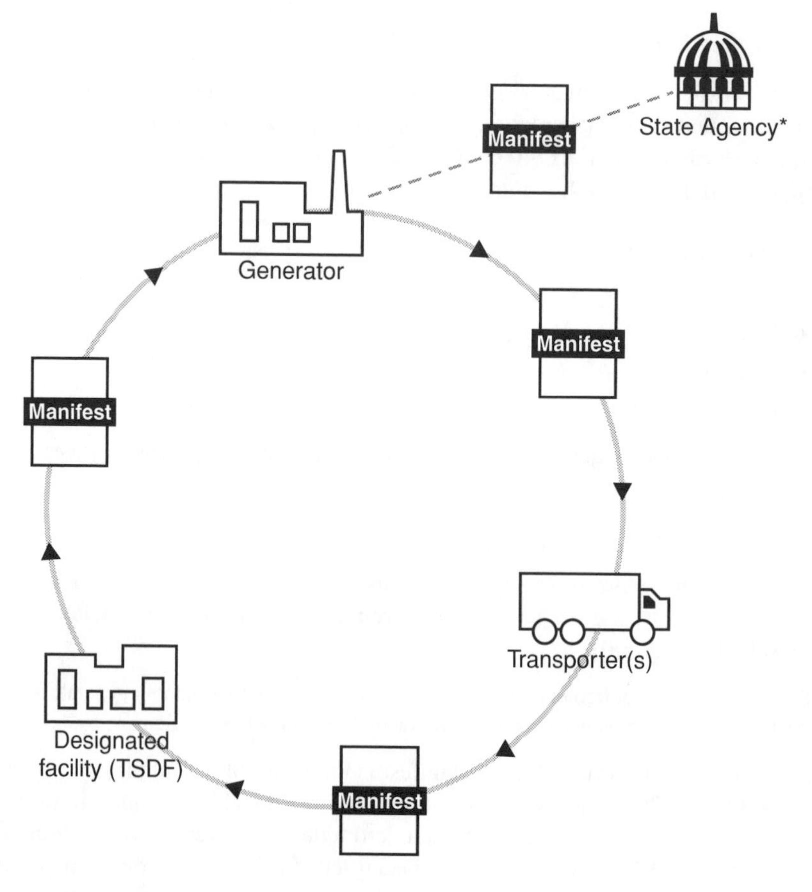

Figure 3-1. The Uniform Hazardous Waste Manifest Cycle for Off-Site Shipments

[40 CFR 262.40(a)]. Generators do not have to submit a copy of the manifest to EPA; however, some states require that a copy of the manifest be sent to the designated state agency.

The designated facility that signs the manifest accepts the responsibility for that shipment and cannot ship the waste back to the generator or any other facility unless the generator or facility is classified as a designated facility (which must be a facility with interim status or a permit). Although a facility and a transporter may accept responsibility for a shipment, the generator retains indefinite liability under §107 of Superfund.

Contractual Relationship Provision. Under the contractual relationship provision [40 CFR 262.20(e)] of the federal program, small-quantity generators do not have to use the Uniform Hazardous Waste Manifest. This conditional exclusion requires the following:

- The waste is reclaimed under a contractual agreement.
- The contractual agreement specifies the type of waste and frequency of shipments.
- The vehicle used to transport the waste to the recycling facility and to deliver regenerated material back to the generator is owned and operated by the reclaimer of the waste.
- The generator maintains a copy of the reclamation agreement for at least three years after the termination of the agreement.

Manifest Acquisition Protocol. The acquisition protocol for the manifest form requires the generator to obtain the manifest from the consignment (disposal) state. However, if that state does not supply or require the use of a particular form, the manifest must be obtained from the generator's state. If the generator's state does not supply or require the use of a particular form, the generator may obtain the manifest from another source (40 CFR 262.21).

Manifest Exception Reporting. A large-quantity generator who does not receive a copy of the signed manifest from the owner or operator of the designated facility within 35 days after the initial transporter accepted the waste must contact the designated facility to determine the status of the waste. If within 45 days the generator has still not received a copy of the manifest, a manifest exception report must be filed with EPA. (Small-quantity generators have up to 60 days before they must file a manifest exception report.) The manifest exception report consists of a copy of the original manifest, a cover letter explaining the efforts taken by the generator to locate the waste, and the results of those efforts (40 CFR 262.42).

Land Disposal Restrictions

There are land disposal restrictions and prohibitions on all hazardous wastes as will be described in Chapter 8. Although the ultimate responsibility for compliance with land disposal restrictions is placed on land disposal facilities, generators are responsible for determining whether their wastes are subject to and meet these restrictions [40 CFR 262.11(d)]. The generator can base the determination about whether a waste is subject to these restrictions on either knowledge of the waste, testing, or both [40 CFR 268.7(a)(1)].

Note: The requirements for the land disposal restrictions are the same for small- and large-quantity generators.

If a generator determines that the waste is subject to the land disposal restrictions [40 CFR 268.7(a)(1)], but does not meet the treatment levels described in Chapter 8 (and is thus prohibited from land disposal), the generator must notify the designated facility in writing of the treatment standards and prohibition on land disposal [40 CFR 268.7(a)(1)].

If a generator determines that a waste subject to the land disposal restrictions may be disposed on the land without further treatment, each shipment of that waste to a designated facility must have a notice and certification stating that the waste meets all of the applicable treatment standards [40 CFR 268.7(a)(2)]. The signed certification must state the following:

> I certify under penalty of law that I personally have examined and am familiar with the waste through analysis and testing or through knowledge of the waste to support this certification that the waste complies with the treatment standards specified in 40 CFR 268 Subpart D. I believe that the information I submitted is true, accurate, and complete. I am aware that there are significant penalties for submitting a false certification, including the possibility of a fine and imprisonment.

Waste Minimization

Small- and large-quantity generators must certify on each manifest (Item No. 16) that a program is in place to reduce the quantity and/or toxicity of the hazardous waste at that generator's facility. The generator must also describe the waste minimization program that is in place, and the results achieved by that program, in the biennial report (or in many states, annual report).

Note: The requirements for waste minimization are the same for small- and large-quantity generators.

Waste minimization is the reduction, to the extent feasible, of hazardous waste that is generated before its treatment, storage, or disposal. Waste minimization is defined as any source reduction or recycling activity that results in (1) reduction of the total volume of hazardous waste, (2) reduction of the toxicity of hazardous waste, or (3) both. Practices that are considered to be waste minimization include recycling, source separation, product substitution, manufacturing process changes, and the use of less toxic raw materials (58 *FR* 31114, May 28, 1993).

Waste Minimization Program. EPA has stated that an effective waste minimization program should include, as appropriate, each of these elements (58 *FR* 31114, May 28, 1993):

Top management support:

- Waste minimization as company policy
- Specified goals for waste minimization
- Commitment to recommendations
- Designated waste minimization coordinator
- Employee training

Characterization of waste generation:

- Periodic waste minimization assessments
- Comprehensive audits of entire waste process to ensure effective minimization
- Comprehensive analysis

Technology transfer program
Program evaluation

Biennial Report

Large-quantity generators must prepare and submit a copy of a biennial report (EPA Form 8700-13A) by March 1 of each even-numbered year (40 CFR 262.41). Many states require an annual report (see Table 10-2). Information required for the biennial report includes the following:

- The generator's EPA identification number.
- The EPA identification number for each transporter used.
- The EPA identification number for each designated facility where hazardous waste was sent.
- A description and accounting of the quantity of hazardous waste generated.
- A report on the efforts undertaken during the year to reduce the volume and toxicity of waste generated and the reductions achieved by the generator's waste minimization program in comparison to previous years.

Recordkeeping

Small- and large-quantity generators must maintain various records. Although each of the recordkeeping requirements were discussed previously in this chapter, the following is a consolidated list:

- Biennial reports [40 CFR 262.40(b)]
- Contingency plan [40 CFR 262.34(c) and (d)]
- Emergency agreements with local authorities [40 CFR 262.34(c) and (d)]
- Land disposal restrictions (40 CFR 268.7)

- Manifest system [40 CFR 262.40(a)]
- Manifest exception reports [40 CFR 262.40(b)]
- Personnel training documentation [40 CFR 262.34(c) and (d)]
- Waste analyses/test results (or other bases used for hazardous waste determination) [40 CFR 262.40(c)]

In general, each of the above records must be maintained for at least three years. However, some records (e.g., contingency plan) must be kept during the active life of the facility.

IMPORTS AND EXPORTS OF HAZARDOUS WASTE

Currently there are four sets of requirements for exports and imports of hazardous waste. These requirements depend on the country of destination/origin. The requirements are as follows:

- Hazardous wastes exported to or imported from Canada are subject to the U.S./Canada bilateral agreement.
- Hazardous wastes exported to or imported from Mexico are subject to the U.S./Mexico bilateral agreement.
- All exports to and imports from countries that are members of the Organization for Economic Cooperation and Development (OECD), excluding Mexico and Canada (which are both members), must comply with the provisions of the OECD decision on transfrontier shipments of hazardous waste.[5] This decision has been codified into 40 CFR part 262, subpart H.
- All other exports to and imports from countries not covered under the above remain subject to 40 CFR part 262, subparts E and F, respectively.

It is important to note that all imports of hazardous waste into the United States are subject to §13 of the Toxic Substances Control Act, which is discussed in Part III of this book.

OECD-Covered Wastes

On April 12, 1996, EPA published a final rule establishing regulations to implement the Organization for Economic Cooperation and Development (OECD) decision (61 *FR* 16289) that requires member countries to establish regulations for hazardous waste exported to or imported from other member countries for recycling. As

[5] OECD member countries include Australia, Austria, Belgium, Canada, Czech Republic, Denmark, Finland, France, Germany, Greece, Hungary, Iceland, Ireland, Italy, Japan, Luxembourg, Mexico, the Netherlands, New Zealand, Norway, Portugal, Spain, Sweden, Switzerland, Turkey, the United Kingdom, and the United States.

required by the OECD decision, the regulations establish a graduated system of controls applicable to wastes when they move across national borders within the OECD for recovery. The level of control is determined by the assignment of the waste on OECD's green, amber, or red list.

Wastes on the green list are subject to basic controls normally applied to international commercial shipments. Amber- and red-listed wastes require special notification to the destination and transit OECD member countries, as well as extra information on tracking forms.

As promulgated by EPA, any person who exports or imports hazardous waste to or from an OECD member country that is subject to federal manifest requirements under 40 CFR parts 262 and 273 or subject to state requirements analogous to 40 CFR part 273 is subject to the OECD export/import requirements.

Green-listed wastes that are sufficiently contaminated or mixed with amber-listed wastes in sufficient quantities to make the waste hazardous under RCRA are subject to amber-list controls. Amber-listed wastes that are contaminated or mixed with red-listed wastes in sufficient quantity to make the waste hazardous under RCRA are subject to red-list controls.[6]

General Requirements. Amber-listed and red-listed wastes exported from the United States to or through an OECD member country must meet basic requirements concerning:

- Notification and consent
- Contracts
- Tracking document
- Reporting

The waste must be destined for recovery operations at a facility that, under applicable domestic law, is operating or is authorized to operate in the importing country [40 CFR 262.82(b)(1)]. The transfrontier movement must be in compliance with applicable international transport agreements and with all applicable international and national laws and regulations in non-OECD countries of transit.

Notification and Consent. Consent must be obtained from the competent authorities (e.g., the national environmental protection agency) of the relevant OECD importing and transit countries before exporting hazardous wastes destined for recovery. There are different requirements for amber-listed, red-listed, and unlisted wastes. There are no notification requirements for green-listed wastes.

Notifications for red, amber, and unlisted waste must include the following information:

[6] Although some OECD-listed wastes are not hazardous under RCRA (e.g., PCBs), they may be subject to other domestic laws (e.g., TSCA) or the national laws and regulations of the countries of import and transit.

- Serial number or other identifier of the notification form.
- Notifier name, EPA identification number, address, telephone, and fax number.
- Importing recovery facility name, address, telephone, and fax number.
- Consignee name, address, telephone, and fax number; a statement as to whether the consignee will engage in waste exchange or store the waste before final recovery and the identification of the recovery operations that will be used for the shipment.
- Intended transporters and/or their agents.
- Country of export, relevant competent authority, and point of departure.
- Countries of transit and relevant competent authorities and points of entry and departure.
- Country of import and relevant competent authorities and points of entry.
- Statement of whether the notification is a single notification or a general notification.
- Projected shipment date.
- OECD waste code and color list, description of waste, estimated quantity, RCRA waste code, and United Nations number.
- The following certification:

I certify that the above information is complete and correct to the best of my knowledge. I also certify that legally enforceable written contractual obligations have been entered into, and that any applicable insurance or other financial guarantees are or shall be in force covering the transfrontier movement.

Name:________________________________

Signature:____________________________

Date:_________________________________

Amber-Listed-Waste Notification Requirements. At least 45 days before the commencement of a movement, the notifier must provide written notification in English of the proposed transfrontier movement to the following office:

Office of Enforcement and Compliance Assurance
Office of Compliance, Enforcement Planning, Targeting and Data Division (2222A)
U.S. Environmental Protection Agency
401 M Street, SW
Washington, DC 20460
Attention: OECD Export Notification

If no objection to such a notification has been lodged by a concerned country (i.e., exporting, importing, and transit) within 30 days after the date of issuance of the Acknowledgment of Consent by the country of import, the shipment may occur (this is known as tacit consent). If written consent is received by all appropriate countries before the 30 days, the shipment may occur as soon as all consents are received.

Red-Listed-Waste Notification Requirements. A shipment of red-listed waste may not occur until written consent is received from the importing and transit countries.

Unlisted-Waste Notification Requirements. For any wastes that are not on any of the OECD lists and are considered hazardous under RCRA (40 CFR 261.3), countries must comply with the procedures for red-listed wastes described above. There are no notification requirements for unlisted wastes not classified as hazardous under RCRA.

Contracts. Transfrontier shipments of red- and amber-listed wastes must be covered under the terms of a valid written contract, chain of contracts, or equivalent arrangements [40 CFR 262.85(a)]. The contracts must specify the name and EPA identification number of the generator, each person who will have physical and legal custody of the waste, and the recovery facility. Contracts also must specify which party listed in the contract will assume responsibility for alternate disposition of the waste if the original disposition listed in the contract cannot occur. In addition, contracts must include provisions for financial guarantees if required by the receiving country.

Tracking Documents. All red- and amber-listed wastes must be accompanied by a tracking document from the initiation of the shipment to its final disposition [40 CFR 262.84(a)]. The tracking document must include all information required for the notification and the following additional information:

- Date the shipment commenced.
- Name, address, and telephone and fax numbers of the primary exporter.
- Company name and EPA identification number of all transporters.
- Description of the transport including the type of packaging.
- Any special precautions to be taken by transporters.
- Appropriate signatures for each custody transfer (e.g., transporter, consignee, and recovery facility).
- The following certification:

I certify that the above information is complete and correct to the best of my knowledge. I also certify that legally enforceable written contractual obligations have been entered into, that any applicable insurance or other financial guarantees are or shall be in force covering the transfrontier movement, and that

1. All necessary consents have been received; or
2. The shipment is directed at a recovery facility within the OECD area and no objection has been received from any of the concerned countries within the 30-day tacit consent period; or

3. The shipment is directed at a recovery facility pre-authorized for that type of waste within the OECD area; such an authorization has not been revoked, and no objection has been received from any of the concerned countries.

(delete sentences that are not applicable)

Name:___________________________________

Signature:________________________________

Date:____________________________________

Reporting. All persons meeting the definition of primary exporter who are involved with transfrontier movements of OECD-covered waste must file an annual report [40 CFR 262.87(a)]. (A *primary exporter* is any person who is required to originate the manifest for a shipment of hazardous waste in accordance with 40 CFR part 262, subpart B, or equivalent state provision which specifies a treatment, storage, or disposal facility in a receiving country as the facility to which the hazardous waste will be sent and any intermediary arranging for the export (40 CFR 262.51). The annual report must be sent no later than March 1 of each year for the previous calendar year to the following address:

Office of Enforcement and Compliance Assurance
Targeting and Data Division (2222A)
Environmental Protection Agency
401 M Street, SW
Washington, DC 20460

The annual report must contain the following information:

- The EPA identification number, name, and mailing and site address of the notifier filing the report.
- The calendar year covered by the report.
- The name and address of each final recovery facility.
- By final recover facility, for each waste exported, a description of the waste, the EPA hazardous waste number, the OECD waste code designation, DOT hazard class, the name and EPA identification number of the transporter, the total amount of hazardous waste shipped, and number of shipments pursuant to each notification.
- The following certification:

I certify under penalty of law that I have personally examined and am familiar with the information submitted in this and all attached documents, and that based on my inquiry of those individuals immediately responsible for obtaining the information, I believe that the submitted information is true, accurate, and complete. I am aware that there are significant penalties for submitting false information including the possibility of fine and imprisonment.

Primary exporters must file an exception report if a signed copy of the tracking document was not received within 45 days from the date it was accepted by the initial transporter; or if a signed copy of the tracking document was not received from the recovery facility within 90 days from the date the waste was accepted by the initial transporter. In addition, an exception report must be filed if the waste is returned to the United States.

HAZARDOUS WASTE GENERATOR COMPLIANCE CHECKLIST[7]

All Generators

Has it been determined that the waste generated at the facility is hazardous? Yes___ No___

Are records of this determination kept? Yes___ No___

Has the amount of *countable* hazardous waste been determined? Yes___ No___

Is the amount generated per calendar month less than 100 kg of nonacutely and less than 1 kg of acutely hazardous waste? Yes___ No___

(If yes, go to "Conditionally Exempt Small-Quantity Generators.")

(If no, go to "Small- and Large-Quantity Generators.")

Conditionally Exempt Small-Quantity Generators

Has more than 1 kg of acute hazardous waste been accumulated? Yes___ No___

(If yes, the CESQG is now considered a large-quantity generator.)

Has more than 1,000 kg of non-acute hazardous waste been accumulated? Yes___ No___

(If yes, the CESQG is now considered a small-quantity generator.)

Has the hazardous waste been treated on-site or sent off-site to a facility that

- Is permitted under 40 CFR part 270? Yes___ No___
- Has interim status under 40 CFR part 270? Yes___ No___
- Is permitted, licensed, or registered by a state? Yes___ No___

[7]Because a state may be authorized to be more stringent or broader in scope than federal law, this checklist should be used only as a guide to the minimum requirements, and individual state requirements also should be consulted

- Beneficially uses or reuses or legitimately recycles the waste? Yes___ No___
- Treats the waste prior to beneficial use, reuse, or reclamation? Yes___ No___
- In the case of universal waste, has it been sent to an authorized universal waste handler or facility? Yes___ No___

Small and Large-Quantity Generators

Note: The following requirements are generally the same for both small- and large-quantity generators. Those noted with an asterisk (*) are additional requirements for large-quantity generators.

EPA ID Number

Has the state or EPA been notified of hazardous waste generation activity? Yes___ No___

Does generator have an EPA Identification number? Yes___ No___

On-Site Accumulation of Hazardous Waste

Is hazardous waste accumulated on-site? Yes___ No___

Is the date upon which accumulation began clearly marked on each container? Yes___ No___

Are the words HAZARDOUS WASTE marked on any container of 110 gallons or less? Yes___ No___

For SQGS, has the waste been accumulated less than 180 days (or 270 days if the facility of choice is over 200 miles away)? Yes___ No___

For SQGS, is the total amount of all hazardous waste less than 6,000 kg? Yes___ No___

(If not, the generator must have a storage permit.)

*For LQGs, is the waste accumulated for less than 90 days? Yes___ No___

Is waste accumulated in a satellite accumulation area? Yes___ No___

- Is the container marked with the words HAZARDOUS WASTE? Yes___ No___
- Are provisions established to move a full container to the regular storage area within 3 days? Yes___ No___

Is the waste managed in containers? Yes___ No___

(If yes, see "Containers.")

Is the waste managed in a tank? Yes___ No___

(If yes, see "Tank.")

Is the waste managed on drip pads? Yes___ No___

(If yes, see "Drip Pads.")

Is the waste managed in containment buildings? Yes___ No___
(If yes, see "Containment Buildings.")

Accumulation Units

Containers

Are all containers in good condition? Yes___ No___

Are containers compatible with the stored contents? Yes___ No___

Are the containers being inspected for leakage or corrosion? Yes___ No___

Are containers always closed unless their contents are being transferred? Yes___ No___

Are containers always handled in such a way as to prevent rupture or leaks? Yes___ No___

Is ignitable or reactive hazardous waste handled? Yes___ No___

*If yes, does the generator locate ignitable or reactive wastes at least 50 feet inside the facility's property line? Yes___ No___

Are the ignitable or reactive wastes separated from sources of ignition? Yes___ No___

*Do containers comply with the air emission control standards? Yes___ No___

Tanks

Are procedures in place to prevent the generation of any extreme heat, explosions, fire, fumes, mists, dusts, or gases? Yes___ No___

Are procedures in place to prevent the placement of hazardous wastes or treatment reagents that may cause rupture, corrosion, or structural failure of the tank? Yes___ No___

Is there at least two feet of freeboard maintained in all uncovered tanks? Yes___ No___

Do continuously fed tanks have a waste-feed cutoff or bypass system? Yes___ No___

Are ignitable, reactive, or incompatible wastes rendered nonignitable, nonreactive, or compatible prior to placement in the tank? Yes___ No___

At least once each operating day, are the waste-feed cutoff and bypass systems, monitoring equipment data, and waste level inspected? Yes___ No___

At least weekly, is the surrounding area of the tank system inspected for visible signs of erosion or leakage? Yes___ No___

At closure, have all hazardous wastes been removed from the tank, containment system, and discharge control systems? Yes___ No___

Note: The following tank requirements are for LQGs only.*

Has a one-time assessment, an integrity test, of the tank system been prepared? Yes___ No___

Were the installation standards for new tank systems followed? Yes___ No___

Does the tank system have appropriate secondary containment? Yes___ No___

Is the tank system inspected daily? Yes___ No___

Was the tank system adequately closed? Yes___ No___

Pertaining to air emission controls, does the tank meet the following criteria?

- Is the tank equipped with a cover (fixed roof) vented to a control device? Yes___ No___
- Is the tank equipped with a fixed roof and an internal floating roof? Yes___ No___
- Is the tank equipped with an external floating roof? Yes___ No___
- Is it a pressure tank designed to operate as a closed system with no detectable emissions at all times? Yes___ No___
- Are emissions from process vents and equipment leaks controlled? Yes___ No___

Drip Pads

Does the drip pad meet the applicable technical standards of 40 CFR part 265, subpart W? Yes___ No___

Are written procedures developed to ensure that wastes are removed from the pad and collection system at least once every 90 days? Yes___ No___

Are records kept documenting that these procedures are followed? Yes___ No___

Is the drip pad cleaned frequently to allow for weekly inspections of the entire drip pad? Yes___ No___

Is the drip pad inspected to ensure that the unit is protective of human health and the environment and thus fit for continued use? Yes___ No___

Is the drip pad inspected weekly and after storms to ensure proper operation and to detect any deterioration or leaks? Yes___ No___

Containment Buildings

Does the containment building meet the technical standards in 40 CFR part 265, subpart DD? Yes___ No___

Has a written description been prepared of the procedures used to ensure that wastes will remain in the unit for no more than 90 days? Yes___ No___

Has documentation been prepared to ensure that these procedures are followed? Yes___ No___

Is the containment building completely enclosed, with four walls, a floor, and a roof? Yes___ No___

Are dust control devices available as necessary to prevent fugitive dust from escaping through building exits? Yes___ No___

Is the containment building's floor free of significant cracks, corrosion, or deterioration? Yes___ No___

Is the containment building inspected at least once a week with all results recorded? Yes___ No___

Personnel Training

Are employees thoroughly familiar with proper waste handling and emergency procedures relevant to their responsibilities Yes___ No___

*Has a formal training program been developed for facility personnel Yes___ No___

*Does the training address the following issues

- Procedures for using, inspecting, repairing, and replacing facility emergency and monitoring equipment Yes___ No___
- Key parameters for automatic waste-feed cutoff systems Yes___ No___
- Communications or alarm systems Yes___ No___
- Response to fires or explosions Yes___ No___
- Responses to groundwater contamination incidents Yes___ No___
- Shutdown of operations Yes___ No___

*Do the facility hazardous waste management personnel have the requisite training documented in their personnel files? Yes___ No___

Has an attempt been made to combine training required under OSHA and HMTA? Yes___ No___

Preparedness and Prevention

Is the facility operated in a manner that minimizes the possibility of fire, explosion, or release? Yes___ No___

Does the facility have basic emergency response and communication equipment? Yes___ No___

Is there adequate aisle space to allow emergency equipment and personnel unobstructed access to any area of the facility? Yes___ No___

Have prior arrangements been made with local emergency organizations? Yes___ No___

**Contingency Plan*

Has a contingency plan been prepared and updated as required? Yes___ No___

Does the plan include a description of actions to be undertaken in case of an emergency? Yes___ No___

Have copies of the plan been distributed to local emergency organizations? Yes___ No___

If an Integrated Contingency Plan has been prepared, have all applicable statutory requirements been followed? Yes___ No___

Emergency Procedures

Is there at least one employee on-site or on call who is able to coordinate an emergency? Yes___ No___

Is the following information posted next to facility telephones:

- Name and telephone number of the designated emergency coordinator? Yes___ No___
- Telephone number of the fire department? Yes___ No___
- Telephone number of other emergency response organizations? Yes___ No___
- Location of fire extinguishers, spill response equipment, and alarms? Yes___ No___

*If implementation of the contingency plan was necessary, has EPA been notified within 15 days of the emergency? Yes___ No___

Manifest System

Is hazardous waste shipped off-site? Yes___ No___

Has a Uniform Hazardous Waste Manifest been obtained? Yes___ No___

Has the manifest been correctly completed? Yes___ No___

(See Chapter 4 for additional transportation requirements.)

Land Disposal Restrictions

Has the land disposal restriction status been determined for each waste? Yes___ No___

If the waste meets the requirements, has the required certification been signed? Yes___ No___

If waste does not meet the requirements, has the required notification been sent to the treatment or disposal facility? Yes___ No___

Waste Minimization

Has the generator certified on the manifest that a waste minimization program is in place? Yes___ No___

*Has the waste minimization program been described in the biennial report? Yes___ No___

Pretransport Requirements

Does the generator package waste for transportation? Yes___ No___

Is the waste packaged in accordance with HMTA requirements? Yes___ No___

(See Chapter 4 for detailed information.)

Are containers leaking, corroding, or bulging? Yes___ No___

Is there evidence of heat generation from incompatible wastes in containers? Yes___ No___

Are containers labeled according to RCRA and HMTA? Yes___ No___

Are containers marked according to HMTA requirements? Yes___ No___

Is each container of 110 pounds or less marked with the following words? Yes___ No___

HAZARDOUS WASTE—Federal law prohibits improper disposal. If found, contact the nearest police or public safety authority or the U.S. Environmental Protection Agency.

Generator's Name and Address____________________

Manifest Document Number____________________

(See Chapter 4 for additional requirements for transportation of hazardous waste.)

Record Keeping and Reports

Are the following records kept?

- Manifests and signed copies from designated facilities Yes___ No___
- Biennial reports Yes___ No___
- Exception reports Yes___ No___
- Test results or other means of waste determination, as required Yes___ No___
- Records documenting compliance with the land disposal restrictions Yes___ No___

For exports of hazardous waste, are the following records kept?

- Notification of intent to export Yes___ No___
- EPA Acknowledgment of Consent Yes___ No___
- Confirmation of delivery Yes___ No___
- Annual export report Yes___ No___
- OECD notification Yes___ No___
- OECD tracking document Yes___ No___

Are all records maintained at least three years after completion or filing? Yes___ No___

4

SHIPPING AND TRANSPORTATION

This chapter outlines the requirements for generators who offer hazardous waste for transportation and transporters of hazardous waste.[1] RCRA requires generators and transporters to comply with requirements established by the U.S. Department of Transportation (DOT). The major law dealing specifically with shipping and transportation is the Hazardous Materials Transportation Act (HMTA), which was enacted in 1974. In addition to the HMTA requirements, RCRA has some other minor requirements for generators and transporters of hazardous waste.

HMTA defines a *hazardous material* as "a substance or material . . . capable of posing an unreasonable risk to health, safety, and property when transported in commerce, and which has been so designated. The term includes hazardous substances, hazardous wastes, marine pollutants, and elevated temperature materials . . ." (49 CFR 171.8). Thus, under HMTA, hazardous wastes are a specified subset of hazardous materials. As such, all RCRA hazardous wastes are automatically hazardous materials under HMTA. For transportation purposes, hazardous waste are treated the same as hazardous materials except for a few additional requirements.

This chapter is divided into two sections: shippers (generators who offer their hazardous waste for transportation) and transporters (carriers who transport hazardous waste). Shippers and transporters share joint responsibility for the safe transportation of hazardous materials.

For example, shippers must correctly

[1] It is important to note that the hazardous materials transportation program administered by DOT focuses on hazardous materials, which includes hazardous waste as a subset.

- identify,
- package,
- mark,
- label, and
- prepare shipments.

Transporters must ensure the following:

- Shipments are acceptable for transport.
- Shipments are properly delivered to the designated facility.
- The shipping documentation (manifest) accompanies the shipment to the designated facility.

SHIPPERS

The basic intent of the HMTA requirements for shippers (generators) is to ensure that the hazardous waste to be shipped is correctly identified, packaged, marked, labeled, and otherwise prepared for transportation. All the applicable requirements under HMTA for a particular shipment depend on the shipping classification of the hazardous waste. The procedure to properly classifying a hazardous waste for transportation under HMTA is described in the following sections and depicted in Figure 4-1.

Hazardous Classification Procedure

The proper classification of a hazardous waste for shipment under HMTA is most important. (The information used to classify the waste as hazardous under RCRA should be sufficient to classify it accurately under HMTA.)

Note: The HMTA classification is in addition to the RCRA classification. After classifying a waste as hazardous under RCRA, HMTA classification is necessary if the hazardous waste will be shipped off-site. Even if a waste is not defined as hazardous under RCRA (e.g., statutory or regulatory exclusion), it may still be classified as a hazardous material under HMTA, subject to transportation controls.

HMTA classification requires a series of steps. The accurate classification of the waste, which depends on the physical/chemical characteristics of the waste, results in the selection of the proper shipping name. Once the waste is properly classified and the proper shipping name determined, identifying the requirements is relatively easy because most of the transportation requirements correspond to the classification of the waste. Most of the applicable requirements are conveniently combined into the Hazardous Materials Table, which is described in the following subsection.

The proper HMTA classification procedure includes determining the following:

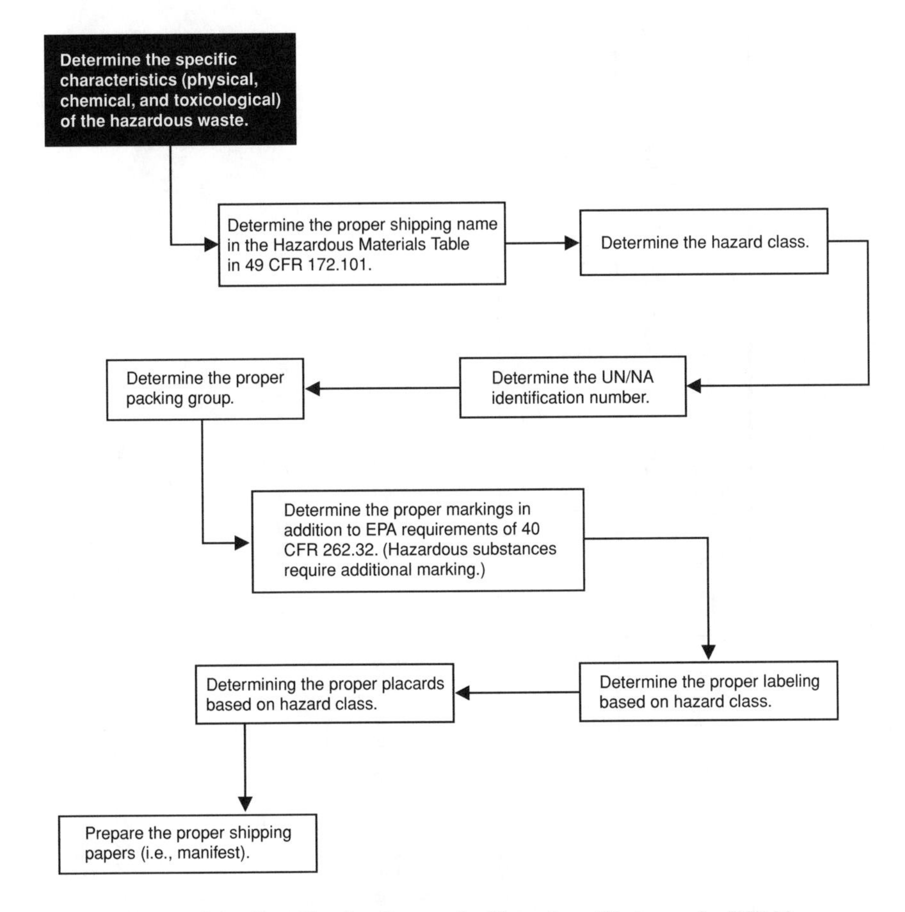

Figure 4-1. Classification Process for Hazardous Wastes under HTMA

- Proper shipping name*
- Hazard class*
- UN/NA identification number*
- Packing group*
- Special provisions

*These elements constitute the *basic shipping description.*

The Hazardous Materials Table. The Hazardous Materials Table (HMT), found in 49 CFR 172.101, alphabetically lists those materials and classes of materials designated as hazardous under HMTA. (The HMT does not differentiate a material versus a waste.) It is therefore the starting point in determining applicable HMTA requirements. The HMT contains information on proper shipping names, hazard classes, identification numbers, and packing groups. Figure 4-2 depicts a sample entry from

(1) Symbol	(2) Hazardous materials description and proper shipping names	(3) Hazard Class or Division	(4) Identification Numbers	(5) Packing Group	(6) Labels Required	(7) Special Provisions	(8) Packaging Authorizations (§173.)		
							Exemptions	Nonbulk Packaging	Bulk Packaging
	Toluene	3	UN1294	II	FLAMMABLE LIQUID	T8	150	202	242

Explanations of Columns

Column 1 Notes the applicability of special regulations, including the following:
+ = The designated proper shipping name and hazard class listed in the HMT must be used.
A or W = The material is subject to regulation only when transported by air (A) or water (W).
D = The proper shipping name as shown is acceptable only for domestic shipments.
I = The proper shipping name as shown is acceptable only for international shipments.

Column 2 Lists the proper shipping name for hazardous materials.

Column 3 Depicts the numerical hazard class or division number of the entry that must be included on the shipping paper (manifest).

Column 4 Depicts the hazard identification number of the entry that must be shown on the shipping paper, portable tanks, and exterior of packages.

Column 5 Depicts the packing group assigned to the entry.

Column 6 Depicts label(s) required for packaging.

Column 7 Lists any special conditions applicable to the entry.

Columns 8a, b, and c Depicts the section number pertaining to the packaging requirements.

Figure 4-2. Sample Entry for the Hazardous Materials Table

the HMT for toluene and contains the explanation codes for each of the columns in the table.

Proper Shipping Name. The *proper shipping name* is the name appearing in Column 2 of the HMT that most accurately describes the waste to be shipped (49 CFR 127.101). (The proper shipping name is only one component of the shipping description.) Only those names listed in roman type (nonitalics) are authorized shipping names, whereas the *italicized* names are to be used primarily as finding aids and cannot be used as the shipping name [49 CFR 172.101(c)].

Because the HMT was originally designed for commodities and not for hazardous waste, determining the proper shipping name for hazardous waste requires additional effort. For example, in addition to the shipping name listed in Column 2, the word WASTE must precede the shipping name for shipments of hazardous waste. For example, spent toluene previously used as a solvent would be labeled WASTE TOLUENE [49 CFR 172.101(c)(9)]. Classification is not always this straightforward, however; because hazardous wastes are often commingled waste streams. HMTA requires the most accurate shipping name to be selected; because many of the names in Column 2 of the HMT are chemical-specific, they cannot be used for mixtures because they are not accurate. Hazardous waste mixtures must follow the hierarchical approach outlined and explained here:

Proper Shipping Name Hierarchy

Listing by the waste's specific chemical name.
Listing by the chemical family name (n.o.s.).[2]
Generic listing by the waste's end-use description.
Generic listing by the n.o.s. end-use description.
Generic listing by n.o.s. class description.

The chemical-specific listings are intended for single-stream wastes. For example, the proper shipping name for spent toluene used in degreasing operations would be WASTE TOLUENE. However, if the toluene was part of a waste solvent mixture, the above shipping name would not be correct. The proper name for the mixture would utilize another description based on the hierarchical approach, which is described below [49 CFR 172.101(c)(12)].

Note: If a non-chemical-specific shipping name is selected, the technical name(s) of the constituents that makes the waste hazardous must be entered in parentheses after the shipping name [49 CFR 172.101(c)(12)(ii) and 172.202(d)]. For example: WASTE COMPOUNDS, cleaning liquid (contains Acetone and Isopropyl alcohol).

If the waste to be shipped is not specifically listed by its chemical name, the chemical family name may be used. But, to repeat, if the specific name is listed, it

[2] *N.O.S.* means "not otherwise specified." *N.O.I.* (not otherwise indexed) and *N.O.I.B.N.* (not otherwise indexed by name) are also acceptable and interchangeable with *N.O.S.*

must be used. For example, for butyl alcohol: BUTYL ALCOHOL. ("Alcohol, n.o.s." is not acceptable.) Examples of chemical family names include

- Chlorate, n.o.s.
- Cyanide solutions, n.o.s.
- Peroxides, inorganic, n.o.s.

The next step of the hierarchy focuses on the waste's end use or purpose—that is, the entry that best describes the purpose of the material before it became a waste (rather than its chemical composition). Examples of end-use descriptions include the following:

- Dye intermediate, liquid
- Engine starting fluid
- Printing ink, flammable

The next step of the hierarchy is the use of generic n.o.s. end-use descriptions, which are less specific than the end-use description just described. Examples of generic n.o.s. end-use descriptions:

- Drugs, n.o.s.
- Esters, n.o.s
- Insecticide, liquid, n.o.s.

The final and least-specific description is based on the generic listing by the n.o.s. hazard classification. Such a listing does not identify the waste itself, but the primary hazard associated with the waste. Examples of generic hazard n.o.s. listings:

- Flammable solid, n.o.s.
- Combustible liquid, n.o.s.
- Oxidizer, n.o.s.

If the waste meets the definition of hazardous under RCRA but does not meet any of the above hierarchical criteria (e.g., soil contaminated with small amounts of hazardous waste, sludges with heavy metals), the proper shipping name may be (49 CFR 173.140) the following:

- Hazardous waste, liquid, n.o.s., or
- Hazardous waste, solid, n.o.s.

However, if either of these names are used, supplemental information must appear in parentheses after the description to indicate the nature of the waste; that information must include the technical names of at least two constituents causing the waste to be hazardous.

Note: Regardless of what proper shipping name is selected, if a plus sign (+) does not appear before the name in Column 1 of the Hazardous Materials Table, the waste's actual characteristics may differ from the entry, and must be so described. This is discussed in more detail under the "Hazard Class" heading that follows.

HMTA requires there be only one proper shipping name per entry. For example, benzene and toluene are each a proper shipping name; however, they may not be used together for the shipping name. In reviewing the "Proper Shipping Name Hierarchy," the most appropriate description for the benzene/toluene mixture is "generic listing by n.o.s. class description." According to the HMT, both toluene and benzene have the designated hazard class of flammable liquid. Mixtures of hazardous materials with generic n.o.s. designations are required to have the technical names of at least two components that most predominantly contribute to the hazards of the mixture. For a mixture of benzene and toluene the proper shipping name would be

WASTE FLAMMABLE LIQUID, n.o.s. (Benzene and Toluene)

For mixtures or solutions containing a hazardous material *and* a nonhazardous material, the mixture or solution may be described using the proper shipping name of the hazardous material, provided [49 CFR 172.101(c)(10)] all of the following conditions are met:

- The packaging specified in Column 8 of the HMT is appropriate for the physical state of the waste.
- The hazard class, packing group, or secondary hazard of the mixture/solution are the same as those of the hazardous mixture itself.
- The word MIXTURE or SOLUTION (as appropriate) is added to the proper shipping name.
- The description does not indicate that the proper shipping name applies only to the pure or technically pure hazardous material.
- There is no change in emergency procedures to be taken.

Hazard Class. Under HMTA, materials (wastes) are defined as hazardous because of their potential danger to human health, property, or the environment during transportation. The hazards of concern are categorized into nine classes, which are further broken down into the divisions shown in Table 4-1. The accurate classification of the hazard(s) exhibited by the waste to be shipped is an important aspect of the HMTA regulations. This classification further determines requirements for packaging, labeling, and any special requirements that must be met by the shipper and carrier. In Column 3 of the HMT, each entry has the "likely" hazard class attributed to that entry. However, the actual hazard class may be different depending on the specific physical, chemical, and biological characteristics of the waste. The presence of

TABLE 4-1. The HMTA Hazard Class and Divisions

Class/Division	Title	Definition	Examples	CFR Citation
Class 1	**Explosives**	—		49 CFR 173.58
Division 1.1	Explosives	Solid or liquid explosives that display a major hazard of mass explosion.	TNT, tritonal, trinitronapthalene	49 CFR 173.58
Division 1.2	Explosives	Explosives that have major hazard of dangerous projectiles.	Rockets and mines with bursting charge	49 CFR 173.58
Division 1.3	Explosives	The major hazard is the release of radiant heat or violent burning, or both, but there is no blast or projection hazard.	Nitrocellulose; sodium salts of aromatic nitro-derivatives; sodium dinitro-*o*-cresolate	49 CFR 173.58
Division 1.4	Explosives	Explosives for which there is a small hazard with no mass explosion and no projection of fragments of appreciable size or range.	Rockets with inert head; primers: cap type	49 CFR 173.58
Division 1.5	Explosives	A material designed for blasting, that is found to be so inactive after undergoing certain prescribed tests that there is very little probability of accidental initiation to explosion or of transition from deflagration to detonation.	Explosives, blasting, Types B and E	49 CFR 173.58
Division 1.6	Explosives	A material that is found to be an extremely insensitive detonating substance after undergoing certain prescribed tests.		49 CFR 173.58

continued

Class 2	**Gases**	—	Nonliquefied compressed gas, liquefied compressed gas, cryogenic liquid, and refrigerant or dispersant gas	49 CFR 173.115
Division 2.1	Flammable Gas	Any material that is a gas at 20°C (68°F) or less and 101.3 kPa (14.7 psi) of pressure that is ignitable at 101.3 kPa (14.7 psi) when in a mixture of 13% or less by volume with air; or that has a flammable range at 101.3 kPa (14.7 psi) with air of at least 12%, regardless of lower explosive limit.	Propane; ethylene, compressed; hydrogen, compressed	49 CFR 173.115
Division 2.2	Nonflammable, Nonpoisonous Gas	Gas that is hazardous by virtue of being under pressure at 280 kPa (41 psi) at 20°C (68°F).	Helium, compressed heptafluoropropane, and sulfur hexafluoride	49 CFR 173.115
Division 2.3	Poisonous Gas	A material that is known to be so toxic to humans and animals as to pose a hazard to health during transportation or is presumed to be so toxic to humans because when tested on laboratory animals it has a LC_{50} value less than 5,000 mL/m^3.	Phosgene, organic phosphate mixed with compressed gas, oxygen difluoride	49 CFR 173.115
Class 3	**Flammable and Combustible Liquids**	Flammable Liquid—A liquid with a flash point of 60°C (140°F) or less. Combustible Liquid—A liquid not meeting the definition of any other hazard class with a flashpoint greater than 60°C (140°F) and below 93°C (200°F).	Benzene, diesel, methyl, isobutyl ketone	49 CFR 173.120
Class 4	**Flammable Spontaneously Combustible Solids**	—		

TABLE 4-1. Continued

Class/Division	Title	Definition	Examples	CFR Citation
Division 4.1	Flammable Solids	Solids that are liable to undergo strongly exothermic decomposition due to high heat during transportation or readily combustible materials that can cause fire through friction.	Nitronaphthalene, paraformaldehyde, picric acid	49 CFR 173.124
Division 4.2	Spontaneously Combustible	Materials that can ignite, without an ignition source, within 5 minutes of exposure to air.	Phosphorus white, molten; sodium hydrosulfite; iron sponge	49 CFR 173.124
Division 4.3	Dangerous When Wet	A material that by contact with water is liable to become spontaneously flammable or to give off flammable or toxic gas.	Barium, calcium, stannic phosphide	49 CFR 173.124
Class 5	**Oxidizers and Organic Peroxides**	—		
Division 5.1	Oxidizers	A material that readily yields oxygen to stimulate the combustion of other substances.	Hydrogen peroxide, potassium perchlorate, calcium permanganate	49 CFR 173.127
Division 5.2	Organic Peroxide	A derivative of hydrogen peroxide in which the hydrogen has been replaced by an organic substance.		49 CFR 173.127
Class 6	**Poisonous and Infectious Materials**	—		
Division 6.1	Poisonous Materials	Liquids or solids known to be so toxic as to create a health hazard during transportation.	Pesticides; cyanides; 1,1,1-trichloroethane	49 CFR 173.134
Division 6.2	Infectious Materials	Materials that can cause human and animal disease.		49 CFR 173.134

continued

Class 7	**Radioactive Materials**	Materials that spontaneously emit radiation capable of penetrating and damaging living tissue.	Uranium hexafluoride, plutonium, cesium	49 CFR 173.389
Class 8	**Corrosives**	A liquid or solid that can cause visible destruction or irreversible damage to living tissue or can severely corrode steel.	Hydrochloric acid, sodium hydroxide, sulfuric acid	49 CFR 173.136
Class 9	**Miscellaneous**	Materials that present a hazardous threat during transportation, but do not meet any other hazard class.	Hazardous waste, n.o.s.; marine pollutant, elevated-temperature material	49 CFR 173.140

a plus sign (+) in Column 1 of the HMT before an entry of a shipping name indicates that this particular entry must use the hazard class shown whether or not the waste meets the definition of the hazard class. An entry that does not have aplus sign (+) may not necessarily have the properties listed in the description. For example, Waste Acetonitrile, flammable liquid, may be more accurately classified as a combustible liquid because the flash point is higher. The absence of a plus sign (+) signifies that the entry's hazard class may differ, depending on the actual characteristics of the specific waste to be shipped, and the waste must be classified accordingly.

Wastes with Multiple Hazard Classes. Hazardous waste can often exhibit more than one of the designated HMTA hazard classes or divisions. A classification hierarchy has been established for such materials that exhibit multiple hazards [49 CFR 173.2a(a)]. A hazardous waste that exhibits more than one hazard or division must use the hazard class ranked highest in the following list:

Class 7—Radioactive Materials

Division 2.3—Poisonous Gas

Division 2.1—Flammable Gas

Division 2.2—Nonflammable Gas

Division 6.1—Poisonous Liquid

Division 4.2—Pyrophoric

Division 4.1—Flammable Solids (self-reactive)

Class 3—Flammable Liquid, Class 8—Corrosives, Division 4.1—Flammable Solids, Division 4.2—Spontaneously Combustible, Division 4.3—Dangerous When Wet, Division 5.1—Oxidizers, and Division 6.1—Poisonous (other than those in Packing Group I, which appears in Column 5 of the HMT)

Class 3—Combustible Liquids

Class 9—Miscellaneous

However, if a hazardous waste has two hazard classes *and* those are Hazard Classes 3 and 8, or Divisions 4.1, 4.2, 4.3, 5.1, or 6.1, then the Precedence of Hazard Table must be used to determine the appropriate hazard class for the shipment. This table is found in 49 CFR 173.2a(b).

In addition, because of the unique hazards presented by some classes and divisions, the above hierarchy does not apply to the following:

- Class 1—Explosives
- Division 5.2—Organic Peroxides
- Division 6.2—Infectious Materials

These classes/divisions require special description and handling.

Hazardous Substances. Hazardous substances are those substances designated under §101(14) (listed in 40 CFR part 302) of the Comprehensive Environmental Response, Compensation, and Liability Act (CERCLA, or Superfund). (See Chapter 12 for more information on Superfund hazardous substances.) It is important to note that every hazardous waste designated under RCRA is automatically classified as a hazardous substance under CERCLA. Other wastes not specifically classified as hazardous under RCRA may still be a hazardous substance under CERCLA, such as PCBs, asbestos, and radionuclides. However, for a substance to be designated as a *hazardous substance* under HMTA, the following criteria must be met (49 CFR 172.101):

- It is listed as a hazardous substance in Appendix A of 49 CFR 172.101.
- The quantity, in *one* package, equals or exceeds the reportable quantity (RQ) listed in Appendix A of 49 CFR part 172.

Wastes that meet the HMTA definition of a hazardous substance must have the letters RQ placed in front or at the end of the shipping description on the container and the manifest.

If the proper shipping name for a material that is a hazardous substance does not identify the hazardous substance by name, one of the following three options must be added to the shipping description (in parentheses) after the shipping name [49 CFR 172.203(c)]:

- The name of the hazardous substance as listed in Appendix A to 49 CFR 172.101.
- The EPA hazardous waste code.
- For characteristic hazardous wastes, the words
  — "EPA—Ignitability"
  — "EPA—Corrosivity"
  — "EPA—Reactivity"
  — "EPA—Toxicity"

Identification Number. The identification numbers are found in Column 4 of the HMT. The four-digit number contains a two-letter prefix, which is based on the United Nations (UN) identification system for international shipments and the North America (NA) identification system for North American shipments [49 CFR 172.101(e)]. For example, the identification number for toluene is UN1294.

Packaging

RCRA requires all generators of hazardous waste to transport their wastes in packaging that complies with HMTA shipping and packaging regulations contained in 49 CFR parts 173, 178, and 179. As with most other HMTA requirements for hazardous waste, the selection of the proper packaging hinges upon the determination of the proper shipping name. To determine the appropriate packaging, consult Columns 5 and 8 of the HMT.

In Column 5, the appropriate packing group is specified (Packing Group I, II, or III), which specifies the degree of danger posed by the material for shipping purposes. Packing Group I presents the greatest danger; Packing Group III, the least danger. Column 8 contains three entries—Exceptions, Nonbulk Packaging, and Bulk Packaging—and directs the shipper to the specific regulatory sections in 40 CFR part 173, *Preparation of Hazardous Materials for Transportation*, which governs the selection of appropriate packaging.

The shipper should also be aware of the package conformance markings. Manufacturers and testers of packages and packaging must mark every appropriate package as to its conformance to United Nations Standards (49 CFR 178.503). The required information includes the following:

- UN packaging symbol
- Packaging identification code
- Designation code for the Packing Group for which the container was tested
- Designation for specific gravity for which the packaging design type was tested
- Test pressure in kilopascals (kPa) for packaging designed for liquids
- If appropriate, minimum container thickness
- Year and month of manufacture
- Country's standards to which the packaging was tested and certified
- Name, address, or symbol of the manufacturer or approval agency certifying compliance with the UN standard

Based on the required packing group, the shipper selects a package appropriate for the physical state of the waste and the amount to be shipped. The authorized packages for each packing groups are listed in 49 CFR part 173, subpart E. Because the hazardous waste management facility may require certain containers, they should be consulted before selecting a container.

Labeling

The shipper is responsible for labeling packages of hazardous waste. Labels are color-coded and diamond-shaped and provide symbolic representations of the hazards exhibited by the shipment. (Labeling is different from marking, which is described in the next section.) The required labels, which are listed in Table 4-2 and depicted in Figure 4-3, appear in Column 6 of the HMT. The label should be affixed to the package near the marked, basic shipping description (49 CFR 172.406). The label may not be located on the bottom. Labels must be durable, clearly visible, not obscured by markings, and displayed on a background of contrasting color or utilize an outer border.

TABLE 4-2. Required Labels for Packages of Hazardous Waste

Hazard Class/Division	Label Name
1.1	EXPLOSIVE 1.1
1.2	EXPLOSIVE 1.2
1.3	EXPLOSIVE 1.3
1.4	EXPLOSIVE 1.4
1.5	EXPLOSIVE 1.5
1.6	EXPLOSIVE 1.6
2.1	FLAMMABLE GAS
2.2	NONFLAMMABLE GAS
2.3	POISONOUS GAS
3 (flammable liquid)	FLAMMABLE LIQUID
4.1	FLAMMABLE SOLID
4.2	SPONTANEOUSLY COMBUSTIBLE
4.3	DANGEROUS WHEN WET
5.1	OXIDIZER
5.2	ORGANIC PEROXIDE
6.1 (packing groups I & II)	POISON
6.1 (packing group III)	KEEP AWAY FROM FOOD
6.2	INFECTIOUS SUBSTANCES
8	CORROSIVE
9	CLASS 9

If a hazardous waste to be shipped has more than one hazard class, more than one label may be required. Column 6 of the HMT prescribes cases in which multiple labels are required (49 CFR 172.402).

Marking

The shipper is responsible for proper marking of each package of hazardous waste for transportation (49 CFR 172.300). *Marking* means placing on the outside of a shipping container one or more of the following (see Figure 4-4):

- Basic shipping description
- Instructions
- Caution
- Weight

Marking also includes any required specification marks on the inside or outside shipping container; these are separate from labeling and placarding (49 CFR 172.301).

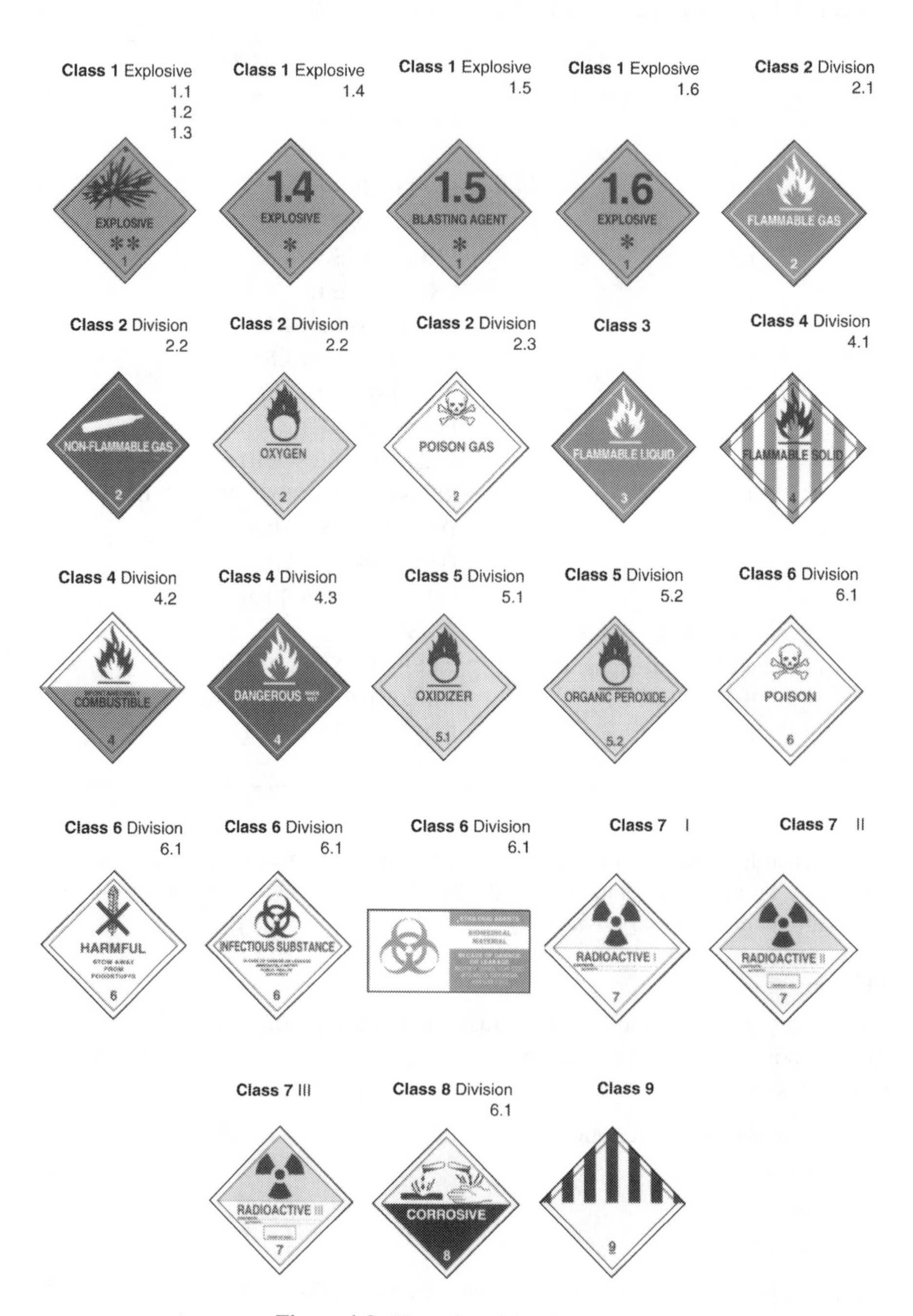

Figure 4-3. Hazardous Materials Labels.

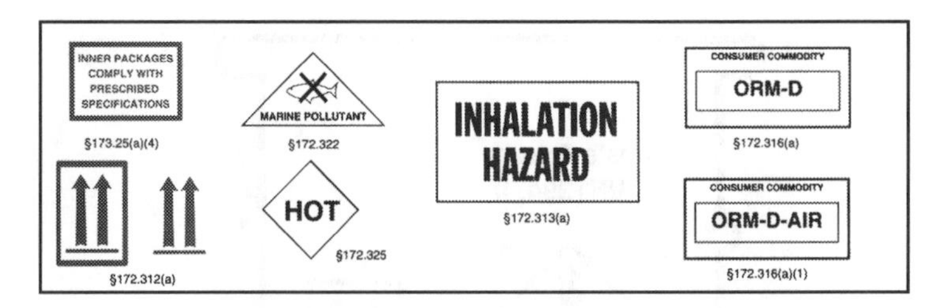

Figure 4-4. Hazardous Materials Package Markings

Markings must be (49 CFR 172.304)

- durable and in English;
- printed directly on packages or by using labels or signs affixed to packages; and
- printed on a background of sharply contrasting color, unobscured by labels or attachments, and located away from other displays (e.g., advertisements) that could reduce their effectiveness.

There are additional marking requirements for marine pollutants (49 CFR 172.322), radioactive wastes (49 CFR 172.310), and poisonous wastes (49 CFR 172.313). There are different marking requirements for bulk and nonbulk packages.

Marking for Nonbulk Packaging. As depicted in Figure 4-5, markings on nonbulk packages (e.g., <450 liters for liquids and <400 kg for solids) of hazardous waste must include the following information (49 CFR 172.301 and 172.324):

- The proper shipping name as shown in Column 2 of the HMT.
- The proper identification number with the prefix UN or NA as shown in Column 4 of the HMT.
- For n.o.s. identification, the required technical name in parentheses.
- If the proper shipping name does not identify the hazardous waste by name, one of the following descriptions must be shown in parentheses:
  — name of the hazardous substance;
  — RCRA waste code; or
  — for characteristic hazardous wastes, the letters EPA followed by the characteristic (e.g., reactivity, ignitability, toxicity, corrosivity).
- If the hazardous waste is a mixture or solution and the proper shipping name does not indicate the components that render it hazardous, the name of the component(s) must be shown in parentheses in association with the proper shipping name.
- For hazardous substances, the letters RQ.
- If the package contains polychlorinated biphenyls (PCBs), the proper shipping name, identification number, and designation RQ must appear (RQ, POLYCHLORINATED BIPHENYLS, UN23152) as well as the PCB label.
- The name and address of the shipper.

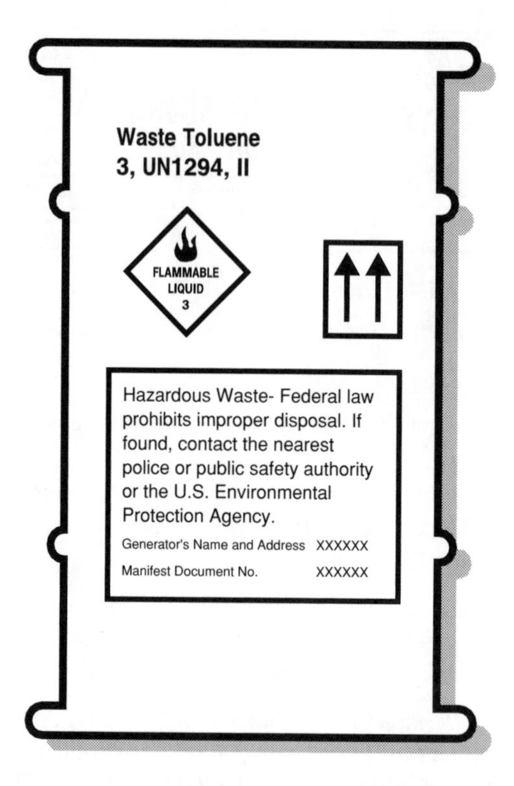

Figure 4-5. Example of Nonbulk Package Marking for Hazardous Waste (Toluene)

In addition to HMTA requirements, there are additional marking requirements under RCRA for hazardous waste [40 CFR 262.32(b), 262.34(a)(2), and 262.34(a)(3)]. A generator accumulating hazardous waste in compliance with 40 CFR 262.34 must mark the container with the words HAZARDOUS WASTE and with the date upon which each period of accumulation begins. In addition, before offering hazardous waste for transportation off-site, a generator must mark each container of 110 gallons or less with the specified information depicted in Figure 4-6.

Marking for Bulk Packaging. The marking requirements for bulk packaging are essentially the same as for nonbulk packaging except for the number of markings. Whereas nonbulk marking requires only one set of markings, bulk containers require markings on each side and end for containers of greater than 3,785 liters (1,000 gallons) (although for containers greater than 450 liters (119 gallons) but less than 3,785 liters, the markings need be only on opposite sides [49 CFR 172.302)].

Shipping Papers

A *shipping paper* is a document used to identify the shipment being offered for transportation and to document its delivery to the designated facility. The Uniform

Figure 4-6. Hazardous Waste Label

Hazardous Waste Manifest, EPA Form 8700-22, must be used for shipments of hazardous waste (49 CFR 172.205 and 40 CFR 262.20).[3]

The Uniform Hazardous Waste Manifest, as mentioned in Chapter 3, is a component of a joint program between DOT and EPA. The manifest is a multicopy form with copies for each participant involved in a shipment of hazardous waste, and it must identify the type and quantity of waste, the generator, the transporter, and the facility designated to receive the waste. (A continuation sheet, EPA Form 8700-22A, must be used when there are more than four waste materials.) The standard federal manifest has white numbered boxes (which must be filled out by all users of a manifest) and shaded lettered boxes (for optional information that a state may require).

The shipper is responsible for properly preparing the manifest, which must accompany the waste at all times while in transit.[4] When the waste reaches its final

[3] There is an exception to this requirement for a contract recycling agreement. See 40 CFR 262.20(c) for additional details.

destination, the receiving designated facility must sign the manifest, provide a signed copy to the transporter and send a signed copy directly to the generator to confirm receipt. The transporter must retain copies of all manifests for at least three years.

The following information must be printed on the manifest by the shipper (generator):

- Basic shipping description of the hazardous waste in sequence
- Total quantity and unit of measure
- Required additional descriptions (described in the next subsection), if applicable
- Shipper certification
- Emergency contact number
- Emergency response information

The basic shipping description must be written in the specified sequence. For example:

proper shipping name—	hazard class—	identification number—	packing group
WASTE SULFURIC ACID, SPENT	8	UN1832	II

Required Additional Descriptions. HMTA regulations require additional descriptions for certain hazardous substances and wastes [40 CFR 172.203(c)]. These descriptions include the following:

- For hazardous wastes listed as hazardous substances in the Appendix to 49 CFR 172.101, if there is a reportable quantity in *one* package, add the letters RQ either before or after the basic shipping description.
- For hazardous substances, if the proper shipping name does not include the name of the hazardous substance, add the name(s) of the hazardous substance, the EPA waste code, or the appropriate EPA characteristic waste.
- If the shipping name includes "n.o.s.," the technical name(s) of the waste must be written with the basic description: for example,

  WASTE OXIDIZING SUBSTANCE LIQUID, CORROSIVE, N.O.S., 5.1, UN3098, I (PEROXYMONOSULFURIC ACID)

- If the material is a mixture or solution, add technical names of at least two materials that contribute most to the hazard of the material after the hazard class: for example,

  WASTE FLAMMABLE LIQUID, N.O.S., 3, UN1993, I (ACETONE, TOLUENE)

[4] A transporter need comply with requirements only in the state of origin and the consignment (destination) state. A transporter is not bound to comply with the state manifest requirements of the states through which the transporter travels (49 *FR* 10492, April 20, 1984).

- In cases of poisonous wastes, regardless of the assigned hazard class, if the name of the constituent that causes the waste to be poisonous (e.g., meeting the definition of Division 6.1 PG I or II, or Division 2.3) is not included in the proper shipping name, the constituent must appear as part of the basic shipping description. Furthermore, if a waste is poisonous (as defined), but it is not obvious in the basic shipping description, the word POISON must appear in the basic shipping description [49 CFR 172.203(m)].

Emergency Contact Number. Shippers must provide an emergency response telephone number on the manifest, and the number must be displayed as such (e.g., **EMERGENCY CONTACT: XXX-XXXX**). The telephone number may be the number of any person or organization capable of, and accepting responsibility for, providing emergency response and accident information 24 hours a day, 7 days a week. It may be an employee, designee of the shipper, or a private organization (e.g., CHEMTREC)[5] (49 CFR 172.604).

Emergency Response Information. Each manifested shipment of hazardous waste must have specific emergency response communication information. (*Emergency response information* means information that can be used to mitigate an accident involving hazardous materials.) Although it is the shipper's responsibility to prepare the emergency response information, the transporter is responsible for ensuring this information is present with the shipment [49 CFR 172.602(c)(1)]. The information may be a material safety data sheet (MSDS) or a copy of the appropriate page from DOT's *North American Emergency Response Guidebook*, provided the seven pieces of information are addressed. The seven specific pieces of emergency response information that must be included as part of the emergency communication information are as follows:

- Basic shipping description
- Immediate hazards to health
- Fire and explosion risks
- Immediate precautions to be taken in the event of an accident or release
- Immediate methods to be taken for handling fires
- Initial methods for handling spills or leaks in the absence of fires
- Preliminary first-aid measures

Manifest Completion. Each manifest comes with its own instructions printed on the reverse side of the form. The generator must fill in the necessary information in Items 1 through 16, the transporter completes Items 17 and/or 18, and the designated facility completes Items 19 and 20. The required information on the manifest includes the following:

[5] CHEMTREC, the Chemical Transportation Emergency Center, is operated by the Chemical Manufacturers Association in Washington, DC. Information on their services can be obtained by calling 1-800-262-8200.

- Generator's (shipper) EPA identification number
- Generator's name and mailing address
- Manifest document number
- Transporter's company name and EPA identification number
- Designated facility's name, address, and EPA identification number
- Weight of each waste and number and type of containers
- Special handling instructions
- Certification by the generator for waste minimization
- Signatures by the appropriate parties

In addition to that required by the federal regulations, some states require other information that appears in the shaded boxes labeled A through K.

Placards

Placards are large, diamond-shaped, color-coded signs that are placed on the outside of transport vehicles indicating the hazards of the cargo as depicted in Figure 4-7. Placards are a joint responsibility between the shipper and carrier. Shippers who offer a hazardous waste for transport by highway must provide the carrier with the required placards unless the carrier has the appropriate placards or the carrier's vehicle is already placarded [49 CFR 172.500(a)]. The carrier is responsible for affixing the placards to the transport vehicle.

All motor vehicles, rails cars, and freight containers carrying *any* hazardous waste meeting the hazard class identified in the top of Table 4-3, or carrying more than 1,000 lbs. of *any* hazardous waste in a hazard class listed in the bottom half of Table 4-3, must be placarded with the specified placard. Also, any vehicles containing any quantity of materials described as "Poison-Inhalation Hazard" must display a Class 6 POISON or Division 2.3 POISON GAS placard in addition to any other placards required.

Placards must be displayed on both sides, the front, and the rear of the vehicle. Placards must be securely attached or in a holder, positioned so that the words or number are read horizontally from left to right, unobstructed from vehicle parts, located at least three inches from other markings, and located so that dirt and other debris from the tires will not cover them.

Transport vehicles carrying one or more PCB transformers or 99.4 lbs. of PCB liquid with a concentration of 50 ppm or greater must be marked with the special PCB label shown in 40 CFR 761.40. (See Chapter 17 for further information on PCBs.)

Training

Each employee involved in handling hazardous materials for transportation must be trained in accordance with 49 CFR 172.704.[6] Each employee must be provided initial training and recurrent training (i.e., every 2 years) to ensure knowledge of proper

[6] Training conducted by employers to comply with OSHA's Hazard Communication Standard may use these portions in lieu of the DOT Hazmat Training where appropriate [49 CFR 172.704(b)].

Table 4-3. HMTA Placarding Tables

If the transport vehicle contains any hazardous waste with the hazard class or division:	The transport vehicle must be placarded on each side and each end with the following:
1.1	EXPLOSIVE 1.1
1.2	EXPLOSIVE 1.2
1.3	EXPLOSIVE 1.3
2.3	POISON GAS
4.3	DANGEROUS WHEN WET
6.1 (PG I, inhalation hazard only)	POISON
7 (Radioactive Yellow III label only)	RADIOACTIVE

If the transport vehicle contains one or more hazardous wastes at 454 kg (1,001 lb) with the hazard class or division:	The transport vehicle must be placarded on each side and each end with the following:
1.4	EXPLOSIVE 1.4
1.5	EXPLOSIVE 1.5
1.6	EXPLOSIVE 1.6
2.1	FLAMMABLE GAS
2.2	NONFLAMMABLE GAS
3	FLAMMABLE
Combustible Liquid	COMBUSTIBLE
4.1	FLAMMABLE SOLID
4.2	SPONTANEOUSLY COMBUSTIBLE
5.1	OXIDIZER
5.2	ORGANIC PEROXIDE
6.1 (PG I or II, other than PG I inhalation hazard)	POISON
6.1 (PG III)	KEEP AWAY FROM FOOD
8	CORROSIVE
9	CLASS 9

loading, unloading, handling, storing, and transportation procedures for hazardous materials and emergency response procedures for hazardous material transportation incidents. This training, which may be conducted in-house or by an outside organization, must include the following:

- General awareness/familiarization training
- Function-specific training
- Safety training
- Testing

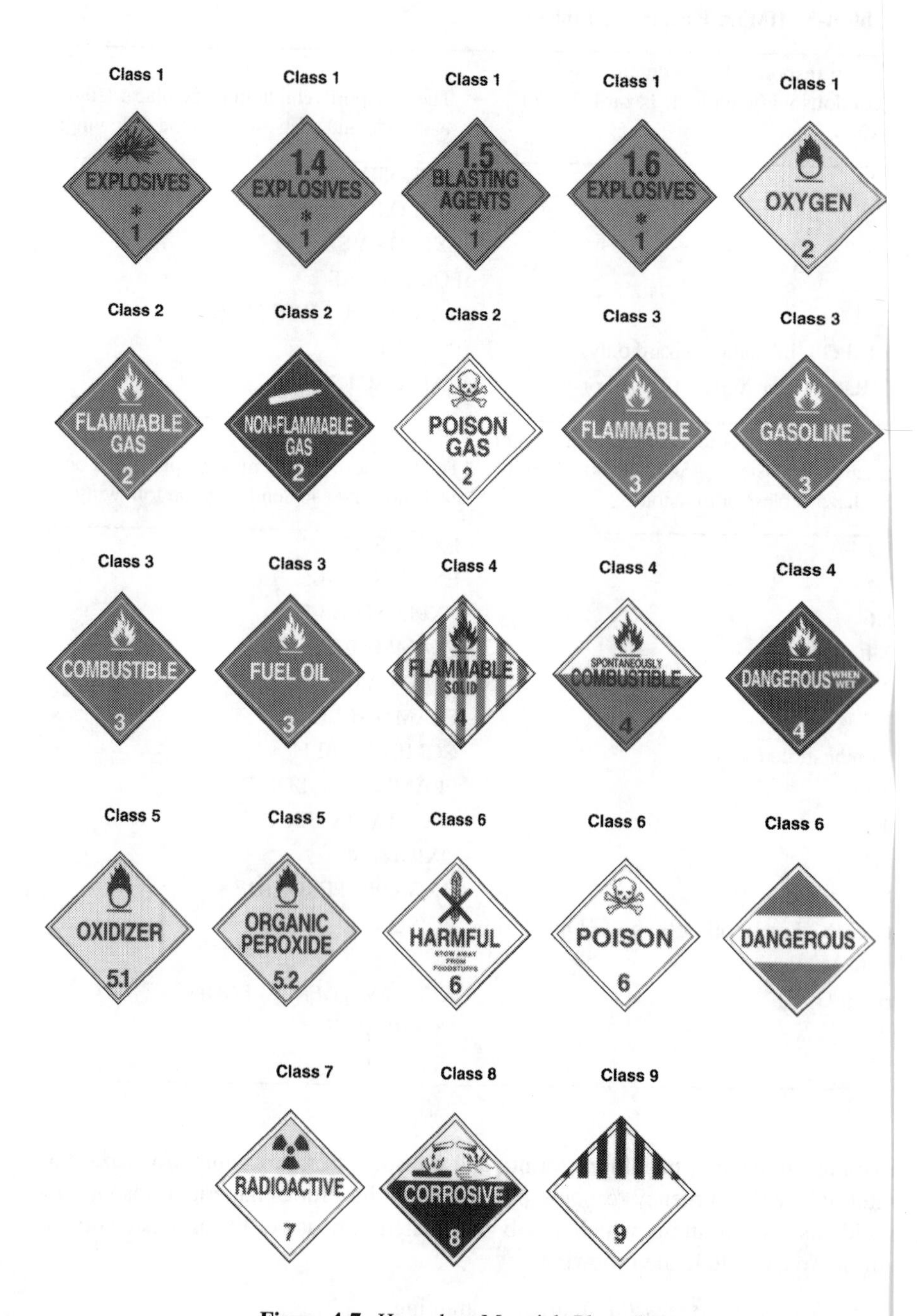

Figure 4-7. Hazardous Materials Placards

General Awareness/Familiarization Training is designed to familiarize employees with the HMTA requirements for shippers and transporters and to ensure that employees recognize and identify hazardous materials [40 CFR 172.704(a)(1)].

Function-Specific Training addresses the specific HMTA requirements that are applicable to the employee's job function [40 CFR 172.704(a)(2)].

Safety Training addresses emergency response information, measures to protect employees from exposure to hazardous materials, and accident prevention measures [40 CFR 172.704(a)(3)].

Testing on each of the above assesses employees' proficiency [40 CFR 172.702(d)].

Recordkeeping. Each employer must establish and maintain records documenting the current training status of each employee. These records must cover at least the previous two years; they must be maintained as long as the employee is employed and at least 90 days after the employee's termination. The records must include the following [49 CFR 172.704(d)]:

- Employee's name
- Date of the most recently completed training
- Description or copy of training materials used
- Name and address of the trainer
- Certification that the employee was trained in compliance with HMTA requirements

TRANSPORTERS

Under RCRA, a transporter of hazardous waste must comply with applicable HMTA requirements under 49 CFR, subchapter C [40 CFR 263.10(a)]. According to RCRA, a *transporter* is any person engaged in the off-site transportation of hazardous waste by air, rail, highway, or water (40 CFR 260.10). Correspondingly, under HMTA, a transporter is known as a carrier. A *carrier* is a person engaged in the transportation of passengers or property by land or water, as a common, contract, or private carrier, or civil aircraft (49 CFR 171.8).

General Transporter Requirements Under RCRA

The transportation requirements apply to any transporter of hazardous waste except for on-site movements (i.e., totally within private property). A transporter who imports waste into the customs territory of the United States or places waste of different HMTA shipping descriptions into a common container is also required to comply with the generator requirements of 40 CFR part 262 [40 CFR 263.10(c)].

Under RCRA, transporters of hazardous waste must comply with the following:

- EPA identification number
- Manifest system
- Transfer facilities
- Delivery of shipments
- Releases of hazardous waste

EPA Identification Number. A transporter may not transport hazardous waste without having an EPA identification number. This number can be obtained by submitting EPA Form 8700-12 to the state, if authorized, or to the EPA Regional Office (40 CFR 263.11). Upon receipt of a complete form, EPA (or the state) will issue the transporter a unique identification number. Transporters with either multiple terminals or vehicles should possess only one identification number. Because the EPA identification number is used for a physical location, the number typically is obtained for a company's headquarters.

Manifest System. Shipments of all hazardous waste, except for conditionally exempt small-quantity generator wastes (and certain contractually controlled waste shipments under a tolling arrangement), must be accompanied by a Uniform Hazardous Waste Manifest (40 CFR 263.20). Although the shipper is responsible for properly completing the manifest, the transporter may not accept the shipment unless the manifest is accurate, complete, and signed by the generator. The initial transporter who accepts the shipment must sign (i.e., handwritten) and date the manifest and present a copy of the signed manifest to the generator before leaving the property (49 CFR 172.204). The manifest must accompany the shipment at all times (except for railroad and water shipments). The manifest must have at least three copies—one for the transporter and two for the designated facility—unless more than one transporter will be used, in which case more copies will be necessary.

A transporter who delivers the shipment to another transporter or to the designated facility must obtain the date of delivery and a handwritten signature of the transporter or the designated facility, retain a copy of the signed manifest, and give the original and remaining copies of the manifest to either the next transporter or the designated facility. Each transporter must retain a copy of the signed manifest for three years (40 CFR 263.22).

Transfer Facilities. A transporter is allowed to store a manifested shipment of hazardous waste at a transfer facility without a permit for up to 10 days during the normal course of transportation (40 CFR 263.12). (It should be noted that this provision differs widely among states.) *Transfer facility* means any transporter-related facility, including loading docks, parking areas, storage areas, and other similar areas where shipments of hazardous waste are held during the normal course of transportation (40 CFR 260.10).

Delivery of Shipments. The entire shipment of hazardous waste must be delivered to either the designated facility or the designated alternate facility listed in Items 9 and 15 on the manifest, respectively. If the waste cannot be delivered to the designated facility or, if listed, the alternate facility, the transporter must contact the generator immediately, obtain instructions on where to transport the waste, and revise the manifest accordingly (40 CFR 263.21).

Releases of Hazardous Waste. In the case of a spill by a transporter, the appropriate authorities must be notified immediately, and the transporter must attempt to contain the spill (40 CFR 263.30). If a government official (state, federal, or local) determines that immediate cleanup is required to protect human health or the environment, the official may require appropriate action, such as removal of the waste by transporters who do not have an EPA identification number, manifest, or permit (40 CFR 263.31). A person may receive an emergency EPA identification number or permit from an authorized state or EPA Regional Office.

The transporter must notify the National Response Center (1-800-424-8802) if the spill is large enough to be a reportable quantity (49 CFR 171.15). The transporter should be able to determine quickly if the spill is a reportable quantity by verifying whether a container held a reportable quantity, which would be noted on the manifest and container markings.

Note: Although under HMTA a "hazardous substance" is determined on a *per container* basis, spills from multiple containers not defined as a "hazardous substance" under HMTA could meet or exceed the Superfund reportable quantity, and would have to be reported immediately to the National Response Center (40 CFR 302.6). (See Chapter 12.)

There are additional reporting requirements under the Emergency Planning and Community Right-to-Know Act (40 CFR 355.40). (See Chapter 12 for a detailed presentation of the spill-reporting requirements under federal laws.)

General Transporter Requirements Under HMTA

Transporters of hazardous waste are subject to the HMTA regulations for hazardous materials transportation because hazardous waste is merely a subset of hazardous material.[7] Basically, all "for-hire" transporters and all "in-house" transporters are subject to the same requirements. In general, the following are the basic requirements for hazardous waste transporters under HMTA:

- Driver qualifications
- Shipment handling
- Determining the acceptability of a shipment
- Shipment routing

[7] Federal regulations allow generators to transport their own hazardous waste to the designated facility, provided their EPA identification number is amended to include transportation; the applicable regulations under RCRA and HMTA are followed for hazardous waste transportation; and, if appropriate, the financial responsibility and liability requirements under the Federal Motor Carrier Act are met. Some states, however, are more restrictive.

- Incident reports

Driver Qualifications. Transporters must ensure that their employees engaged in receiving, processing, or transporting hazardous waste are thoroughly instructed in the applicable HMTA regulations relevant to their job functions (49 CFR 174.7, 175.20, 176.13, and 177.800).

Shipment Handling. Shipments of hazardous waste must be properly handled to minimize the possibility of a release or the accidental mixing of incompatible wastes (49 CFR 117.848). Before hazardous wastes are accepted and loaded, the transporter must determine whether the wastes are compatible for shipment. DOT has established requirements for segregation to ensure safe handling during transportation. The Segregation Table for Hazardous Materials lists materials by hazard class or division that may or may not be loaded together (49 CFR 177.848). Hazardous wastes must be loaded, transported, and stored in accordance with this table.

Determining the Acceptability of a Shipment. A transporter must ensure that a shipment of hazardous waste is prepared for transportation in accordance with 49 CFR, parts 171, 172, and 173 [49 CFR 174.3, 175.3, 176.3, and 177.801(a)], including the following criteria:

- The manifest is prepared in the proper format and is accurate and complete.
- The manifest matches the shipment.
- Damaged or leaking materials are not loaded.
- The shipment is properly blocked and braced to prevent movement and/or damage while in transit.
- Proper placards and identification numbers are displayed.

Shipment Routing. If a transporter is transporting hazardous waste in a vehicle that is required to be marked or placarded, DOT routing requirements must be followed (49 CFR 397.9). The route taken by the operator must be planned carefully in advance. Unless there are no practicable alternatives, the vehicle must not be operated on routes that pass through or near the following:

- Heavily populated areas
- Places where crowds are assembled
- Tunnels
- Narrow streets or alleys

Incident Reports. Transporters are required to file reports of any accident or incident involving the spillage or unintentional release of any hazardous waste from its packaging (40 CFR 171.15 and 171.16). At the earliest practicable moment, each transporter who transports hazardous waste must notify the National Response Center (1-800-424-8802) after any incident that occurs during the course of transportation (including loading, unloading, and temporary storage) in which, as a direct result of hazardous waste, any of the following occurs:

- A person is killed.
- A person is injured such that admittance to a hospital is required.
- Estimated property damage exceeds $50,000.
- The general public is evacuated for a period lasting one or more hours.
- One (or more) major transport arteries is (are) closed for one or more hours.
- Fire, breakage, spillage, or suspected contamination occurs involving a shipment of radioactive material or etiological agents (for etiological agents, reporting to the Center for Diseases Control at 404-633-5313 may be made in lieu of the National Response Center).
- A situation occurs that the transporter believes should be brought to the attention of authorities, even though none of the above criteria is met.

Within 30 days of the date of discovery for each incident that occurs during the course of transportation (including loading, unloading, or temporary storage) in which, as a direct result of the hazardous waste, any of the circumstances warranting an immediate telephone report have occurred, or if there has been an unintentional release of hazardous waste from a package (including a tank), the transporter must prepare and submit a written report (in duplicate) on DOT Form F-5800.1. If the report involves a release of a hazardous waste, a copy of the manifest also must be included as part of the report (40 CFR 171.16). The report must be sent to the following party:

Information Systems Manager
Research & Special Programs Administration
U.S. Department of Transportation
Washington, DC 20590

HAZARDOUS WASTE SHIPPING/TRANSPORTATION COMPLIANCE CHECKLIST

Shipper

Hazardous Classification Procedure

Are the physical and chemical characteristics of the waste accurately identified? Yes___ No___

Has the proper shipping name been selected? Yes___ No___

Does a plus (+) sign appear in Column 1 of the Hazardous Materials Table? Yes___ No___

(If no, does the waste meet the suggested hazard class?) Yes___ No___

Has the hazard class or division been identified? Yes___ No___

Has the UN or NA identification number been identified? Yes___ No___

Has the packing group been determined? Yes___ No___

Does the waste meet the definition of a HMTA hazardous substance? Yes___ No___

Packaging

Have the proper packaging requirements been determined? Yes___ No___

Has the proper packaging been obtained? Yes___ No___

Labeling

Has the proper labeling been identified and affixed? Yes___ No___

Marking

Is the package adequately marked with the following?

- Proper shipping name Yes___ No___
- Shipper's address Yes___ No___
- UN/NA identification number Yes___ No___
- Orientation arrows Yes___ No___
- Hazardous waste designation Yes___ No___
- The letters RQ if it is a hazardous substance Yes___ No___
- Technical name(s) of hazardous components if they are not listed Yes___ No___

Shipping Papers

Has a uniform hazardous waste manifest been obtained? Yes___ No___

Are there enough copies for all participants in the shipment? Yes___ No___

Is a continuation sheet necessary (i.e., more than four wastes)? Yes___ No___

Has a unique manifest document number been listed? Yes___ No___

Has each waste's basic shipping description (name, hazard class, identification number, packing group) been noted on the manifest? Yes___ No___

Has the word WASTE been included as part of the basic shipping description? Yes___ No___

Does the waste meet the definition of a hazardous substance? Yes___ No___

If so, has the designation RQ been added? Yes___ No___

Has the total quantity and container type been listed for each waste? Yes___ No___

Are any additional descriptions required for the particular waste stream? Yes___ No___

Has the EPA ID for the generator, each transporter, and the designated facility been listed? Yes___ No___

Have the names and addresses of the generator and designated facility been listed? Yes___ No___

Has the emergency response contact been included? Yes___ No___

Has the emergency response information been included? Yes___ No___

Has the manifest been signed and dated by hand? Yes___ No___

Has a copy of the manifest been maintained? Yes___ No___

Are copies of the manifests retained for three years? Yes___ No___

Were all manifests signed and dated? Yes___ No___

Were the handwritten signature and date of acceptance from the initial transporter obtained? Yes___ No___

Is one copy of the manifest signed by the generator and transporter retained? Yes___ No___

Do return copies of the manifest include a facility owner/operator signature and date of acceptance? Yes___ No___

If the copy of the manifest from the facility was not returned within 45 days, was a manifest exception report filed? Yes___ No___

If yes, did it contain:

- a legible copy of the manifest? Yes___ No___
- a cover letter explaining efforts to locate the waste and the results of those efforts? Yes___ No___

Placards

Have the proper placards been obtained and available to the transporter? Yes___ No___

Transporter

General Requirements

Does the transporter have an EPA ID number to cover transportation? Yes___ No___

Is the waste being imported into the United States? Yes___ No___

If yes, have the generator requirements been followed? Yes___ No___

Are hazardous wastes of different shipping description in a single container? Yes___ No___

If yes, have the generator requirements been followed? Yes___ No___

Have the appropriate routes been determined? Yes___ No___

Shipping Papers

Does the shipment match the information contained in the manifest? Yes___ No___

Does all required information appear on the manifest? Yes___ No___

Has the manifest been signed and dated? Yes___ No___

Has a signed and dated manifest been maintained? Yes___ No___

Are all shipments of hazardous wastes accomplished by manifest? Yes___ No___

If export of hazardous waste(s) out of the United States is intended, are the date of exit and name and address of the receiving facility indicated on the manifest? Yes___ No___

Is all waste shipped to either the designated facility listed on the manifest or the alternate facility (when applicable) or the next designated transporter? Yes___ No___

Are copies of manifests and shipping papers retained for the required three-year period? Yes___ No___

Shipment Preparation

Has the shipment been properly identified, packaged, marked, labeled, and not leaking? Yes___ No___

Have the proper placards been affixed to the vehicle? Yes___ No___

Do the number of containers match the number listed on the manifest? Yes___ No___

Are potentially incompatible wastes segregated from each other? Yes___ No___

Is the shipment properly secured, blocked, and braced? Yes___ No___

Is the condition of containerization safe for transportation (intact, nonleaking, corrosion free, not fuming, not damaged, properly sealed, proper lining, and so forth)? Yes___ No___

Is the emergency response information readily available? Yes___ No___

Is the emergency response contact information readily available? Yes___ No___

Has the shipper marked each container of 110 gallons or less with the following words and information displayed in accordance with the requirements? Yes___ No___

HAZARDOUS WASTE—Federal law prohibits improper disposal. If found, contact the nearest police or public safety authority or the U.S. Environmental Protection Agency.

Generator's Name and Address ______________________________

Manifest Document Number______________________________

Transfer Facilities

Is hazardous waste stored at transfer facilities? Yes___ No___

If yes, are manifested shipments stored in authorized packaging? Yes___ No___

Is all hazardous waste stored at a transfer facility shipped off-site within 10 days? Yes___ No___

Spills/Incidents

Has the transporter been involved in a discharge of hazardous wastes? Yes___ No___

If yes, was the National Response Center (800-424-8802), state, and principal office of the transporter notified? Yes___ No___

Has the transporter obtained an Emergency Identification Number from EPA for the cleanup operation? Yes___ No___

Was a written report submitted to DOT within 10 days following the discharge? Yes___ No___

5

GENERAL STANDARDS FOR HAZARDOUS WASTE MANAGEMENT FACILITIES

All hazardous waste management facilities must be designed, constructed, and operated in a manner that is protective of human health and the environment. Because there are significant differences among the types of hazardous waste generated, there are also major differences among the types of waste management facilities. As a result of these differences, there are two major sets of standards: general standards applicable to all hazardous waste management facilities and standards unique to a specific type of hazardous waste management unit. This chapter outlines the general standards for owners and operators of hazardous waste management facilities (the unit-specific standards are addressed in Chapter 6). These general standards are divided into sections addressing the following:

- General facility standards
- Groundwater monitoring
- Closure/post-closure activities
- Financial responsibility requirements

Any person who treats, stores, or disposes of hazardous waste is considered to be an owner or operator of a treatment, storage, or disposal facility (TSDF, also referred to as a hazardous waste management facility or a designated facility) and is subject to the requirements of 40 CFR part 264 or 265, unless excluded by 40 CFR 264/265.1.

Treatment, storage, and disposal are thus defined in 40 CFR 260.10:

Treatment is any method, technique, or process, including neutralization, designed to change the physical, chemical, or biological character or composition of any hazardous waste so as to neutralize such waste, or so as to recover energy or material resources from the waste, or so as to render such waste nonhazardous or less hazardous; safer to transport, store, or dispose of; or amenable to recovery, amenable to storage, or reduced in volume.

Storage is the holding of hazardous waste for a temporary period, at the end of which the hazardous waste is treated, disposed of, or stored elsewhere.

Disposal is the discharge, deposit, injection, dumping, spilling, leaking, or placing of any solid waste or hazardous waste into or on any land or water so that such solid waste or hazardous waste or any constituent thereof may enter the environment or be emitted into the air or discharged into any waters, including groundwaters.

As stated in 40 CFR 264.1(g) and 265.1(c), the following persons or activities are excluded from the requirements of 40 CFR, parts 264 and 265:

- Facilities approved by the state to handle conditionally exempt small-quantity generator waste (< 100 kg/mo) exclusively.
- A totally enclosed treatment facility as defined in 40 CFR 260.10. It is important to note that the exemption is for the unit itself; effluent is not excluded. In addition, a totally enclosed treatment facility must meet the following conditions, according to OSWER Directives 9432.01(83) and 9432.02(84):
  - — Be completely contained on all sides.
  - — Pose negligible potential for escape of constituents to the environment.
  - — Be connected directly to a pipeline or similar totally enclosed device to an industrial production process, which produces a product, by-product, or intermediate material that is reused in the process.
- A generator accumulating hazardous waste in compliance with 40 CFR 262.34.
- A farmer disposing of waste pesticides in compliance with 40 CFR 262.51.
- The management of hazardous wastes in an elementary neutralization unit, defined as a container, tank, transport vehicle, or vessel for neutralizing waste that is hazardous solely because of corrosivity characteristic (45 *FR* 767074, November 17, 1980).
- Wastewater treatment units that manage hazardous wastewater, meet the definition of a tank, and have a discharge subject to Clean Water Act pretreatment standards or permitting requirements.

  Note: Only the wastewater treatment unit itself is excluded. Any sludge generated and removed from the treatment unit is subject to regulation as a hazardous waste (45 *FR* 767074, November 17, 1980).

- A person engaged in the immediate treatment or containment of a discharge of hazardous waste.

- A transporter storing manifested waste at a transfer facility in compliance with 40 CFR 262.30.
- The addition of absorbent material to waste or the addition of waste to absorbent material.
- A universal waste handler or transporter as defined in 40 CFR part 273.

Although these persons or activities are generally excluded from parts 264 and 265, there are specific regulatory references to these parts for some of these persons or activities. For example, generators accumulating hazardous waste onsite must comply with specified sections of part 265. (See 40 CFR 262.34 for these references.)

The basic difference between 40 CFR part 265 (interim status standards) and 40 CFR part 264 (final operating standards) is that the interim status standards were written for facilities treating, storing, or disposing of hazardous waste when the RCRA regulations first went into effect on November 19, 1980. Congress wanted EPA to establish interim standards that would allow facilities to continue to operate as though they had a permit while EPA developed the more stringent part 264 standards for new and existing facilities. A facility received interim status by filing a RCRA §3010 notification (EPA Form 8700-12, *Notification of Hazardous Waste Activity*) by August 26, 1980, and a Part A permit application by November 19, 1980. (Refer to Chapter 7 for a more detailed discussion of interim status.) An owner or operator who qualified for and obtained interim status remains subject to the part 265 standards until the final administrative disposition of the facility's permit application is made. Another important distinction is that the part 265 standards are self-implementing, whereas the part 264 standards are used to establish the minimum requirements, in the form of conditions, of a facility's permit as discussed in Chapter 7.

GENERAL FACILITY STANDARDS

The general facility standards are applicable to all RCRA hazardous waste management facilities unless the facility is specifically excluded. The general facility standards are found in subpart B of parts 264 and 265 and are essentially the same for both permitted and interim status facilities. Any differences are noted in the following text. The general facility standards are separated into the following sections:

- Notification and record-keeping requirements
- General waste handling requirements
- Preparedness and prevention
- Contingency plan
- Emergency procedures
- Manifest system

Notification and Record Keeping

This section addresses the requirements for the following:

- EPA identification number
- Required notices
- Record keeping
- Operating log
- Biennial report

EPA Identification Number. Owners or operators of a hazardous waste management facility must have an EPA identification number before they can accept hazardous waste from off-site sources or generate, treat, store, or dispose of hazardous waste. The EPA identification number is obtained by submitting EPA Form 8700-12 (40 CFR 264/265.11).

Required Notices. Before transferring ownership or operational control of a hazardous waste management facility during its operating life, or of a land disposal facility during its post-closure care period, the owner or operator must notify the new owner or operator in writing of the requirements of Subtitle C of RCRA. However, even if the old owner or operator fails to comply with this requirement, the new owner or operator must comply with all the applicable RCRA regulations [45 *FR* 33179, May 19, 1980, and 40 CFR 264.12(c) and 265.12(b)].

Only owners or operators of permitted units must report to the EPA Regional Administrator every six months, indicating any instances during the reporting period in which an air emission control device exceeded or operated outside of its design specifications for longer than 24 hours without being corrected (40 CFR 264.1065).

Importing Hazardous Waste. Hazardous waste management facilities receiving waste from a foreign source must notify the appropriate EPA Regional Office of the shipment at least 4 weeks before receiving the shipment. Subsequent shipments from the same foreign source containing the same waste do not require notification [40 CFR 264/265.12(a)].

When importing hazardous wastes into the United States, the importer, who must be a U.S. citizen, must comply with §13 of the Toxic Substances Control Act (TSCA). This section requires all importers of *chemical substances* (see Chapter 16 for a definition) to certify either that the shipment is in compliance with all applicable rules and regulations under TSCA, or that it is not subject to TSCA. Hazardous wastes that are not subject to TSCA under §13 (i.e., materials that are not included under the definition of chemical substance) are limited to active pesticide ingredients, drugs, cosmetics, and nuclear materials. In determining whether a material is in compliance with TSCA, there are two primary concerns: Every chemical component in the waste stream must be an "existing chemical," meaning that the chemical must be included in EPA's TSCA Chemical Substances Inventory; and none of the chemical components of the waste stream is prohibited from importation. Regulatory information pertaining to the import status of chemical substances under TSCA can be obtained from EPA's TSCA assistance hotline at 202-554-1404.

Importing OECD-Covered Waste. Facilities that have arranged to receive OECD-covered hazardous waste under 40 CFR part 262, subpart H (see Chapter 3 for a detailed discussion), must provide a copy of the tracking document bearing all required signatures to the notifier, EPA, and the competent authorities of all other concerned countries within 3 working days of receipt of the shipment.[1] The original of the signed tracking document must be maintained for at least 3 years [40 CFR 264/265.12(a)(2)].

Record Keeping. All records required to be maintained for the various regulations must be made available to any EPA-designated inspector upon request. Unless otherwise specified, all records must be maintained for at least three years after the recorded activity ceases. In addition, a copy of all records containing waste locations and their quantities must be submitted at closure, if applicable.

Operating Log. An owner or operator must maintain a written operating record (log) at the facility until closure (40 CFR 264/265.73). The contents of the operating record must include the following:

- Description of waste received and the method, quantity, and date of treatment, storage, or disposal.
- Location and quantity of all hazardous waste at the facility (landfills require a map).
- Record and results of waste analyses.
- Details of any instances requiring implementation of the contingency plan.
- Records and results of inspection.
- Results of required groundwater monitoring analyses.
- Information pertaining to the operation, inspection, and monitoring of any required air emission control devices.
- Waste minimization statement if applicable.
- The information contained in the notice, certification, and demonstration as applicable to comply with the land disposal restrictions.
- Closure cost estimates.
- Post-closure cost estimates (land disposal facilities only).
- Notices to generators (part 264 only).

[1] The EPA address designated to receive the OECD tracking document is as follows: The Office of Compliance Assurance, Office of Compliance, Enforcement Planning, Targeting and Data Division (2222A), Environmental Protection Agency, 401 M Street, SW, Washington, DC 20460.

In addition to the above requirements for the operating log, the following records must be maintained by the facility:

- Closure plan (40 CFR 264/5.112)
- Contingency plan (40 CFR 264/5.51 & 53)
- Groundwater monitoring (40 CFR 264.97(j) & 265.93-94)
- Inspection schedule (40 CFR 264/5.15)
- Land disposal restrictions (40 CFR 268.7 & 19)
- Manifest system (40 CFR 264/5.71-77)
- Personnel training (40 CFR 264/5.16)
- Post-closure plan (40 CFR 264/5.118)
- Waste analysis (40 CFR 264/5.13)

Biennial Report. The owner or operator must submit a biennial report to the designated state agency (or EPA) before March 1 of each even-numbered year (40 CFR 264/265.75). It is important to note that most *states* require annual reporting.) The required contents of the report are as follows:

- The facility's EPA identification number.
- The EPA identification number of each generator that sent waste to the facility.
- The quantity and description of each hazardous waste received.
- The method of treatment, storage, or disposal for each waste.
- The most recent closure and post-closure cost estimate.
- A signed statement certifying the accuracy of the information.

General Waste Handling Requirements

This section addresses the following waste handling requirements:

- Security
- Personnel training
- General waste analysis
- Unstable waste handling
- Facility inspections
- Location standards

Security. An owner or operator must either establish a security system or upgrade an existing system to prevent unknowing entry and minimize unauthorized entry of any person or livestock into the active portion of a facility. Either a 24-hour surveillance system or an access barrier around the active portion of a facility is required. A fence with locked gates is considered an access barrier. Signs must be prominently displayed along the periphery of the active portion, warning that entering the active portion is potentially hazardous (40 CFR 264/265.14).

Personnel Training. Each facility must establish a training program for appropriate facility personnel (40 CFR 264/265.16). *Facility personnel* means all persons who work at or oversee the operations of a hazardous waste facility, and whose actions or failure to act may result in noncompliance with the requirements of 40 CFR parts 264 or 265 (40 CFR 260.10). The purpose of the training requirements is to reduce the potential for errors that might threaten human health or the environment by ensuring that facility personnel acquire expertise in the areas to which they are assigned. As shown in Table 5-1, in addition to the RCRA requirements, the Occupational Safety and Health Administration (OSHA) has established training requirements for facility personnel involved in hazardous waste operations. There are also training requirements under HMTA if the facility is involved in shipping hazardous waste as described in Chapter 4. This training, however, may be combined with the training described as follows.

RCRA Training Requirements These are personnel training requirements under RCRA:

- Facility personnel must successfully complete a training program that ensures the facility's compliance with the requirements of RCRA.
- The training program must be directed by a person trained in hazardous waste management procedures.
- The training program must be designed to ensure that facility personnel are able to respond adequately during an emergency situation.
- Facility personnel must successfully complete the program within six months of their assignment.
- Facility personnel must take part in an annual review of the training program.

The content and format of the program are unspecified in the regulations except that the following subjects must be addressed:

- Procedures for using, inspecting, repairing, and replacing facility emergency and monitoring equipment.
- Key parameters for automatic waste-feed cutoff systems.
- Communications or alarm systems.
- Response to fires or explosions.
- Responses to groundwater contamination incidents.
- Shutdown of operations.

EPA accepts the use of on-the-job training as a substitute for or supplement to formal classroom instruction (45 *FR* 33182, May 19, 1980). However, the job title,

TABLE 5-1. Summary of Training Requirements for Personnel Involved in Hazardous Waste/Materials Handling

Statute	Applicable Staff	Training Subjects	Training Requirements
RCRA	Hazardous waste handlers and emergency response personnel	• Procedures for using, inspecting, repairing, and replacing facility emergency and monitoring equipment. • Key parameters for automatic waste-feed cutoff systems. • Communications or alarm systems. • Response to fires or explosions. • Responses to groundwater combination incidents. • Shutdown of operations.	No specific initial training requirements; annual update.
OSHA (HAZWOPER)	Hazardous waste handlers and emergency response personnel	• A program for developing and implementing new response technologies (e.g., foams, absorbents, neutralizers). • At facilities where employees will handle drums or containers, a special material handling program must be developed to ensure that employees handle drums properly. • Employees are to be informed of the degree and nature of any safety and health hazards specific to the work site.	No specific initial training requirements; annual update.
OSHA (Hazard Communication Standard)	Hazardous chemical handlers	• Employees must be trained to recognize hazards and protect themselves. • Employees must be trained on how they can access information to protect their health.	No specific initial training requirements; annual update.
HMTA	Hazardous materials handlers	• Proper loading, unloading, storing, and transporting hazardous materials. • Emergency spill response.	No specific initial training requirements; periodic updates.

employee name, training program content, schedule of training taken, and techniques used for training must be described in the training records, which must be maintained at the generator's facility; this training is subject to approval during the permitting process. [40 CFR 265.16(d)].

OSHA Training Requirements. On March 6, 1989 (54 *FR* 92921), OSHA promulgated a final rule to protect the health and safety of employees engaged in hazardous waste operations. Only some of these regulations [20 CFR 1910.120(p)] are applicable to hazardous waste management facilities. Guidance on health and safety is also contained in the *Occupational Safety and Health Guidance Manual for Hazardous Waste Site Activities,* NIOSH/OSHA/USCG/EPA, 1985.

The major elements of the OSHA regulations regarding employees at RCRA hazardous waste management facilities include the following:

Safety and Health Program. Each employer must develop and implement a written safety and health program for its employees involved in hazardous waste operations. The program must identify, evaluate, and control safety and health hazards for the purpose of employee protection.

Hazard Communication. Employees, contractors, and subcontractors are to be informed of the degree and nature of any safety and health hazards specific to the work site.

Medical Surveillance Program. A medical surveillance program must be instituted for employees involved in hazardous waste operations. Examinations are to be conducted before an assignment, annually, upon an employee's termination, and as soon as possible if an employee has symptoms that may indicate an exposure.

Decontamination Program. Decontamination procedures must be developed and implemented for employees to minimize contact between contaminated personnel and equipment and uncontaminated equipment and personnel.

New Technology Program. A program for developing and implementing procedures must be instituted for the introduction of effective new technologies and equipment developed for the improved protection of employees. These new technologies may include use of foams, absorbents, adsorbents, neutralizers, and so forth.

Material Handling Program. At facilities where drums or containers will be handled by employees, a special material handling program must be developed. The purpose of this program is to ensure that employees handle drums properly.

Training Program. A training program that includes both initial (24 hours) and refresher (8 hours annually) training must be provided to employees before they engage in hazardous waste operations that could expose them to hazardous substances or safety or health hazards. A written certificate attesting they have successfully completed the program is required.

Emergency Response Plan. In addition to the health and safety programs, each employer must establish an emergency response plan [29 CFR 1910.120(p)(8)]. Many of the requirements may already be addressed by the facility's program for preparedness and prevention, its contingency plan, and its emergency procedures. The emergency response plan must be part of the required written safety and health plan, and it must address the following:

- Pre-emergency planning and coordination with outside parties.
- Personnel roles, lines of authority, and communication.
- Emergency recognition and prevention.
- Safe distances and places of refuge.
- Site security and control.
- Evacuation routes and procedures.
- Decontamination procedures.
- Emergency medical treatment and first aid.
- Emergency alerting and response procedures.
- Critique of response and follow-up.
- Personal protective equipment and emergency equipment.

The employees must be trained specifically for the emergency response procedures before an employee is called upon to perform in an emergency [29 CFR 1910.120(p)(8)(iii)]. This training must include the elements of the emergency response plan discussed in this bulleted list, standard operating procedures, and personal protective equipment to be worn during emergency operations.

General Waste Analysis. An owner or operator of a hazardous waste management facility must obtain a detailed analysis of representative samples of any waste before that waste is managed. Beyond the general waste analysis requirements, the regulations for each specific unit (e.g., surface impoundments, incinerators, land treatment area) addressed in Chapter 6 include additional waste analysis requirements that are appropriate for that unit [40 CFR 264/265.13(a)]. The general waste analysis requirements distinguish between two types of hazardous waste management facilities: on-site and off-site facilities.

Waste Analysis Plan. Each facility must prepare and maintain a waste analysis plan to implement the waste analysis requirements. The objective of a waste analysis plan is to describe the procedures that will be undertaken to obtain sufficient information about a waste to ensure that a facility will handle the waste in accordance with its permit or interim status requirements. The waste analysis plan establishes the hazardous waste sampling and analysis procedures that the facility must follow. These objectives are the same for both on-site and off-site facilities. However, because on-site facilities are better acquainted with the waste generation process and its characteristics, off-site facilities are required to conduct more frequent checks on wastes than on-site facilities.

At a minimum, the waste analysis plan must contain the following elements:

- Procedures to ensure that the waste from off-site sources is the waste described in the manifest.
- Sampling methods.
- Testing and analytical methods.
- Parameters to be analyzed.
- Frequency of waste reevaluation or spot-check analysis for facilities receiving off-site shipments.
- Acceptance and rejection criteria for unacceptable wastes received from off-site sources.

Handling Requirements for Unstable Waste. Ignitable, reactive, and incompatible wastes must be handled in a way that prevents accidental ignition or reaction. While ignitable or reactive wastes are being handled, smoking and operations involving open flames (e.g., welding activities) must be limited to specially designed locations. NO SMOKING signs must be conspicuously placed wherever there is a potential hazard from ignitable or reactive wastes (40 CFR 264/265.17).

Facility Inspections. An owner or operator must develop and follow an inspection schedule written specifically for that facility based on its critical processes, equipment, and structures, and on the potential for failure and the rate of any processes that may promote deterioration. The owner or operator must inspect the facility for the following [40 CFR 264/265.15(a)]:

- Monitoring equipment, if applicable
- Safety and emergency equipment
- Security devices
- Operating and structural equipment
- Malfunction or deterioration (e.g., inoperative sump pump, leaking fitting, eroding dike, corroded pipes or tanks)
- Operator error
- Discharges (e.g., leaks from valves or pipes, joint breaks)

If the owner or operator finds any malfunctioning equipment or structures, the equipment or structure must be repaired or replaced immediately [40 CFR 264/265.15(c)]. All inspections must be noted by date and time of inspection, name of inspector, description of observations, and the date and nature of any necessary repairs or corrective actions. The inspection information must be maintained in the facility's operating record for at least three years [40 CFR 264/265.15(d)].

Location Standards. Although there are no site-specific location standards based solely on hydrogeologic considerations, the groundwater protection standards, as well as general design and operating requirements, contain performance standards that implicitly involve hydrologic and geologic factors. Current regulations do not provide the legal basis to deny a RCRA permit based on sensitive locations, such as vulnerable groundwater formations, although a RCRA §7003 order can be used if there *may* be an imminent threat to health or the environment. (See Chapter 10 for further explanation of this topic.) In addition, EPA's omnibus permit authority under §3005(c)(3) of RCRA may be used as necessary to protect human health and the environment when a facility is sited in a poor location (e.g., involving unstable terrain, karst terrain, high population densities, complex and vulnerable hydrogeology, and near drinking-water recharge zones).

Currently, the only location restrictions are the following:

- The facility must be at least 200 feet from an active (during the last 10,000 years) Holocene fault.
- Facilities in a 100-year floodplain must be designed to prevent washout from 100-year floods.

Note, though, that facilities are allowed in a 100-year floodplain only if one of the three following conditions is met:

- The facility is protected, using dikes or equivalent measures, from washout during a 100-year flood.
- All hazardous wastes can be removed to safe ground before flooding.
- It can be demonstrated that no adverse effects to human health or the environment will occur should flood waters reach the hazardous wastes.

New facilities in active fault zones are prohibited. However, only those facilities located in certain political jurisdictions listed in Appendix VI to 40 CFR part 264 are required to demonstrate compliance with these standards. These jurisdictions include areas in Alaska, Arizona, Colorado, Hawaii, Idaho, Montana, New Mexico, Utah, Washington, and Wyoming and all of the states of California and Nevada.

In addition to satisfying the floodplain and seismic standards, owners of operators of facilities on federally owned lands must consider Executive Order 11990 (Protection of Wetlands) if a facility failure could impact wetlands. Other location areas may be subject to review from other federal laws, state laws, or local ordinances; those locations include critical habitat for threatened and endangered species, predominantly poor or minority locations (the environmental justice program), historic preservation areas, and culturally significant areas.

For further information concerning current location standards, consult EPA's *Permit Writer's Guidance Manual for Hazardous Waste Land Storage and Disposal Facilities,* February 1985.

Preparedness and Prevention

A facility must be operated and maintained in a manner that will minimize the possibility of any fire, explosion, or unplanned sudden or nonsudden release. Subpart C of parts 264 and 265 of 40 CFR, requires a facility to address the following:

- Required equipment
- Aisle space
- Outside assistance
- Access to communication equipment

Required Equipment. The facility must have certain equipment, including an alarm system, a communication device to contact emergency personnel (e.g., telephone, radio, air horn), portable fire extinguishers, fire control equipment, and an adequate firefighting water supply system in the form of hoses, hydrants, or automatic sprinkler systems. This equipment must be routinely tested and maintained in proper working order (40 CFR 264/265.32).

Aisle Space. There must be adequate aisle space to allow the unobstructed movement (i.e., deployment and evacuation) of emergency equipment and personnel to any area of the facility. The regulations do not specify the aisle space; this should be determined upon consultation with local emergency organizations (40 CFR 264/265.34).

Outside Assistance. Facilities must make prior arrangements with local emergency organizations and personnel for an emergency response. The arrangements should include notification of the types of waste handled, a detailed layout of the facility, facility contacts, and specific agreements with various state and local emergency response organizations. Refusal of any state or local authorities to enter into such arrangements must be documented and noted in the facility's operating record (40 CFR 264/265.37).

Access to Communication Equipment. Whenever hazardous waste is handled physically, all personnel involved in the operation must have immediate access to an internal alarm or emergency communication device to summon assistance directly or through visual or voice contact with another employee. If, however, there is only one employee on the premises, the employee must have direct, immediate access to communication equipment (40 CFR 264/265.33).

Contingency Plan

Each facility must prepare a contingency plan designed to minimize hazards in the case of a sudden or nonsudden release, fire, explosion, or similar emergency as presented previously in Table 3-3. On June 5, 1996 (61 *FR* 28642), the National Response Team announced the Integrated Contingency Plan (ICP). The intent of the

ICP is to provide a mechanism for consolidating multiple plans (e.g., RCRA, SPCC, OSHA's HAZWOPER, etc.) under various federal statutes into one functional emergency response plan. Thus, one "super" emergency response plan, following the ICP guidance, may be prepared instead of multiple emergency response plans. (See Table 3-4 for the basic elements of the ICP as suggested by the National Response Team.)

The provisions of the plan must be carried out immediately whenever there is a fire, explosion, or a release of hazardous waste or constituents (40 CFR 264/265.51). A copy of the contingency plan must be maintained at the facility and submitted to all local police, fire departments, hospitals, and emergency response teams (40 CFR 264/265.52). The plan must also be submitted by interim status facilities to EPA upon request. Permitted facilities would already have submitted the plan as part of their permit application. The contents of the plan must include the following:

- A description of the planned response actions that will be undertaken in the event of an emergency.
- Information on any arrangements with local and state organizations to provide emergency response support when needed.
- A list of current names, addresses, and phone numbers of all persons qualified as emergency coordinators. There must be at least one employee on-site or close by and on call who is designated *emergency coordinator.*
- A list of all available emergency equipment located at the facility, including the location and a physical description of each item and a brief outline of the equipment's capabilities.

The contingency plan must be amended if any of the following conditions applies:

- Whenever there are revisions to applicable interim status regulations.
- Whenever there are revisions to a facility's permit.
- If the plan fails in an emergency.
- If there are changes in the person(s) qualified to act as facility emergency coordinator.
- If there are changes in the facility design, construction, operation, maintenance, or other circumstances that materially increase the potential for fire, explosion, or releases of hazardous waste or change the response requirements in an emergency situation.

Emergency Procedures

In accordance with 40 CFR 264/265.56, whenever there is an imminent or an actual emergency situation, the emergency coordinator must do the following:

- Immediately activate the facility alarm system, notify all facility personnel, and, if needed, notify appropriate state or local agencies.
- Institute measures to prevent the spread of fires and explosions in other wastes at the facility.
- Immediately assess the possible hazards to the environment and to human health outside the facility.
- Immediately after the emergency, provide for the treating, storing, or disposing of any contaminated material as a result of the emergency.
- In the event of a release, fire, or explosion, identify the source, character, and amount of materials involved. If any hazardous substances are released into the environment constituting a Superfund reportable quantity, the owner or operator must immediately contact the National Response Center at 800-424-8802. (See Chapter 12 for further explanation of this topic.)

The owner or operator must submit a written report to EPA, within 15 days, for any incident that requires implementation of the contingency plan [40 CFR 264/265.56(j)]. The written report must include the following:

- Name, address, and telephone number of the owner or operator and the facility.
- Date, time, and type of accident (e.g., fire, explosion).
- Name and quantity of materials involved.
- The extent of injuries.
- An assessment of actual or potential hazards to human health or the environment.
- Estimated quantity and disposition of recovered material following the incident.

Emergency Coordinator. There must be one employee either on the facility premises or on call to serve as the designated emergency coordinator (40 CFR 264/265.55). The emergency coordinator is responsible for coordinating all emergency response measures and must be familiar with all aspects of the facility's contingency plan, operations, location of wastes, and physical layout. Also, the emergency coordinator must have the authority to commit any resources needed to carry out the contingency plan.

Manifest System

The manifest requirements are applicable only to facilities accepting hazardous waste from offsite sources. A *facility representative* must sign and date each copy of the manifest received from an off-site source, providing one copy to the transporter immediately. Within 30 days, a signed copy of the manifest must be sent directly to the generator, and a copy of the manifest must be retained for at least three years in the facility's files (40 CFR 264/265.71). If a shipment is initiated from a hazardous waste management facility, the facility must comply with the part 262 generator standards, which require that the receiving facility be a *designated facility.*

Manifest Discrepancies. Upon discovery of a significant discrepancy involving information contained in the manifest, the owner or operator must reconcile it with the generator or transporter verbally. If the discrepancy is not resolved within 15 days after receipt of the waste, the owner or operator must notify EPA in writing and describe the discrepancy (40 CFR 264/265.72). A *manifest discrepancy* is a significant difference between the type or quantity of waste received and the type or quantity of waste described on the manifest. *Significant discrepancies* include variations of 10 percent or more by weight of bulk waste, any variation in piece count (e.g., one drum), or a difference in waste description [40 CFR 264/265.72(a)].

Unmanifested Waste Report. If a facility receives a shipment of unmanifested waste that is not excluded by 40 CFR 261.5, the owner or operator must prepare and submit to EPA an unmanifested waste report (EPA Form 8700-13B; 40 CFR 264/265.76).

GROUNDWATER MONITORING

The goal of the RCRA groundwater monitoring program is to determine whether a land disposal unit has adversely impacted the quality of underlying groundwater. This is accomplished through field sampling, laboratory analysis, and statistical evaluation of the data. If the results of statistical tests indicate the groundwater has been impacted, additional monitoring, and possibly cleanup, will be required.

Groundwater monitoring must be conducted at RCRA-regulated hazardous waste management units (and radioactive mixed waste management units) where hazardous waste is stored or disposed of in or on the land. Such units include interim status and permitted surface impoundments, landfills, and land treatment units and permitted waste piles. Groundwater monitoring may also be required for certain miscellaneous units regulated under 40 CFR part 264, subpart X.

RCRA groundwater monitoring falls into two general categories: monitoring at interim status units and monitoring at permitted units. As shown in Figure 5-1, the interim status regulations under 40 CFR part 265 establish a two-stage groundwater program designed to detect and characterize the release of hazardous waste or constituents at interim status units, whereas the 40 CFR part 264 regulations establish a three-stage program designed to detect, evaluate, and clean up groundwater contamination from permitted land disposal units. At some point, even interim status units will have to comply with the part 264 groundwater regulations. Each of these programs is discussed separately in the following pages.

Groundwater Monitoring at Interim Status Units

Owners or operators of interim status hazardous waste management facilities with land treatment areas, landfills, waste piles, and/or surface impoundments must have

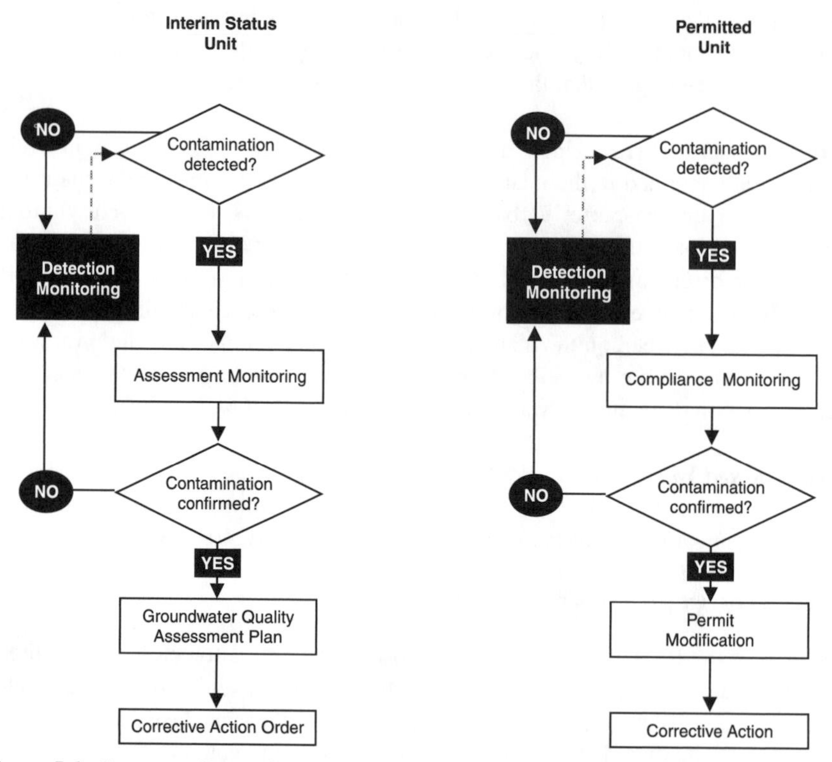

Figure 5-1. Summary Comparison of Interim Status and Permitted Groundwater Monitoring Programs.

a groundwater monitoring program capable of evaluating the impact of the facility on the quality of the groundwater underlying the site [40 CFR 265.90(a)].

The groundwater monitoring program is to be carried out during the *active life* of the hazardous waste management unit, which includes the closure and post-closure period. The post-closure period uses 30 years as a baseline, but this time period can be reduced or extended by EPA as appropriate [40 CFR 265.117(a)(1)].

The Part 265 groundwater monitoring program for interim status units consists of the following:

- Groundwater monitoring system
- Sampling and analysis plan
- Detection monitoring phase
- Assessment monitoring phase

Some facilities may qualify for a partial or complete waiver [40 CFR 265.90(c)] of the monitoring requirements if the owner or operator can demonstrate that there is a low potential for hazardous waste (or waste constituents) to migrate from the unit via the underlying groundwater to surface water or water-supply wells.

Note: Only a very small percentage of interim status facilities would be expected to qualify for such a waiver of all monitoring requirements based on the underlying hydrogeology.

The owner or operator claiming such a waiver must keep a detailed written demonstration, certified by a qualified geologist or geotechnical engineer, at the facility [40 CFR 265.90(c)].

Pursuant to 40 CFR 265.90(k), the written demonstration must establish the following:

- The potential for migration of hazardous waste or hazardous waste constituents from the facility to the underlying groundwater by an evaluation of
  - —A water balance of precipitation, evapotransportation, runoff, and infiltration.
  - —Unsaturated zone characteristics (i.e., geologic materials, physical properties, and depth to groundwater).
- The potential for hazardous waste or hazardous waste constituents that enter the underlying groundwater to migrate to a water supply well or surface water, by an evaluation of
  - —Saturated zone characteristics (i.e., geologic materials, physical properties, and rate of groundwater flow).
  - —The proximity of the facility to water-supply wells or surface water.

An additional waiver from the groundwater monitoring program is made for surface impoundments used solely to neutralize corrosive wastes. Such a waiver is based not on hydrogeologic factors but on the premise and documentation that only corrosive wastes will be added to the impoundment, and that the neutralization occurs so rapidly that the waste is no longer corrosive if it migrates out of the impoundment. This waiver also must be certified by "a qualified professional" and maintained at the facility [40 CFR 265.90(e)].

Groundwater Monitoring System. A groundwater monitoring system must be installed in a manner that will yield samples of sufficient quality to meet the technical and regulatory requirements of RCRA.[2] The monitoring wells must be cased in a manner that maintains the integrity of the monitoring well bore hole. This casing must be screened or perforated and, if necessary, packed with sand or gravel to enable sample collection at depths where appropriate aquifer flow zones exist. The annular space (the space between the bore hole and the well casing) above the selected sampling depth must be sealed with a suitable material, such as bentonite slurry or grout [40 CFR 265.91(a)].

If a facility has more than one hazardous waste land-based unit, a separate groundwater monitoring system is not necessarily required, provided that the system is adequate to detect any discharge from any of the units. The *waste management*

[2] See *RCRA Ground-Water Monitoring: Draft Technical Guidance* (EPA/530-R-93-001), U.S. Environmental Protection Agency, Office of Solid Waste, November 1992.

area in this situation would be described by an imaginary line that surrounds all of the units [40 CFR 265.91(b)].

Upgradient Wells. A minimum of one groundwater monitoring well must be installed hydraulically upgradient from the waste management area. The number, location, and depth of the well(s) must be sufficient to yield samples that are representative of the background groundwater quality in the uppermost aquifer and are not affected by the facility [40 CFR 265.91(a)(1)].

Downgradient Wells. A minimum of three monitoring wells must be installed hydraulically downgradient of the waste management area. Their number, location, and depth must ensure the immediate detection of any statistically significant amounts of hazardous wastes or hazardous waste constituents that migrate from the facility to the uppermost aquifer [40 CFR 265.91(a)(2)].

In general, the required minimum number of upgradient and downgradient wells (four) typically is not sufficient to detect contamination. There are many conditions that can complicate the detection of contaminants in the groundwater. Large multicomponent waste management areas consisting of several landfills, surface impoundments, or land treatment zones would require a greater number of monitoring wells than the minimum.[3] Units underlain by complex geological conditions also may have to use more monitoring wells. The minimum number of wells would be adequate only for a small unit in which the contaminants and hydrogeology were simple and well documented.

Sampling and Analysis Plan. An owner or operator must develop and follow a groundwater sampling and analysis plan for each groundwater monitoring system [40 CFR 265.92(a)]. This plan must be maintained at the facility. Owners or operators also must simultaneously prepare and maintain an outline of a *groundwater quality assessment program*, which is discussed below [40 CFR 265.93(a)].

The contents of the sampling and analysis plan must include procedures and techniques for:

- Sample collection
- Sample preservation
- Sample shipment
- Analytical procedures
- Chain-of-custody control

The chain-of-custody program should include sample labels, sample seals, a field log book, a chain-of-custody record, sample analysis request sheets, and a laboratory log book. A detailed description of each element, including sample forms, is

[3] See *Waste Management Area (WMA) and Supplemental Well (SPW) Guidance*, U.S. Environmental Protection Agency, Office of Solid Waste, Washington, DC, July 1993.

available from EPA's *Test Methods for Evaluating Solid Waste: Physical/Chemical Methods,* SW-846.

In accordance with 40 CFR 265.93(a), an *outline* of the groundwater quality assessment program must describe a more comprehensive program than the detection monitoring program. This outline will form the basis of the groundwater quality assessment program if hazardous waste or constituents enter the underlying groundwater. The outline must describe procedures that can determine whether hazardous waste or constituents have entered the groundwater and the concentration of such hazardous waste or constituents.

Detection Monitoring Phase. *Detection monitoring*, the first stage in the interim status groundwater monitoring program, is a program developed to determine whether a land disposal unit has released hazardous waste or constituents into the underlying groundwater in quantities sufficient to cause a significant change in groundwater quality, as shown in Figure 5-2.

Facilities not qualifying for a waiver from the groundwater monitoring program must install a basic detection monitoring system [40 CFR 265.92(b)]. For one year, the facility owner or operator must conduct quarterly sampling of wells upgradient and downgradient of the facility to account for seasonal variation. The owner or operator must analyze these samples for the following:

- Drinking water suitability
- Groundwater quality
- Indicators of groundwater contamination

In addition, the elevation of the groundwater surface at each monitoring well must be determined each time a sample is obtained to determine if horizontal and vertical flow gradients have changed since the initial site characterization [40 CFR 265.92(e)].

The first year of monitoring is intended to establish baseline information (background data) on the underlying groundwater for future statistical comparison.

Alternate Groundwater Monitoring Program. If an owner or operator assumes or knows that a statistically significant increase (or pH decrease) in one or more of the specified indicator parameters could occur, the owner or operator may install, operate, and maintain an *alternate groundwater monitoring program* [40 CFR 265.90(d)], which allows facilities suspected or known to be discharging hazardous waste or waste constituents to groundwater, to enter immediately into the assessment phase rather than delay the assessment program a year by doing detection monitoring and background comparisons. An owner or operator who chooses the alternate groundwater monitoring program must submit to EPA a groundwater assessment plan as outlined in 40 CFR 265.93(d)(3).

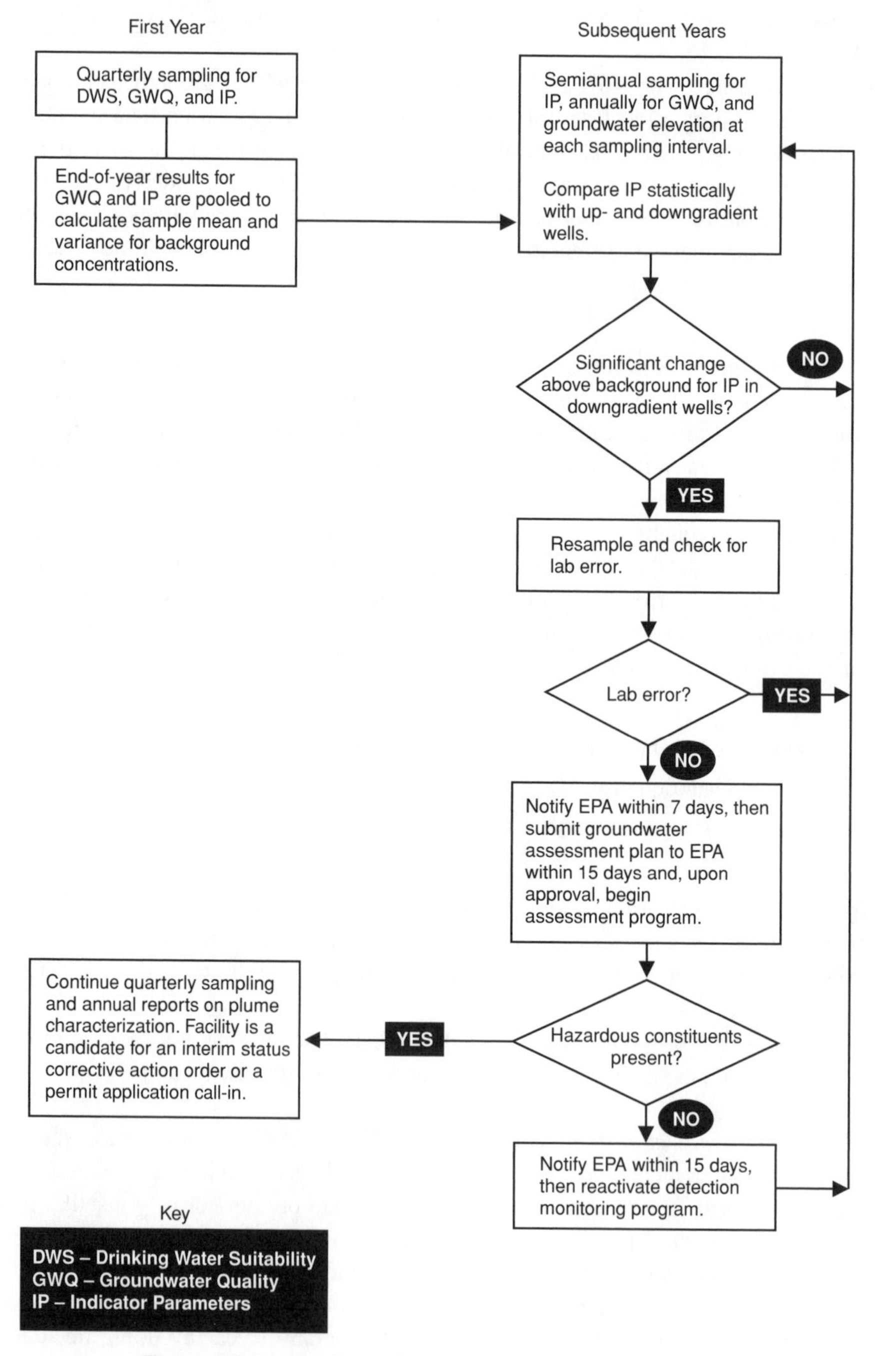

Figure 5-2. Interim Status Groundwater Monitoring Program

Monitoring Parameters. To determine if a release has occurred after the first year of monitoring (detection monitoring), the owner or operator each quarter thereafter must compare monitoring data obtained from downgradient wells to background concentration levels (established at the end of the required first-year quarterly up- *and* downgradient monitoring) for drinking water suitability, groundwater quality, and indicator parameters.

Parameters of Drinking Water Suitability. The following parameters are used to assess the suitability of the aquifer as a drinking water supply [40 CFR 265.92(b)(1)]:

- Arsenic
- Barium
- Cadmium
- Chromium
- Coliform bacteria
- 2,4-D
- Endrin
- Fluoride
- Gross alpha
- Gross beta
- Lead
- Lindane
- Mercury
- Methoxychlor
- Nitrate
- Radium
- Selenium
- Silver
- Toxaphene
- 2,4,5-TP (Silvex)

Within 15 days of completing each quarterly analysis, the owner or operator must report to EPA the concentrations of each drinking water suitability parameter [40 CFR 265.94(a)(2)(i)]. The report must note if any of these constituents exceed the levels established in Appendix III to 40 CFR part 265. It is important to note that after the first year, the owner or operator is not required to analyze for drinking water suitability. However, the first year's data will have to be submitted if an assessment plan is required [40 CFR 265.93(b)(2)].

Parameters of Groundwater Quality. The following parameters are used to assess the suitability of groundwater for other nondrinking purposes 40 CFR 265.92(b)(2)]:

- Chloride

- Iron
- Manganese
- Phenols
- Sodium
- Sulfate

During the first year, the owner or operator must establish the initial background concentrations based on quarterly measurements. After the first year, the owner or operator must analyze samples of both upgradient and downgradient wells annually for groundwater quality. Although there is no requirement for the statistical evaluation of these parameters, the data are to be used as a basis for comparison if a groundwater quality assessment program is implemented as required by 40 CFR 265.93(b)(3).

Indicator Parameters. The following parameters are used as gross indicators of whether contamination has occurred [40 CFR 265.92(b)(3)]:

- pH
- Specific conductance
- Total organic carbon
- Total organic halogen

During the first year, for each quarterly sampling, at least four replicate measurements must be obtained for each sample from the upgradient well(s); then initial background arithmetic mean and variance must be determined by pooling the replicate measurements for the respective indicator parameter concentrations. For the downgradient wells, initial background concentrations must be obtained based on quarterly sampling [40 CFR 265.92(c)(2)]. After background values have been established from the first year, both upgradient and downgradient indicator parameter values are subsequently compared with the initial background values of the upgradient well only.

In subsequent years of monitoring, all monitoring wells must be sampled for indicator parameters semiannually. A mean and variance, based on four replicate measurements, must be determined for each of the indicator parameters.

By March 1 of each calendar year, the concentrations of each of the indicator parameters (as well as the groundwater elevations) must be reported to EPA for each monitoring well. Each monitoring well must also be compared statistically both with its initial background arithmetic mean and variance for each indicator parameter derived from the first year's quarterly sampling of the upgradient well, as well as with subsequent means and variances of each indicator parameter derived from all wells, including the upgradient well [40 CFR 265.93(b)]. The comparison must consider each of the wells individually in the monitoring system.

Statistical Comparisons. Pursuant to 40 CFR 265.93(b), the owner or operator must use the Student's t-test to determine statistically significant changes in the concentrations of an indicator parameter in groundwater samples compared to the initial background concentration of that indicator parameter.[4] The comparison must consider individually each of the wells in the monitoring system.

For three of the indicator parameters (specific conductance, total organic carbon, and total organic halogen) a single-tailed Student's t-test must be used at the 0.01 level of significance for increases over background. The test for pH uses a two-tailed test because both increases and decreases in pH can be significant.

Significant Changes. If the comparison for any of the *downgradient wells* shows a significant increase (or any significant change in pH), the affected wells must be resampled and the samples must be split into two (duplicates) for a qualitative check for possible laboratory error. The samples may be split into four replicates and another t-test run, but this is not required [40 CFR 265.93(c)(2)].

If this resampling indicates that the change was due to laboratory error, the owner or operator should continue with the detection monitoring program [40 CFR 265.93(d)]. However, if resampling shows that the significant change was not a result of laboratory error, the owner or operator must notify EPA within 7 days that contamination may have occurred. Within 15 days of that notification, the owner or operator must submit to EPA a groundwater quality assessment plan, based on the required assessment outline, that has been certified by a qualified geologist or geotechnical engineer [40 CFR 265.93(d)(1)].

If the comparison for the *upgradient wells* shows a significant increase (or pH decrease), the owner or operator must submit this information to EPA without a qualitative laboratory check pursuant to 40 CFR 265.94(a)(2)](ii). The groundwater quality assessment phase is not triggered when there is a significant change in upgradient wells. The owner or operator must, however, submit this information to EPA [40 CFR 265.93(c)(1)].

If a facility never detects a statistically significant increase in any of the wells for the indicator parameters and an assessment plan is never done, detection monitoring for groundwater quality and indicator parameters continues on an annual and a semiannual basis, respectively, through closure and throughout the post-closure period.

Assessment Monitoring Phase. If a significant change in water quality is discovered during the interim status detection monitoring phase, the owner or operator must undertake a more comprehensive groundwater monitoring program called *assessment monitoring.*

Because the parameters used in detection monitoring are nonspecific, a statistically significant change in a parameter may not necessarily signify leakage into the underlying groundwater. Thus, determining whether hazardous waste constituents have indeed migrated into the groundwater is the first step in the assessment monitoring phase. If this has occurred, the owner or operator must determine the vertical

[4] See EPA's Unified Guidance on the *Statistical Analysis of Ground-Water Monitoring Data at RCRA Facilities.*

and horizontal concentration profiles of all hazardous waste constituents in the plume(s) emanating from the waste management area. In addition, the owner or operator must establish the rate and extent of contaminant migration [40 CFR 265.93(d)(3)].

The information developed during assessment monitoring is used by EPA to evaluate the need for corrective action at the facility. If corrective action is necessary, EPA may issue an enforcement order compelling corrective action under §3008(h) of RCRA before or in conjunction with the issuance of a permit (see Chapters 9 and 10 for further information).

Owners or operators required to conduct plume characterization activities for the assessment program are required to have a *written* assessment monitoring plan. According to 40 CFR 265.93(d)(3), the required elements of the plan include the following:

- Number, location, and depth of wells
- Sampling and analytical methods to be used
- Evaluation procedures of groundwater data
- A schedule of assessment monitoring implementation

If the assessment confirms that hazardous constituents have entered the groundwater, the owner or operator must continue quarterly assessments until final closure of the facility or until a permit is issued [40 CFR 265.93(d)(7)]. If the initial assessment confirming contamination was performed during post-closure, no subsequent quarterly monitoring is required [40 CFR 265.93(d)(7)(ii)]. However, this type of facility would be a likely candidate for a post-closure permit or an interim status corrective action order. In applying for a post-closure permit, the owner or operator would be required to conduct a comprehensive groundwater characterization program to generate the information required for the permit application found in 40 CFR 270.14(c)(8).

By March 1 of each calendar year, the owner or operator must submit to EPA a report containing the results of the facility's assessment program, which must include the calculated (or measured) rate of migration of the hazardous waste or hazardous waste constituents for the reporting period.

If this assessment confirms that no hazardous constituents have entered the groundwater, EPA must be notified within 15 days of the determination with a written report. The owner or operator may then reinstate the detection monitoring program or enter into a consent agreement with EPA to follow a revised protocol designed to avoid false triggers [40 CFR 265.93(d)(6)].

Groundwater Monitoring at Permitted Units

Whereas the principal objective of the part 265 groundwater monitoring program is to identify and assess releases from interim status land disposal units, the objectives of the groundwater monitoring program for permitted units are to characterize any leachate and to ensure that corrective action is taken to prevent leachate migration beyond the waste management area. To achieve this goal, the regulations establish a three-step program designed to detect, evaluate, and clean up groundwater contamination arising from leaks or discharges from regulated units, as outlined in Figure 5-3. (A *regulated unit* is any land disposal unit that accepted waste after July 26, 1982.) *All* regulated units are subject to the part 264 groundwater monitoring requirements [40 CFR 265.90(a)]. Thus, even interim status units will at some point have to comply with or meet the part 264 groundwater monitoring requirements.

An interim status facility must comply with the part 265 requirements until the final administrative disposition is made concerning the facility's permit. When a permit is issued for a unit, the owner or operator is then subject to the conditions of the permit, which are based on the part 264 standards. Thus, an interim status unit that is a regulated unit follows the part 265 groundwater monitoring program only until a permit is issued (40 CFR 264.3).

The groundwater protection requirements of part 264 define a general set of responsibilities that the owner or operator must meet; however, the specific requirements are tailored to the individual facility through its permit. The permit provisions concerning groundwater are based on 40 CFR 264.90 through 264.100 and 40 CFR 270.14(6), (7), and (8).

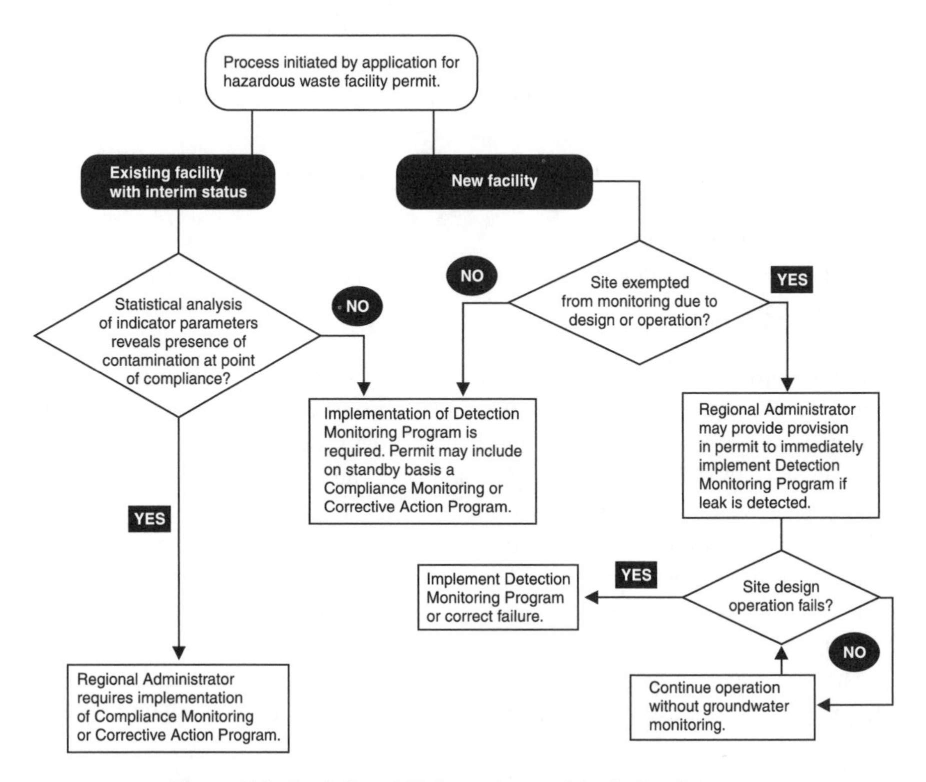

Figure 5-3. Basic Part 264 Ground water Monitoring Program

The nature of the monitoring program in the permit will depend on the information available at the time of permitting. The key question is whether a regulated unit has begun to leak. For new units, this is not an issue. For existing units, there should be a reliable base of information available to determine whether hazardous constituents have entered the groundwater.

Variances. There are four variances from the part 264 groundwater monitoring requirements available to owners and operators [40 CFR 264.90(b)]. These variances, requested through the submittal of the Part B permit application (see Chapter 7 for further information), involve the following conditions:

- EPA finds that a regulated unit is an engineered structure; does not receive or contain free liquids or wastes containing free liquids; is designed and operated to exclude liquid, precipitation, and other runon and runoff; has both inner and outer layers of containment enclosing the waste; and has a leak detection system built into each containment layer. Also, the owner or operator will provide continuous operation and maintenance of the leak detection systems throughout the facility's "active life" and, to a reasonable degree of certainty, will not allow hazardous constituents to migrate beyond the outer containment layer before the end of the post-closure period [40 CFR 264.90(b)(2)].
- EPA finds, pursuant to 40 CFR 264.280(d), that the treatment zone of a land treatment unit does not contain hazardous constituents above background levels.
- EPA finds that there is no potential for migration of liquid from regulated units to the uppermost aquifer during the facility's active life [40 CFR 264.90(b)(4)].
- The unit is an enclosed waste pile and is operated in compliance with 40 CFR 264.250(c).

General Requirements. The facility must have a sufficient number of wells representing (1) the background water quality (that is, groundwater that is not affected by leakage from a waste management unit) and (2) the water quality passing the point of compliance [40 CFR 265.97(a)]. (The point of compliance is addressed in the discussion on detection monitoring later in this section.)

If the facility has more than one regulated unit, separate groundwater monitoring systems are not necessarily required, provided that the system is adequate for all units.[5] The *waste management area* would be described by an imaginary line circumscribing all the regulated units (40 CFR 264.97(b)].

The sampling program must ensure a reliable indication of groundwater quality. The program also must include chain-of-custody control, sample collection proce-

[5] See *Waste Management Area (WMA) and Supplemental Well (SPW) Guidance*, U.S. Environmental Protection Agency, Office of Solid Waste, Washington, DC, July 1993.

dures, sample preservation and shipment, and analytical procedures [40 CFR 264.97(d)].

Statistical Analysis. To determine if a regulated unit is affecting the groundwater quality, statistical analysis of the monitoring results is required. Whereas the interim status groundwater monitoring program requires the use of the Student's t-test, the part 264 program provides five options. An owner or operator must choose, for each of the chemical parameters and hazardous constituents listed in the permit, one or more of the statistical methods described in the following list. In determining which statistical test is appropriate, the owner or operator should consider the theoretical properties of the test, the data available, the site hydrogeology, and the fate and transport characteristics of potential contaminants at the unit. EPA will review and, if appropriate, approve the proposed statistical methods and sampling procedures when issuing the permit.

EPA has identified the following statistical methods that an owner or operator may use (40 CFR 264.97(h)]:

- A parametric analysis of variance followed by multiple comparisons procedures to identify statistically significant evidence of contamination. The method must include estimation and testing of the contrasts between each compliance well's mean and the background mean levels for each constituent.
- An analysis of variance based on ranks followed by multiple comparison procedures to identify statistically significant evidence of contamination. The method must include estimation and testing of the contrasts between each compliance well's median and the background median levels for each constituent.
- A tolerance or prediction interval procedure in which an interval for each constituent is established from the distribution of the background data, and in which the level of each constituent in each compliance well is compared to the upper tolerance or prediction limit.
- A control chart approach that gives control limits for each constituent.
- Another statistical test method submitted by the owner or operator and approved by the EPA Regional Administrator.

For further information concerning the requirements and applicability of each test for groundwater monitoring, consult EPA's Unified Guidance for the *Statistical Analysis of Ground-Water Monitoring Data at RCRA Facilities*.

Detection Monitoring Phase. Detection monitoring, the first stage of the part 264 groundwater monitoring program, requires sampling for *indicator* parameters to determine if the unit is leaking.

Indicator parameters, established in the permit, may include specific conductance, total organic carbon, waste constituents, or reaction products. The selected parameters, established in the permit, are determined by considering the following:

- The types, quantities, and concentrations of constituents in the waste to be managed in the regulated unit.
- The mobility, stability, and persistence of waste constituents and reaction products in the underlying unsaturated zone.
- The detectability of the parameters.
- The concentration or values and coefficients of variation of the parameters in the groundwater background.

Detection monitoring is implemented at facilities where no hazardous constituents are known to have migrated from the facility to the groundwater. Applicants who are seeking permits for new facilities and for interim status facilities that have not triggered the assessment phase would generally qualify for the detection monitoring phase. Facilities that are in the assessment program of part 265 would not start with detection monitoring but rather with compliance monitoring [40 CFR 264.91(a)(1)].

The detection monitoring program, like its interim status counterpart, is based on one year of background groundwater monitoring. The background, or upgradient, wells must be sampled quarterly for one year. The permittee then routinely monitors for a selected set of indicator parameters specified in the permit rather than the four indicator parameters used in the part 265 program.

The number and types of samples (to be specified in the permit) collected to establish background levels must be appropriate for the form of statistical test employed [40 CFR 264.97(g)].

Values from the upgradient wells are used to establish background levels during the first year of monitoring only. However, the downgradient wells (at the point of compliance) must continue to be monitored after the first year. The regulations do not explicitly require continued monitoring of the upgradient wells, but the permit may require this practice.

In accordance with 40 CFR 264.98(h), if it is determined that there is significant change above background levels for any of the indicator parameters, the owner or operator must do the following:

- Notify EPA within seven days of those indicators that have changed.
- Run a complete Appendix IX (see Appendix C of this book) scan of all of the monitoring wells to determine the chemical composition of the leachate.
- Establish a background level for each Appendix IX constituent found at each well.
- Submit a Class 3 permit modification application to EPA within 90 days to establish a compliance monitoring program (see Chapter 7).
- Submit an engineering feasibility plan for a corrective action program.

The owner or operator may, within 90 days, attempt to demonstrate that a source other than the regulated unit is the cause of the increase, or that there is an error in sampling or analysis. EPA must be notified while the owner or operator is attempting to make these demonstrations [40 CFR 264.98(i)].

Compliance Monitoring Phase. The goal of the compliance monitoring program is to ensure that the leakage of hazardous constituents into the groundwater does not exceed designated levels. The framework for a compliance monitoring program is established by incorporating a groundwater protection standard into the permit.

Appendix IX Constituents. The Appendix IX constituents, found in Appendix IX of 40 CFR part 264 (and Appendix C of this book), are substances specifically designed for monitoring releases to groundwater at permitted facilities. The Appendix IX list, based on the Appendix VIII list, was developed to select constituents that are appropriate for characterizing potential releases to the underlying groundwater. This list, used specifically for the RCRA groundwater monitoring program, was promulgated on July 9, 1987 (52 *FR* 25942).

Groundwater Protection Standard. The Groundwater protection standard (GWPS), established in the permit, is a standard that places a constituent-based limit on the leachate from a regulated unit, a limit measured at the point of compliance. In other words, the GWPS is used to determine if and when corrective action is required. As depicted in Figure 5-4, the *point of compliance* is a vertical plane located at the hydraulically downgradient limit of the waste management area that extends down

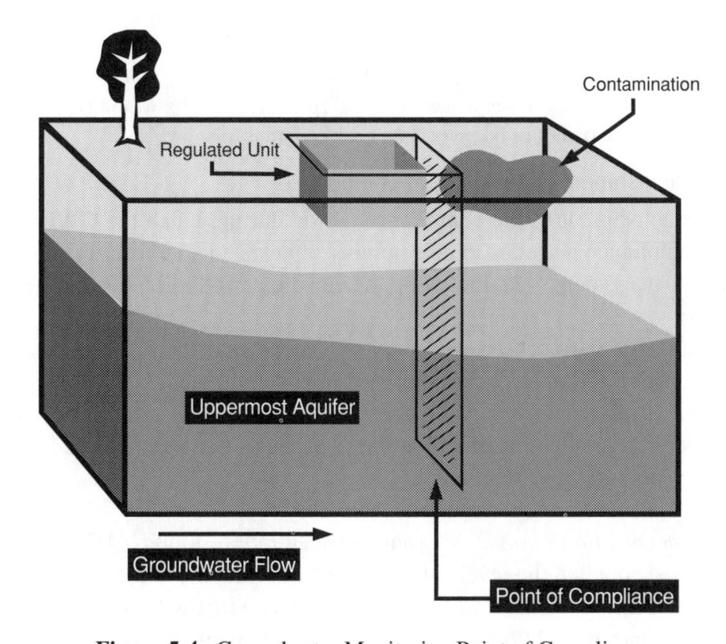

Figure 5-4. Groundwater Monitoring Point of Compliance

into the groundwater underlying the regulated unit. The point of compliance is specified in the permit (40 CFR 264.95).

EPA will establish the GWPS in the facility's permit when hazardous constituents have entered the groundwater from a regulated unit. A GWPS is not established at a regulated unit where groundwater contamination has not been detected (40 CFR 264.92).

The GWPS consists of the following four elements, each of which is specified in the permit:

- A listing of selected Appendix VIII (Part 261) hazardous constituents that could reasonably have been derived from the waste at the facility and that are present in the groundwater. The burden of demonstrating that a particular constituent could not reasonably be derived from the waste, or is incapable of posing a substantial present or potential hazard to human health or the environment, lies with the owner or operator in accordance with 40 CFR 264.93(a) and (b).
- The establishment of concentration limits for each hazardous constituent listed. Where possible, concentration limits must be based on well-established numerical concentration limits so as to prevent degradation of water quality unless the owner or operator can demonstrate that a higher limit will not adversely affect public health or the environment [40 CFR 264.94(a)]. The concentration limits are established using one of the following:
  — Maximum contaminant levels (MCLs)
  — Alternate concentration limits (ACLs)
  — Background levels

  Note: Background levels are generally not used because the GWPS is implemented only when contamination is detected. Thus, ACLs and MCLs would typically be less stringent for the owner or operator than background levels.
- The establishment of the point of compliance.
- The establishment of the compliance period during which the GWPS applies. The compliance period is to be a number of years equal to the active life of the waste management area, including the closure period (40 CFR 264.99).

The compliance period, however, may extend beyond the 30-year post-closure period if corrective action has been initiated but not completed. If the owner or operator is engaged in corrective action, the compliance period is extended until the owner or operator can demonstrate that the GWPS has not been exceeded for three consecutive years [40 CFR 264.96(c)].

Alternate Concentration Limits. Alternate concentration limits (ACLs) generally may be established when the levels of hazardous constituents in the groundwater are found above background. An ACL may be established if it will not pose a substantial or potential hazard to human health or the environment, and if the ACL will not be exceeded at the point of compliance. The ACL demonstration must justify all claims regarding the potential effects of groundwater contaminants on human health and the environment. In terms of human considerations, the regulations require assessments of toxicity, exposure pathways, and exposed populations [40 CFR 264.94(b)].

There are three basic policy guidelines established for ACLs at facilities with "usable" groundwater [OSWER 481.00-6c]:

- Groundwater contaminant plumes should not increase in size or concentration above allowable health or environmental exposure levels.
- Increased facility property holdings should not be used to allow a greater ACL.
- ACLs should not be established so as to contaminate off-site groundwater above allowable health or environmental exposure levels.

The following information must be considered when establishing ACLs [40 CFR 264.94(b)]:

- Potential adverse effects on groundwater quality, in view of these factors:
  - —The physical and chemical characteristics of the waste in the regulated unit, including its potential for migration.
  - —The hydrogeologic characteristics of the facility and surrounding land.
  - —The quantity of groundwater and the direction of groundwater flow.
  - —The proximity and withdrawal rates of groundwater users.
  - —The current and future uses of groundwater in the area.
  - —The existing quality of groundwater, including other sources of contamination and their cumulative impact on the groundwater quality.
  - —The potential for health risks caused by human exposure to waste constituents.
  - —The potential damage to wildlife, crops, vegetation, and physical structures caused by exposure to waste constituents.
  - —The persistence and permanence of potential adverse effects.
- Potential adverse effects on hydraulically connected surface water quality, in view of these factors:
  - —The volume and physical and chemical characteristics of the waste in the regulated unit.
  - —The hydrogeologic characteristics of the facility and surrounding land.
  - —The quantity and quality of groundwater and the direction of groundwater flow.
  - —The patterns of rainfall in the region.
  - —The proximity of the regulated unit to surface waters.
  - —The current and future uses of surface waters in the area and any water quality standards established for those surface waters.
  - —The existing quality of surface water, including other sources of contamination, and the cumulative impact on surface water quality.

—The potential for health risks caused by human exposure to waste constituents.
—The potential damage to wildlife, crops, vegetation, and physical structures caused by exposure to waste constituents.
—The persistence and permanence of potential adverse effects.

Requirements During Compliance Monitoring. A facility in the compliance monitoring phase must continue the compliance monitoring program for the active life of the facility, including closure [40 CFR 264.99(a)(4) and 264.96]. During post-closure, the facility may switch back to detection monitoring, but this will occur only if the levels in compliance monitoring have consistently reached background. Otherwise, the facility must remain in compliance monitoring.

In accordance with 40 CFR 294.99(a), during the compliance monitoring program, the owner or operator must

- Determine whether the regulated units are in compliance with the groundwater protection standard.
- Determine the concentration of constituents at the point of compliance as determined by the permit, but at least semiannually.
- Determine the groundwater flow rate and direction at least annually.
- Conduct a complete Appendix IX scan at least annually to determine if there are any new constituents at the point of compliance.

According to 40 CFR 264.99(i), if it is determined that the groundwater protection standard is being exceeded at any monitoring well during compliance monitoring, the owner or operator must

- Notify EPA in writing within seven days.
- Submit a Class 3 permit modification application to EPA within 180 days to establish a corrective action program.
- Prepare a plan for a groundwater monitoring program that will demonstrate the effectiveness of the corrective action program.

Corrective Action Phase. The goal of the corrective action phase is to bring the facility back into compliance with its groundwater protection standard (40 CFR 264.100). The elements of the corrective action program include the following:

- Implementation of corrective measures to remove or treat the constituents as specified in the permit.
- The time period for implementing the corrective action program as specified in the permit.
- Termination of the corrective action program only upon demonstration of the facility meeting the groundwater protection standard.

- Submittal of a written report semiannually to EPA on the effectiveness of the corrective action program.

In conjunction with a corrective action program, the owner or operator must implement a groundwater monitoring program at least as effective as the compliance monitoring program (in determining compliance with the groundwater protection standard) to demonstrate the effectiveness of the corrective action program [40 CFR 264.100(d)]. Corrective action will continue until the owner or operator can demonstrate that the groundwater protection standard has not been exceeded for a period of three consecutive years [40 CFR 264.96(c) and 264.100(f)].

Once contamination has been reduced below the concentration limits set in the GWPS, the facility may discontinue corrective action measures and return to the compliance monitoring program [40 CFR 264.100(f)].

CLOSURE AND POST-CLOSURE

All hazardous waste management facilities eventually stop accepting waste, which subsequently requires proper closure of all waste management units. The primary purpose of closure and post-closure care program is to ensure that all hazardous waste management facilities are closed in a manner that, to the extent necessary, (1) protects human health and the environment and (2) controls, minimizes, or eliminates post-closure escape to the ground or atmosphere of hazardous waste, hazardous constituents, leachate, contaminated precipitation runoff, or waste decomposition products (40 CFR 264/265.111). The closure and post-closure requirements are divided into general standards applicable to all hazardous waste management facilities (addressed in the following sections) and technical standards specific to the type of waste management unit (addressed in Chapter 6).

All waste management units are subject to the closure requirements. However, only those land-based units (e.g., landfills, surface impoundments, waste piles) that close with waste in place are subject to the post-closure care requirements.

The first part of this section discusses the closure requirements; post-closure requirements follow.

Closure Requirements

Closure is the act of securing a hazardous waste management facility pursuant to the closure requirements (40 CFR 270.2). It is the period after which wastes are no longer accepted and during which the owner or operator is required to complete all treatment, storage, and disposal operations. The owner or operator may also conduct partial closure of a facility. *Partial closure* is closure of a hazardous waste management unit at a facility that contains other operating hazardous waste management units. Closure of the last unit constitutes *final closure* of the facility.

Closure is implemented through a closure plan, which establishes how a unit will be closed and establishes the cleanup requirements. At some point, all closure plans must be approved by EPA or the state.

Closure Plan. All hazardous waste management facilities must prepare and maintain a written closure plan that outlines the procedures to complete closure (40 CFR 264/254.112). The plan must identify the steps necessary to perform partial or final closure at any time during the facility's active life. (A sample outline of a typical closure plan is shown in Table 5-2.)

The required steps of closure are as follows:

- A description of how each hazardous waste management unit will be closed in accordance with applicable closure performance standards.
- A description of how final closure of the entire facility will be conducted, including a description of the maximum extent of the operation that will remain unclosed during the facility's active life.
- An estimate of the maximum inventory of hazardous wastes on-site at any time during the active life of the facility. The estimate of the maximum inventory of wastes should include (OSWER Directive 9476.00-5) the following:
  — the maximum amount of hazardous wastes, including residues, in all treatment, storage, and disposal units;

TABLE 5-2. Sample Outline of a Typical Closure Plan

Activities to Be Described in Closure Plan	Regulatory Citation (40 CFR)
Facility description	264/265.112(b)
Partial closure activities	264/265.112(b)(1)
Final closure activities based on the maximum extent of operations	264/265.112(b)(2)
Treating, removing, or disposing of the maximum amount of inventory	264/265.112(b)(3)
Facility decontamination	264/265.112(b)(4) and 264/265.114
Final cover*	264/265.112(b)(5), 264/265.228, 264/265.258, 264/265.280, and 264/265.310
Groundwater monitoring*	264/265.112(b)(5)
Ancillary closure activities (e g., leachate management, gas monitoring, runon and runoff control)	264/265.112(b)(5)
Survey plat*	264/265.116
Closure certification	264/265.115
Closure (final and partial) schedule	264/265.112(b)(6)

Source: OSWER Directive 9476.00-5

*May not be required

  - the maximum amount of contaminated soils and residues from drips and spills from routine operations; and[6]
  - if applicable, the maximum amount of hazardous wastes from manufacturing/process areas and raw material/product storage and handling areas.
- A description of the steps needed to remove or decontaminate all hazardous waste residues from any of the facility components, equipment, or structures.
- A sampling and analysis plan for testing surrounding soils to determine the extent of decontamination required to meet closure standards.
- A description of all other required activities during closure, including groundwater monitoring, leachate collection, and precipitation control.
- A schedule for closure of each hazardous waste management unit and for final closure of the facility.
- The expected year of closure.

Unlined storage surface impoundments and waste piles and tank systems without secondary containment that intend to clean close (that is, remove all hazardous waste and residues at the time of closure) must prepare a *contingent closure plan* in addition to a clean closure plan. The contingent closure plan must describe how the closure activities will be undertaken if the owner or operator cannot remove all hazardous waste and constituents from the unit, which must be closed with waste in place, similar to a landfill (i.e., "dirty" closure).

A copy of the closure plan, including all revisions to the plan, must be kept at the facility until closure is certified to be complete [40 CFR 264/265.112(a)]. The plan must be furnished to EPA upon request. In addition, for facilities without *approved* plans, the plan must be provided on the day of a site inspection to any duly designated representative of EPA [40 CFR 265.112(a)]. An owner or operator who intends to clean close a unit must include specific details on how removal and/or decontamination will be demonstrated, details including sampling protocols, schedules, and the cleanup levels that will be used as a standard for assessing whether clean closure is achieved (52 *FR* 8706, March 19, 1987).

Amendments to the Closure Plan. Facilities operating under interim status may amend the closure plan at any time before notification of partial or final closure. A written request for approval of any amendments to an *approved* closure plan (approval of closure plans is discussed later in its own section) must be submitted to EPA [40 CFR 265.112(c)(2) and (3)].

As shown in Table 5-3, the owner or operator of a facility with either an approved or a nonapproved closure plan must amend the plan whenever any of the following applies:

- There are changes to the facility design or operations that affect the closure plan.

[6] Residues means 40 CFR part 261, Appendix VIII, hazardous constituents.

TABLE 5-3. Amending Closure Plans for Interim Status Units

Units without Approved Closure Plans	Units with Approved Closure Plans
Voluntary Amendments	
May amend the plan at any time prior to closure	Must submit a written request to EPA to amend the plan
Mandatory Amendments*	
Must amend the plan at least 60 days prior to facility design and operation	Must submit the modified plan at least 60 days prior to the proposed change in facility design and operation
Must amend the plan no later than 60 days after an unexpected event has occurred that affects the plan	Must submit the modified plan no later than 60 days prior to an unexpected event
Must amend the plan no later than 30 days after an unexpected event that occurs during the closure period	Must submit the modified plan no later than 30 days after an unexpected event that occurs during the closure period

*A closure plan must be amended whenever changes in operating plans or facility design affect the closure plan; whenever there is a change in the expected year of closure (if applicable); or whenever, during closure activities, unexpected events require a modification to the closure plan.

- The expected year of closure changes.
- Unexpected events occur during closure.

The plan must be amended at least 60 days before a proposed change or no later than 60 days after an unexpected change. If there is an unexpected change during closure, the plan must be amended within 30 days [40 CFR 265.112(c)(3)]. For example, if a surface impoundment or waste pile originally intended to clean close but cannot, and instead must close with waste in place, the closure plan must be amended within 30 days.

Notification of Closure. Owners or operators without approved closure plans must notify EPA of the intent to close a hazardous waste management unit by submitting the closure plan, which must be submitted at least 180 days before the expected date on which closure will begin for the first land disposal unit (i.e., surface impoundment, waste pile, land treatment, or landfill unit). An owner or operator of a facility that has only incinerators, tanks, or container storage units must submit the closure plan at least 45 days before the expected closure date, as shown in Table 5-4 [40 CFR 264/265.112(d)].

The *expected date of closure* must be either within 30 days after the date on which any hazardous waste management unit receives the known final volume of hazardous wastes, or, if the hazardous waste management unit will be receiving additional hazardous wastes, no later than one year after the date when the unit

Table 5-4. Closure Notification Requirements

	Facility Type	Notification Requirement
	Permitted and Interim Status Facilities and Approved Plans	
Partial closure	Disposal unit	60 days
	Other units (tank, incineration, container storage)	No notification required
Final closure	Disposal unit	60 days
	Other units	45 days
	Interim Status Facilities without Approved Plans	
Partial closure	Disposal unit	180 days and submittal of closure and post-closure plan
	Other units	No notification required
Final closure	Disposal unit	180 days and submittal of closure and post-closure plan
	Other units	45 days and submittal of closure plan

Source: OSWER Directive 9476.00-5

received the most recent volume of hazardous waste [40 CFR 264/265.112(d)(2)]. However, this date can be extended upon approval by EPA.

If the facility's interim status is terminated for reasons other than the issuance of a permit, the owner or operator must submit a closure plan within 15 days of termination unless the facility is issued either a judicial decree or a compliance order to close [40 CFR 265.112(d)(3)].

Facilities that have *approved* closure plans must notify EPA at least 60 days before closure for any land disposal unit, or 45 days for facilities with only container storage, tanks, or an incinerator [40 CFR 264/265.112(d)].

Closure Plan Approval. Owners and operators of interim status facilities must submit their plans (closure and, if required, post-closure) for review and approval to EPA before final closure of the facility or closure of the first land disposal unit. (A permitted facility was to have included the closure plan as part of the permit application [40 CFR 270.14(13)]. When the permit is issued, the closure plan, which is now a condition of the permit, is the *approved* closure plan.)

Upon receipt of a closure plan, EPA will provide the owner or operator and the public the opportunity to submit written comments or requests for modifications to the plan, and a public hearing may be held if requested [40 CFR 265.112(d)(4)].

EPA must approve, modify, or disapprove a closure plan for interim status facilities (permitted facilities would have had their closure plan approved upon issuance of the permit) within 90 days of its receipt.

If the plan is not approved, EPA must provide a detailed written statement to the owner or operator with the reasons for disapproval. The owner or operator then has 30 days after receiving the written statement to modify the plan or submit a new one. EPA must approve or modify the new or resubmitted plan within 60 days. If EPA modifies this plan, the modified plan becomes the *approved* closure plan [40 CFR 265.112(d)(4)].

There are, however, no explicit provisions under RCRA that allow an owner or operator to appeal the final closure plan issued by EPA. The owner or operator would have to pursue other legal avenues outside of the RCRA regulations to appeal provisions in a final closure plan.

Closure Period. As shown in Figure 5-5 (interim status units) and Figure 5-6 (permitted units), within 90 days after receiving the final volume of hazardous waste or 90 days after approval of the closure plan, whichever is later, all hazardous waste must be treated, disposed of, or removed. The owner or operator has an additional 90 days to complete the decontamination of the facility, dismantling of equipment, and so forth. Thus, closure must be completed within 180 days after the final receipt of hazardous waste or plan approval. A time extension may be granted by EPA (under specified conditions outlined below), but the request must be made no later than 30 days before the expiration of either the initial 90 days or the total 180-day period (40 CFR 264/265.113).

Note: Very few facilities have met the 6-month timeframe, especially for those facilities attempting a clean closure.

The conditions for a time extension during closure include the following:

- The required closure activities will, if necessary, take longer than the applicable 90 or 180 days.
- The facility has the capacity to receive additional wastes.
- There is reasonable likelihood that a person other than the owner or operator will recommence operation at the site.
- Closure of the facility would be incompatible with continued operation of the site.

An approval of a time extension for closure is contingent on whether the owner or operator has taken and will continue to take all steps necessary to prevent threats to human health and the environment.

Delay of Closure. Landfills, surface impoundments, and land treatment facilities may continue to receive nonhazardous waste after the final receipt of hazardous waste [40 CFR 262/265.113(d)] provided there is sufficient capacity, the wastes will

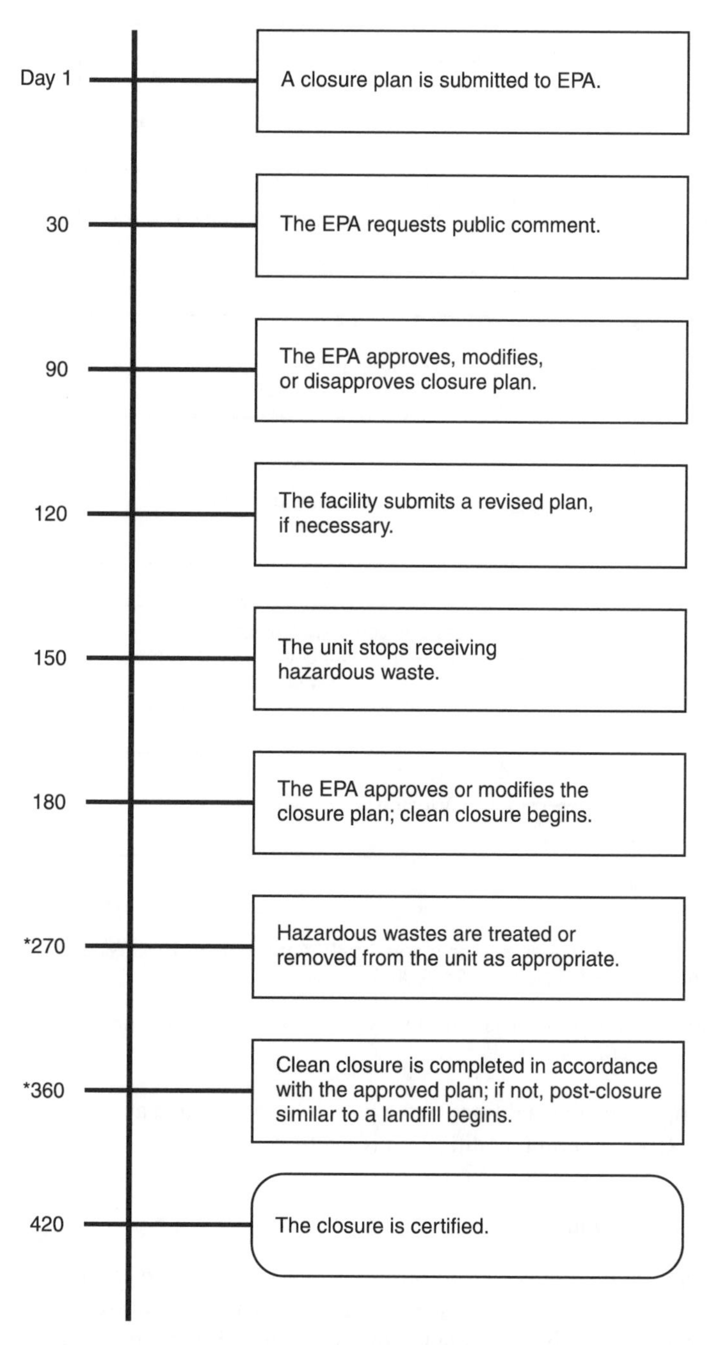

*A longer time period may be granted by EPA pursuant to 40 CFR 265.113.
Source: OSWER Directive 9476.00-5

Figure 5-5. Approximate Time Frame for Clean Closure of Interim Status Units

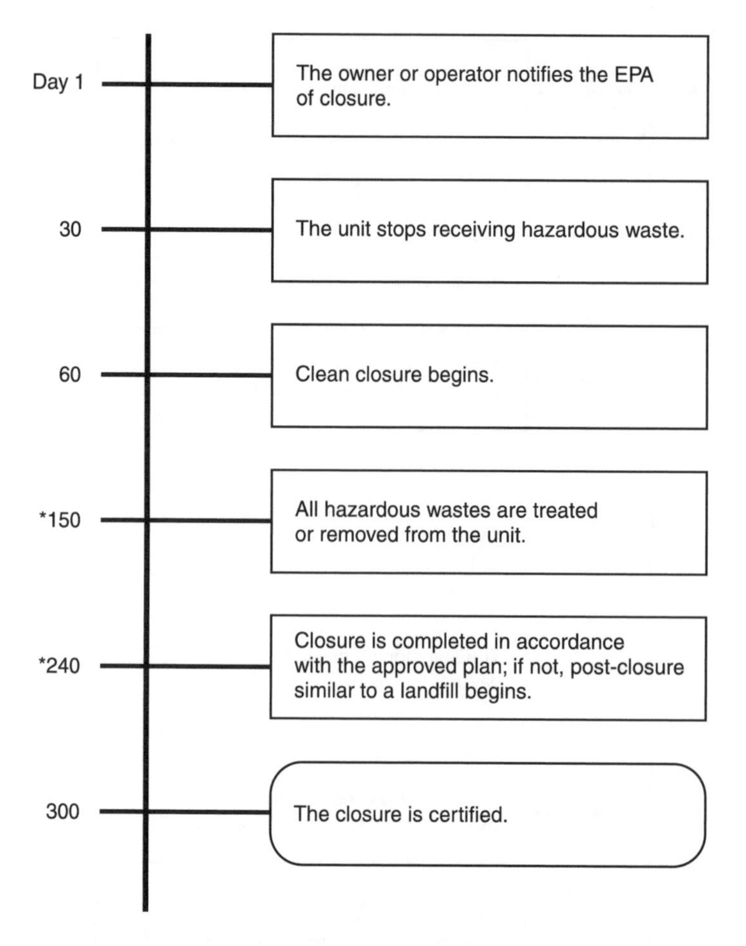

*A longer time period may be granted by EPA pursuant to 40 CFR 265.113.
Source: OSWER Directive 9476.00-5

Figure 5-6. Approximate Time Frame for Clean Closure of Permitted Units

not be incompatible, all conditions of the permit have been and are being met, and a request for a Class 2 permit modification is made.

Cleanup Requirements

During partial or final closure, all hazardous waste and waste residues; contaminated liners, equipment, and structures; and contaminated subsoils must be removed (and properly disposed of) or decontaminated.[7] Otherwise, the unit must be closed

[7] EPA interprets "contaminated subsoils" to include contaminated groundwater (53 *FR* 9944, March 28, 1988).

with waste in place similar to a landfill (40 CFR 264/265.114). During the closure process, all hazardous wastes, waste residues, contaminated subsoils, groundwater, and equipment must be managed as a hazardous waste unless the material is delisted (if it was a listed waste) or the waste does not exhibit a characteristic of hazardous waste. This means that contaminated material that is not (by definition) a hazardous waste must be removed, but may be managed as a nonhazardous industrial (Subtitle D) waste, subject to state rules and regulations (40 CFR 264/265.114).

To satisfy the closure requirements, it may be necessary to create new treatment, storage, or disposal units. There is no exclusion from the permitting requirements just because a facility is subject to the closure requirements, as 40 CFR part 264 standards are applicable to new units added during closure as well as to new operating units. However, a Class 2 permit modification (for tanks, containers, and some waste piles) or a Class 3 permit modification (for landfills, surface impoundments, incinerators, and waste piles) may be sought to create these units. Interim status facilities, however, may add more units if the addition constitutes an allowable change to a facility during interim status. According to 40 CFR 270.72(c), changes in processes or addition of processes may be allowed if a revised Part A permit application and justification are submitted and EPA approves the change. (See Chapter 7 for further explanation.) In addition, a facility will probably become a generator during the closure and cleanup process, subject to the part 262 generator standards, and it must comply accordingly.

Clean Closure. At closure, owners and operators of hazardous waste management units may choose between removing/decontaminating all hazardous wastes and waste residues *(clean closure)* and terminating further regulatory responsibility under RCRA for the unit or closing the unit with hazardous waste or waste residue remaining in place *(dirty closure)* and instituting post-closure care similar to that of landfills.

Previously, owners and operators of interim status facilities attempting a clean closure were required to remove wastes from a unit to the point that wastes were no longer *hazardous.* The criteria for this determination depended on whether the wastes were hazardous because they were listed or because they exhibited a characteristic of hazardous waste. Thus, if a surface impoundment contained only ignitable hazardous waste, the owner or operator could cease the removal of waste if that waste no longer exhibited the ignitability characteristic. At the point when the waste is no longer hazardous, the responsibility of the owner or operator under Subtitle C of RCRA ceased at the time of certification of clean closure (45 *FR* 33203, May 19, 1980). Consequently, the clean-closure standard allowed facilities to be relieved of their RCRA responsibility, even though there may have been contamination remaining (assuming that the contamination was not defined as a hazardous waste).

To address this contamination, EPA promulgated final regulations on March 19, 1987 (52 *FR* 8704), and December 1, 1987 (52 *FR* 45788), that significantly strengthened the clean-closure requirements for interim status units. Those requirements now are nearly identical to the part 264 requirements. An important aspect of these regulations is that they are applicable to *any* land disposal unit that received waste after July 26, 1982, or certified closure after January 26, 1983. Thus, even if a unit was previously clean closed successfully, the facility owner or operator must demonstrate (as discussed in the following subsection) that the unit has met the part 264 clean-closure requirements (52 *FR* 45795, December 1, 1987). For example, on June 13, 1983, an owner of a surface impoundment (which was used to treat ignitable waste only) clean closed the surface impoundment in accordance with the approved closure plan. The owner of the impoundment ceased removing contaminated subsoils at the point where the soil did not exhibit the ignitability characteristic in accordance with 40 CFR 265.228(b). However, the changes to the closure rules (52 *FR* 8704 and 52 *FR* 45788) require the owner either to remove *all* wastes and waste residues (i.e., Appendix VIII hazardous constituents) from the previously closed impoundment to attain a clean closure or to close the impoundment similar to a landfill and commence the 30-year post-closure care period.

Clean-Closure Requirement. Clean closure is achieved when all wastes, residues, and associated equipment are removed or decontaminated. EPA interprets *remove* and *decontaminate* to indicate the amount of removal of decontamination that obviates the need for post-closure care (52 *FR* 8706, March 19, 1987). This means that an owner or operator must remove all hazardous waste or waste residue (i.e., all hazardous constituents) that pose a "substantial present or potential threat to human health or the environment." EPA reviews site-specific demonstrations submitted by the owner or operator to determine if the removal or decontamination is sufficient. The closure demonstration submitted must document that any contaminants left in the soil and/or groundwater, surface water, or atmosphere, in excess of EPA-recommended limits or factors, based on a direct contact scenario (i.e., fate and transport considerations are not allowed) through inhalation or ingestion, will not result in a threat to human health or the environment. *EPA-recommended limits or factors* are those that have undergone peer review by EPA. At the present time, these include maximum contaminant levels, federal water quality criteria, verified reference doses, and carcinogenic slope factors, as shown in Table 5-5. The most current levels associated with these criteria are available from EPA's Integrated Risk Information System.[8] If no EPA-recommended exposure limit exists for a particular constituent, then the owner or operator must either remove the constituent to background levels, submit data of sufficient quality for EPA to determine the environmental and health effects of the constituent in accordance with TSCA (40 CFR parts 797 and 798), or follow closure and post-closure requirements similar to landfills (52 *FR* 8706, March 19, 1987).

Clean Closure of an Interim Status Unit. Previously, it was interpreted that a unit that clean closed was no longer subject to 40 CFR part 265 (i.e., post-closure care). However, on March 28, 1986 (51 *FR* 10716), EPA asserted its opinion that an interim status disposal unit that clean-closes after July 26, 1982, is not relieved from

[8] Contact EPA's Risk Information Hotline at 513-569-7254 for further information on IRIS, or contact IRIS directly through the internet at http://www.epa.gov/iris.

Table 5-5. Clean-Closure Cleanup Levels

Media	EPA-Recommended Limits or Factors
Groundwater	MCLs; * if nonexistent, oral RfDs* for noncarcinogenic constituents, and CSFs* for carcinogens. Carcinogens must be in risk range of 10^{-4} to 10^{-7}, with 10^{-6} as the point of departure.
	If above limits or factors do not exist, may use either natural background levels or conduct testing in compliance with TSCA to develop a health-based standard. If background levels are elevated, may use detection limit.
Surface Water	If surface water is used as a drinking water source, use MCLs; if nonexistent, use oral RfDs and CSFs. Use Ambient Water Quality Criteria (WQC) to determine if protective of the environment (aquatic species).
Soil	Use oral RfDs and CSFs.
Air	National Ambient Air Quality Standards (NAAQS); if nonexistent, use inhalation RfDs and CSFs.

*MCL = maximum contaminant level; RfD = reference dose; CSF = carcinogenic slope factor

the post-closure permit requirement. If the unit is a regulated unit, it is subject to the part 264 groundwater regulations when a post-closure permit is issued.

EPA's rationale for this opinion is that the clean closure standards [40 CFR 265.228(b)] state "that provided the cleanup standards are met, the impoundment is not further subject to this part (part 265)." It is not stated that the unit is excluded from the part 270 permit requirements or the part 264 groundwater requirements. EPA is stating that the congressional intent was for all regulated units to be subject to the part 264 groundwater regulations, regardless of a successful clean closure. However, EPA will allow an interim status unit to be relieved of any further requirements of RCRA if the owner or operator can demonstrate that any further cleanup is unnecessary; otherwise, a unit that has clean closed may be required to obtain a post-closure permit or its equivalent. However, a permit application will trigger the §3004(u) corrective action provision to address releases from solid waste management units (hazardous waste management units are a subset of these units) at the entire facility. (See Chapter 9 for further information.) If a permitted hazardous waste management unit clean closes, its obligations under RCRA end.

Equivalency Demonstration. Owners or operators may demonstrate that an interim status unit was clean closed in accordance with the part 264 requirements either by submitting a part B permit application for a post-closure permit or by petitioning EPA for an equivalency demonstration under 40 CFR 270.1(c)(5). The *equivalency demonstration* is a petition that attempts to demonstrate that a post-closure permit or its equivalent (i.e., further cleanup) is not required because the owner or operator has met the applicable part 264 closure standards for that unit (52 *FR* 45795, December 1, 1987). The demonstration submitted by the owner or operator, at a minimum, must contain sufficient information for identifying the type and location of the unit, the unit boundaries, the waste that had been managed in the unit, and the extent of waste and soil removal or decontamination undertaken at closure. Relevant groundwater monitoring and soil sampling data also should be submitted to demonstrate that any Appendix VIII and Appendix IX constituents that remain at closure are below levels that pose a threat to human health or the environment. The demonstration may use data developed at the time of closure. If insufficient data are available to support the demonstration, new data may have to be collected for the determination (OSWER Directive 9476.00-18).

EPA must make a determination as to whether the unit has met the removal or decontamination requirements within 90 days of receiving a demonstration as outlined in Figure 5-7. EPA must also provide for a 30-day public comment period. If EPA finds that the closure did not meet the applicable cleanup requirements (i.e., hazardous constituents above EPA-recommended limits or factors remain), EPA must provide a written statement of the reasons why the closure was not successful. The owner or operator may submit additional information in support of the demonstration within 30 days after receiving such a written statement. EPA must review this additional information and make a final determination within 60 days. If EPA determines that the closure does not meet the part 264 standards, the owner or operator must submit a Part B permit application containing all the applicable information required in part 270, including groundwater quality data.

Certification of Closure. Within 60 days of the completion of closure for each individually and disposal *unit* at a facility, or within 60 days of closure completion for an incinerator, tank, or container storage *facility,* a certification prepared by the owner or operator and a qualified, independent, registered professional engineer must be submitted to EPA. The owner or operator must certify that the facility or unit was closed in accordance with the approved facility closure plan. Within 60 days after receiving the closure certification, EPA will notify the owner or operator in writing that, under RCRA, said owner or operator is no longer required to maintain financial assurance for closure for that particular unit or facility unless EPA has reason to believe that closure has not been done in accordance with the closure plan (40 CFR 264/265.115).

Closure Notices. At the time of closure certification, the owner or operator must submit to the local zoning authority a survey plat specifying the location and dimensions of landfill cells or other hazardous waste units with respect to permanently displayed benchmarks.

Within 60 days of closure certification, the owner or operator must submit a written record to the local zoning authority and EPA. The record must specify the type, quantity, and location of hazardous waste disposed of in each cell [40 CFR 264/265.119(a)].

Within 60 days of closure certification, a permanent notation must be made in the deed stating the following [40 CFR 264/265.119(b)]:

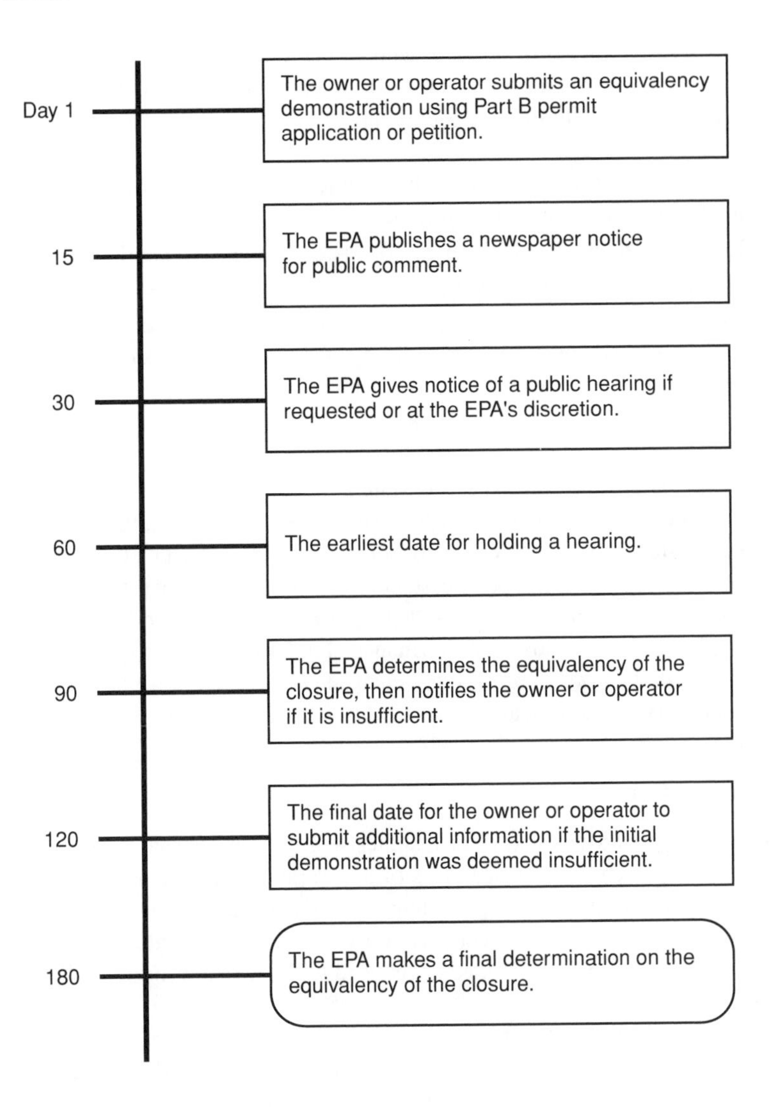

Source: OSWER Directive 9476.00-18

Figure 5-7. Equivalency Demonstration Process for Clean Closure

- Hazardous waste management occurred on the property.
- The property's use is restricted under RCRA.
- The survey plat and other applicable information are available at the local zoning authority.

A certification by the owner or operator that the notification was placed on the deed, and a copy of the deed, must be submitted to EPA.

Dirty Closure. An owner or operator who either chooses not to conduct a clean closure or fails to do so must provide post-closure care similar to that for a landfill (40 CFR 264/265.310), including the following:

- Eliminating all free liquids by either removing the liquid wastes/residues from the impoundment or solidifying them.
- Stabilizing the remaining waste and waste residues to a bearing capacity sufficient to support a final cover.
- Installing a final cover that provides long-term minimization of infiltration into the closed unit, functions with minimum maintenance, promotes drainage, and minimizes erosion.
- Performing post-closure care and groundwater monitoring.

There are additional unit-specific standards (e.g., for surface impoundments, waste piles, tanks) that provide specific instructions. In addition, the general closure performance standard (40 CFR 264/265.111) applies to activities that are not otherwise addressed by the unit-specific standards but are necessary to ensure that the facility is closed in a manner that will protect human health and the environment. For example, under the closure performance standard, an owner or operator can be required to install source control (e.g., leachate collection and runon/runoff control) that is not otherwise specifically required (OSWER Directive 9476.00-13).

Post-Closure Care

Units at which hazardous wastes or residues will remain after closure are subject to post-closure care. Owners or operators of hazardous waste management units must conduct certain monitoring and maintenance activities during the post-closure care period, which begins with the certification of closure and continues for 30 years; however, this date may be extended or reduced by EPA as appropriate (40 CFR 264/265.117).

Except for clean-closed units, *all* land disposal units that close after July 26, 1982, are subject to the post-closure care requirements [40 CFR 270.1(c)]. Thus, an interim status unit that is a regulated unit closing after July 26, 1982, and entering its post-closure period may be required to obtain a post-closure permit or its equivalent at any time during its post-closure care period.[9] If such an interim status facility receives a post-closure permit, it is then subject to both the 40 CFR 264.100 corrective action requirements for groundwater and the §3004(u) provisions (the latter being initiated by the permit application), which may require corrective action for any solid waste management unit (SWMU) at the entire facility, regardless of when the SWMU closed. [See Chapter 9 for further explanation of §3004(u).]

[9] EPA has reinterpreted the need for post-closure permits *per se*. In certain cases, EPA will not require the application or issuance of a post-closure permit provided that post-closure care is implemented through an enforcement order and all of the substantive requirements of post-closure are met (e.g., corrective action, groundwater assessment, public involvement).

Post-Closure Plan. Facilities with land disposal units must prepare and maintain a written post-closure plan at the facility. A facility without an approved post-closure plan must maintain a copy of the most current plan, which must be furnished to EPA upon request (40 CFR 264/265.118).

The post-closure plan must describe the frequency of monitoring and maintenance activities to be conducted after closure of each disposal unit. The plan should include provisions for the kinds of monitoring and maintenance activities that reasonably can be expected during the post-closure care period. It should be noted that any subsequent use of the property may not allow any disturbance of the integrity of the final cover containment system, or monitoring system [40 CFR 264/265.118(b)]. Because of difficulty in predicting what may be required, the plan should include a range of alternatives, thus avoiding a potential permit modification.

The post-closure plan must include the following:

- Inspection schedule.
- A description of the planned groundwater monitoring requirements, including the frequency of sampling.
- A description of the maintenance of the groundwater system (e.g., sedimentation buildup in the wells, protection and repair off casings).
- Evaluation and response to possible groundwater contamination.
- Security measures.
- Information and documentation concerning the cap and final cover or other containment systems.
- A description of the planned maintenance activities for the cap, containment, and monitoring equipment.
- Record keeping and reporting (e.g., groundwater analysis results).
- The name and address of the facility contact person overseeing post-closure care.

Amendments to the Post-Closure Plan. The plan may be amended by the owner or operator at any time during the active life or post-closure care period. However, an owner or operator with an *approved* post-closure plan must submit a written request to EPA for authorization to modify the approved plan. The plan must be amended whenever operating plans or facility designs affect the post-closure plan, whenever any event occurs that affects the plan, and at least 60 days before a change in operations or 60 days after an unexpected event. If the plan needs to be amended and the facility has a permit, a Class 2 permit modification must be sought [40 CFR 264/265.118(d)].

Approval of the Post-Closure Plan. The post-closure plan for an interim status facility (a permitted facility's post-closure plan is already approved) must be submitted at least 180 days before the owner or operator expects to begin partial or final closure of the first hazardous waste disposal unit. If an owner or operator intended to clean close but cannot, and must instead close as a landfill, that party must submit a post-closure plan or the contingent post-closure plan (if required) within 90 days of making that determination [40 CFR 265.118(e)].

Upon receipt of a post-closure plan, EPA will provide the owner or operator and the public the opportunity to submit written comments or requests for modifications to the plan. EPA must approve, modify, or disapprove a post-closure plan within 90 days of its receipt. EPA may hold a public hearing if one is requested.

If EPA does not approve the plan, a detailed written statement must be provided to the owner or operator with the reasons for disapproval. The owner or operator then has 30 days after receiving the written statement to modify the plan or submit a new one. EPA must approve or modify the new or resubmitted plan within 60 days. If EPA modifies this plan, the modified plan becomes the *approved* post-closure plan [40 CFR 265.118(f)]. All post-closure activities must conform to the approved post-closure plan.

Post-Closure Certification. Within 60 days after the post-closure care period for each hazardous waste disposal unit is completed, the owner or operator must submit, by registered mail, a certification stating that post-closure care was performed in accordance with the approved post-closure plan. The certification must be signed by both the owner and operator and an independent, qualified, registered professional engineer (40 CFR 264/265.120).

FINANCIAL RESPONSIBILITY REQUIREMENTS

The financial responsibility requirements were established to ensure that funds are available to pay for properly closing a facility, to maintain post-closure care at facilities that close with waste in place, and to compensate third parties for bodily injury and property damage caused by sudden and nonsudden accidents related to the facility's operation. Federal- and state-owned facilities are exempt from these requirements [40 CFR 264/265.140(c)]. However, this exclusion does not cover county or municipally owned facilities (45 *FR* 33199, May 19, 1980).

Two programs are addressed under the financial responsibility regulations: financial assurance for closure/post-closure and liability coverage.

Financial Assurance for Closure/Post-Closure

The purpose of the financial assurance requirements is to ensure that there are sufficient funds for the proper closure and post-closure care of a hazardous waste management facility should the facility close with waste or contamination remaining.

The owner or operator must have a detailed written estimate for facility closure and post-closure (if applicable) in current dollars. These cost estimates are used to determine the level of financial assurance required. The closure cost estimate, based on the closure plan, must equal the cost of final closure at the point during the active life when closure would be the most expensive and must be based on actual third-party contractor costs, not on the use of facility personnel. The costs may not take into account value from any potentially salvageable materials, equipment, or wastes (40 CFR 264/265.142).

An owner or operator must maintain financial assurance for closure until an acceptable certification of closure is made. The same requirement applies to the post-closure care period.

The key activities to be included in the closure cost estimate include the following:

- Management of the facility's entire inventory of hazardous waste by
  - transportation to an off-site TSDF, or
  - on-site treatment or disposal.
- Removal and/or decontamination of facility equipment
- Monitoring activities
- Final cover installation
- Maintenance of a facility security program
- Survey plat preparation
- Closure certification

The post-closure cost estimate, like that for closure, must be in current dollars; it must be based on the annual cost required for proper post-closure maintenance according to the post-closure plan. The annual cost is then multiplied by the established post-closure care period (i.e., 30 years). Cost estimates must also be based on the actual cost of hiring a third party to conduct post-closure activities, and not on using facility personnel (again, the same conditions as for closure). Facilities with land-based units that intend to clean close are not required to maintain a post-closure cost estimate [40 CFR 264/265.144(a)].

The minimum, key activities to be included in the post-closure cost estimate include the following:

- Monitoring of groundwater and leachate as required
- Leachate management
- Routine maintenance of the monitoring systems and other equipment
- Filing post-closure notices
- Maintenance of a security program
- Post-closure certification

The closure and post-closure cost estimate must be updated to reflect any changes that would affect the costs. For example, adding new or closing existing waste management units may require new cost estimates. The closure and post-closure cost estimate must also be revised yearly to account for inflation [40 CFR 264/265.142(b) and 264/265.144(b)]. This revision is based on the annual inflation factor, a value that can be obtained either from EPA's RCRA/Superfund hotline

(800-424-9346) or by dividing the latest annual implicit price deflator (IPD) for the gross domestic product by the previous annual deflator. (The deflators are published by the U.S. Department of Commerce in its *Survey of Current Business.*) The costs estimates must be revised according to the following formula:

$$\text{IPD current year} \div \text{IPD previous year} = \text{inflation factor}$$

$$\text{Inflation factor} \times \text{previous year cost estimate} = \text{updated cost estimate}$$

The cost estimate must be updated within 30 days after the close of the firm's fiscal year (for facilities using the financial test or corporate guarantee) or within 60 days of the anniversary of the date the first cost estimate was made (for facilities using other financial mechanisms).

Financial Assurance for Closure and Post-Closure. Each owner or operator must establish financial assurance for the proper closure and post-closure care (if applicable) of the facility and must choose from the following options:

- Trust fund
- Surety bond
- Letter of credit
- Insurance
- Financial test
- Corporate guarantee

These options may be used individually or in combination. The owner or operator may substitute another form of financial assurance at any time, provided that EPA approves of the substitution [40 CFR 264/265.143(g) and 264/265.145(f)].

Trust Fund. A *trust* is an arrangement in which one party, the grantor, transfers money to another party, the trustee, who manages the money for the benefit of one or more beneficiaries. For the purposes of this section, the facility owner or operator is the grantor, the financial institution is the trustee, and EPA is the beneficiary. These trusts are irrevocable; they cannot be altered or terminated by the owner or operator without the consent of EPA and the financial institution. The trust is established when the trust agreement is signed by the grantor and the financial institution.

For permitted facilities, payments must be made annually over the term of the initial RCRA permit (i.e., 10 years). For interim status facilities, payments are made annually over 20 years or the remaining operating life of the facility, whichever is shorter. The operating life of the facility is determined by using the expected year of closure, which should be identified in the closure plan. If the facility subsequently receives a permit, the pay-in period is adjusted accordingly [40 CFR 264/265.143(a)].

After the pay-in period is complete, more payments must be made if the closure cost estimate increases or inflation factoring shows that a greater amount is needed.

However, if the amount in the trust fund is greater than needed for closure, the owner or operator may submit to EPA a written request to release excess money. After partial or final closure has begun, a request for reimbursement for closure expenditures may be submitted to EPA using itemized bills.

Surety Bond. A *surety bond* is a bond guaranteeing performance of an obligation, such as guaranteeing payment into a trust fund or guaranteeing the performance of closure and/or post-closure. If an owner or operator uses a surety bond or a letter of credit, a standby trust fund (essentially the same as the trust fund) must be established. In most cases, a standby trust fund is established with an initial nominal fee agreed on by the owner or operator and the trustee. Further payments into this fund are not required until the standby trust is funded by a surety company as required. The surety company must be listed as an acceptable surety in *Circular 570* of the U.S. Department of the Treasury.

If an administrative order is issued to compel closure or if closure is required by an appropriate court order, the amount equal to the penalty sum must be placed into the standby trust fund within 15 days of issuance.

Letter of Credit. A *letter of credit* is an agreement by the issuing institution that it will make available to the beneficiary (EPA) a specific sum of money during a specific time period on behalf of its customer (the facility owner or operator). The letter of credit must be irrevocable and issued for one year. The letter must have an automatic extension unless the issuing institution notifies both EPA and the facility owner or operator at least 120 days before the expiration date that the letter will not be extended. The letter of credit must also establish a standby trust fund. In addition, the letter of credit must be increased within 60 days of whenever the current cost estimate increases.

If the facility fails to perform final closure in accordance with the approved closure plan, EPA may draw upon the letter of credit to pay for proper closure.

Insurance. Insurance is basically a contract through which one party guarantees monies to another party (usually a prespecified amount) to perform the closure in return for premiums paid.

An insurance policy must be obtained for the face amount (the total money the insurer is obligated to pay under the policy) at least equal to the cost estimate of closure and post-closure. Also, the face value must be increased if the cost estimate increases.

Financial Test. The financial test is a means that may be used to provide financial assurance. An owner or operator may choose between the following alternatives:

Alternative 1

- The firm must have a tangible net worth of at least $10 million.

- Net working capital in the United States must be at least six times the amount of the closure and post-closure cost estimates or 90 percent of total assets located in the United States.
- At least two of the following must be satisfied:
  - —liabilities to net worth less than 2,
  - —assets to liabilities ratio greater than 1.5, or
  - —net income to liabilities ratio greater than 0.1

Alternative 2

- The firm must have a tangible net worth of at least $10 million.
- Tangible net worth must be at least six times the amount of the closure and post-closure cost estimates.
- At least 90 percent of total assets are located in the United States.
- The current bond rating for the most recent bond issuance is Aaa, Aa, A, or Baa (Moody's) or AAA, AA, A, or BBB (Standard & Poor's).

Corporate Guarantee. An owner or operator may use a corporate guarantee to ensure adequate financial coverage for closure and post-closure of another entity. This guarantee is used primarily by parent firms to guarantee the obligations of their subsidiaries. The following criteria must be met to satisfy the requirements of the corporate guarantee:

- The guarantor must be a direct corporate parent (owns at least 50 percent of the voting stock), a corporate grandparent (indirectly owns more than 50 percent through a subsidiary), a sibling corporation (shares the same parent corporation), or a firm with a substantial business relationship with the owner/operator.
- The guarantor must meet the financial test described in the previous subsection.
- The guarantor must perform the required closure/post-closure care activities or establish a trust fund should the owner/operator fail to carry out the closure/post-closure requirements.

Liability Coverage

Owners and operators must demonstrate the existence of sufficient liability coverage during the operating life of a facility for bodily injury and property damage to third parties resulting from facility operations. Under the liability coverage regulations, owners and operators of all hazardous waste management facilities are required to demonstrate, on an owner/operator (per firm) basis, adequate liability coverage (40 CFR 264/265.147). This means that even if a firm has multiple hazardous waste management facilities, the coverage stays the same. Liability coverage is required until an acceptable certification of closure is made. Liability is generally not required for the post-closure care period.

Liability coverage is required for nonsudden accidental occurrences in the amount of $3 million per occurrence and $6 million annual aggregate, exclusive of legal defense costs. An *accidental occurrence* is an accident that is neither expected nor intended. A *nonsudden accidental occurrence* is an accidental occurrence that takes place over a period of time and involves continuous or repeated exposure, such as a hazardous waste leaching into groundwater. Liability coverage is also required for sudden accidental occurrences in the amount of $1 million per occurrence with an annual aggregate of at least $2 million, exclusive of legal defense costs. A *sudden accidental occurrence* is an accidental occurrence that is not continuous or repeated, such as a fire or an explosion. An owner or operator may combine coverage for sudden and nonsudden accidental occurrences [40 CFR 264/265.147(b)]. The coverage must be at least $4 million per occurrence ($1 million sudden plus $3 million nonsudden) and $8 million annual aggregate ($2 million sudden plus $6 million nonsudden).

First-dollar coverage is required; that is, the amount of any deductible must be covered by the insurer, with right of reimbursement from the insured. Liability coverage must be continuously provided for a facility until the certification of closure is received by EPA. The financial mechanisms authorized to establish liability coverage options include a financial test, corporate guarantee, liability insurance, surety bond, letter of credit, or a trust fund.

HAZARDOUS WASTE TREATMENT, STORAGE, AND DISPOSAL FACILITIES: GENERAL STANDARDS COMPLIANCE CHECKLIST[10]

General Facility Standards

Does the facility have an EPA identification number? Yes___ No___

Are all facility records maintained for at least three years? Yes___ No___

Has the facility imported hazardous waste from a foreign source? Yes___ No___

If yes, was a notice filed with the EPA Regional Administrator four weeks in advance of the initial shipment? Yes___ No___

Was the waste an OECD-covered waste? Yes___ No___

If yes, was the tracking document signed and sent to EPA, the importer, and the competent foreign authority? Yes___ No___

Is a written operating log maintained at the facility? Yes___ No___

Is the following noted in the operating log?

- Description and management of waste received Yes___ No___
- Location and quantity of all hazardous waste at the facility Yes___ No___

[10] Because states are authorized to be more stringent and/or broader in scope than the federal program, this checklist should be used only as a guide and individual state requirements should also be consulted.

- Record and results of waste analyses Yes___ No___
- Details of any instances requiring implementation of the contingency plan Yes___ No___
- Records and results of inspection Yes___ No___
- Results of required groundwater monitoring analyses Yes___ No___
- Documentation pertaining to the land disposal restrictions Yes___ No___
- Closure and post-closure cost estimates Yes___ No___

Has the biennial report been completed and submitted? Yes___ No___

General Waste Handling Requirements

Has an adequate security system been established for the active portion of the facility? Yes___ No___

Has the facility established a training program for appropriate personnel? Yes___ No___

Is a training review given annually? Yes___ No___

Are the following records kept?

- Job title and written job description of each position Yes___ No___
- Description of the type and amount of introductory and continuing training Yes___ No___
- Documentation that training has been given to employees Yes___ No___

Are these records maintained at the facility? Yes___ No___

Does the training program contain the following?

- Procedures for using, inspecting, repairing, and replacing emergency equipment Yes___ No___
- Key parameters for automatic waste-feed cutoff systems Yes___ No___
- Communications or alarm systems Yes___ No___
- Response to fires or explosions Yes___ No___
- Responses to groundwater contamination incidents Yes___ No___
- Shutdown of operations Yes___ No___

Has an OSHA required training program, which addresses the following issues, been implemented?

- Safety and health Yes___ No___
- Hazard communication Yes___ No___
- Medical surveillance Yes___ No___
- Decontamination procedures Yes___ No___
- New technologies Yes___ No___
- Material handling procedures Yes___ No___
- Pre-emergency planning and coordination with outside parties Yes___ No___

- Personnel roles, lines of authority, and communication Yes___ No___
- Emergency recognition and prevention Yes___ No___
- Safe distances and places of refuge Yes___ No___
- Site security and control Yes___ No___
- Evacuation routes and procedures Yes___ No___
- Decontamination procedures Yes___ No___
- Emergency medical treatment and first aid Yes___ No___
- Emergency alerting and response procedures Yes___ No___
- Critique of response and follow-up Yes___ No___
- Personal protective equipment and emergency equipment Yes___ No___

Does the facility have a written waste analysis plan? Yes___ No___

Does the plan include the following information?

- Parameters for which each waste will be analyzed Yes___ No___
- Rationale for the selection of these parameters Yes___ No___
- Test methods used to test for these parameters Yes___ No___
- Sampling method used to obtain sample Yes___ No___
- Frequency with which the initial analysis will be reviewed or repeated Yes___ No___

For off-site facilities only: Does the plan include the following?

- Waste analyses that generators have agreed to supply Yes___ No___
- Procedures that are used to inspect and analyze each shipment Yes___ No___
- Rejection criteria for an unacceptable shipment Yes___ No___

Has a written inspection schedule been prepared for the facility? Yes___ No___

Are records of inspections maintained in the inspection log? Yes___ No___

If yes, does it include the following?

- Date and time of inspection Yes___ No___
- Name of inspector Yes___ No___
- Notation of observation Yes___ No___
- Date and nature of repairs or remedial action Yes___ No___

Is ignitable or reactive waste handled? Yes___ No___

If yes, is waste separated and confined from sources of ignition or reaction? Yes___ No___

Are NO SMOKING signs posted in hazardous areas where ignitable or reactive wastes are handled? Yes___ No___

Preparedness and Prevention

Is the facility equipped with the following?

- Easily accessible internal communications or alarm system Yes___ No___
- Telephone or two-way radio to call emergency response personnel Yes___ No___
- Portable fire extinguishers, fire control equipment, spill control equipment, and decontamination equipment Yes___ No___
- Water of adequate volume for hoses, sprinkler, or water spray system Yes___ No___

Is this equipment tested and maintained as necessary to assure its proper operation? Yes___ No___

Is there sufficient aisle space to allow unobstructed movement of equipment? Yes___ No___

Have arrangements been made with the local authorities? Yes___ No___

If a local authorities declined to enter into an agreement, is there documentation? Yes___ No___

Contingency Plan and Emergency Procedures

Does the facility have a contingency plan? Yes___ No___

If yes, does it contain the following information?

- Actions to be taken in response to emergencies Yes___ No___
- A description of arrangements with police, fire, and hospital officials Yes___ No___
- A list of names, addresses, phone numbers, of emergency coordinators Yes___ No___
- A list of all emergency equipment at the facility Yes___ No___
- An evacuation plan for facility personnel Yes___ No___

Is a copy of the contingency plan maintained at the facility? Yes___ No___

Has a copy been supplied to local police and fire departments? Yes___ No___

Is the plan an Integrated Contingency Plan? Yes___ No___

If so, have all statutory requirements been met? Yes___ No___

Is there an emergency coordinator on-site or on call at all times? Yes___ No___

Manifest System

Does the facility receive hazardous waste from off-site? Yes___ No___

If yes, are copies of all manifests retained? Yes___ No___

Has the facility received any shipments that were inconsistent with the manifest? Yes___ No___

If yes, has the discrepancy been resolved with the generator and the transporter? Yes___ No___

If no, has the appropriate authority been notified? Yes___ No___

Has any nonexcluded waste been received that was not accompanied by a manifest? Yes___ No___

If yes, has an unmanifested waste report been submitted? Yes___ No___

Groundwater Monitoring

Interim Status Facilities

Has a groundwater sampling and analysis plan been prepared? Yes___ No___

Does the plan include the following?

- Sample collection procedures Yes___ No___
- Sample preservation procedures Yes___ No___
- Sample shipment procedures Yes___ No___
- Analytical procedures Yes___ No___
- Chain-of-custody controls Yes___ No___

Has a groundwater monitoring system been installed that will yield groundwater samples of sufficient quality to detect a potential releases? Yes___ No___

Have at least three downgradient wells been installed? Yes___ No___

Has at least one upgradient well been installed? Yes___ No___

Has the facility gathered one year's worth of background data for the following parameters?

- Drinking water suitability Yes___ No___
- Groundwater quality Yes___ No___
- Indicators of groundwater contamination Yes___ No___
- Groundwater elevation Yes___ No___

Have statistical comparisons been made with indicators of groundwater contamination compared to the first year's baseline data? Yes___ No___

Has a statistical increase (or decrease with pH) occurred? Yes___ No___

If yes, has the facility entered the assessment monitoring phase? Yes___ No___

Has a groundwater assessment monitoring plan been prepared? Yes___ No___

Permitted Facilities

Has the facility installed a sufficient number of wells representing background water quality? Yes___ No___

Has the facility installed a sufficient number of wells representing the water quality passing the point of compliance? Yes___ No___

Has an authorized, appropriate statistical test been selected? Yes___ No___

Has a groundwater sampling and analysis plan been prepared? Yes___ No___

Does the plan include the following?

- Sample collection, preservation, and shipment procedures Yes___ No___
- Chain-of-custody controls Yes___ No___

Has the facility gathered one year's worth of background data of the parameters specified in the facility's permit? Yes___ No___

Have statistical comparisons been made with the specified parameters compared to the first year's baseline data? Yes___ No___

Has a statistical increase occurred? Yes___ No___

If yes, has EPA been notified? Yes___ No___

Has a complete Appendix IX scan been run for all monitoring wells? Yes___ No___

Has a Class 3 permit modification for compliance monitoring been applied for? Yes___ No___

Has an engineering feasibility plan for corrective action been submitted to EPA? Yes___ No___

Has a groundwater protection standard been established? Yes___ No___

Has the facilities entered the compliance monitoring phase? Yes___ No___

Has the groundwater protection standard been exceeded? Yes___ No___

If yes, has EPA been notified? Yes___ No___

Has a Class 3 permit modification for corrective action been applied for? Yes___ No___

Has a corrective action plan been prepared? Yes___ No___

Closure and Post-Closure

Has a closure plan been prepared? Yes___ No___

If clean closure is planned, has a contingent closure plan been prepared? Yes___ No___

Has the closure plan been approved by EPA? Yes___ No___

Does the closure plan address the following?

- Description of how to achieve closure Yes___ No___
- Estimate of maximum inventory of waste on-site Yes___ No___
- Description of other closure activities such as groundwater monitoring Yes___ No___
- Schedule of closure dates and time required to conduct closure Yes___ No___

Has the closure plan been modified to address any facility or operation modifications? Yes___ No___

Has EPA been notified that closure is about to begin? Yes___ No___

For facilities without approved closure plans, has the plan been submitted to EPA? Yes___ No___

Has the facility removed or decontaminated all waste, waste residues, subsoils, and liners and equipment? Yes___ No___

If not, have contingencies been made to close the unit similar to a landfill? Yes___ No___

Has closure been certified? Yes___ No___

Has a survey plat been prepared and provided to the local land authority? Yes___ No___

Has a notation on the deed been made? Yes___ No___

Is the facility subject to post-closure care (i.e., dirty closure)? Yes___ No___

Has a post-closure plan been prepared? Yes___ No___

Have there been any modification to the facility or operations necessitating modification to the post-closure plan? Yes___ No___

Has the post-closure plan been approved by EPA? Yes___ No___

Has the post-closure period been satisfied? Yes___ No___

Has the completion of post-closure been certified? Yes___ No___

Financial Assurance

Have the closure cost estimates been determined? Yes___ No___

Have the post-closure cost estimates been determined? Yes___ No___

Are the cost estimates based on hiring a third-party contractor? Yes___ No___

Have any changes been made to the facility necessitating a cost update? Yes___ No___

Have the costs been updated, taking into account annual inflation? Yes___ No___

Has an appropriate mechanism to cover financial assurance been obtained? Yes___ No___

Has an appropriate mechanism to cover liability been obtained? Yes___ No___

6

TECHNICAL STANDARDS FOR HAZARDOUS WASTE MANAGEMENT UNITS

In addition to the general standards applicable to all hazardous waste management facilities, there are specific technical operating requirements for each type of hazardous waste management unit. Generally, it is not the waste management activity per se, that is regulated; the unit in which the activity occurs is regulated. For example, a facility may employ many different treatment processes occurring in tanks; however, the facility would generally comply with the same standards for all of the tanks regardless of which process is being employed.

The following hazardous waste management units are addressed in this chapter:

- Container storage units
- Tank systems
- Surface impoundments
- Waste piles
- Land treatment areas
- Landfills
- Incinerators
- Thermal treatment units
- Chemical, physical, and biological treatment units
- Miscellaneous units
- Containment buildings
- Underground injection wells

This chapter describes the technical requirements for hazardous waste management under 40 CFR part 265 (interim status units) and part 264 (permitted units).

Because the interim status standards are self-implementing, they are applicable to all interim status units. The part 264 standards, however, are used to establish the conditions of an operating permit. Therefore, each specific permit may differ from the requirements presented in this chapter because each permit is unique. Chapter 7 of this book presents more detailed information concerning permit conditions.

The major requirements summarized here include both parts 264 and 265. Any difference between the standards of those parts is noted in the text.

CONTAINER STORAGE UNITS

Containers used to store hazardous waste must always be closed and sealed except when necessary to add or remove waste, inspect or repair equipment located inside the container, or vent gases or vapors to an air emission control device. A container must not be opened, handled, or stored in a way that might cause the contents to leak. If a container leaks or is in poor condition, the contents of that container must be placed into a sound container (40 CFR 264/265.171). Containers holding ignitable or reactive wastes must be stored at least 50 feet from the facility's property boundary (40 CFR 264/265.176).

The owner or operator must inspect the area(s) where containers are stored at least weekly for leaks and corrosion or other indications of potential container failure, and record the results of these inspections in the facility's operating log (40 CFR 264/265.174).

The waste must be compatible with the container and may not be placed in an unwashed container if it previously held an incompatible waste. Incompatible wastes may not be placed into the same container. A container storing waste that is incompatible with other materials must be separated from these materials by a physical structure (e.g., dike or wall) or removed from the area (40 CFR 264/265.172).

Containers that will manage hazardous waste having a volatile organic concentration greater than 500 ppmw must comply with the air emission control requirements. (Containers that have a capacity less than or equal to 0.1 m^3, approximately 26 gallons, are excluded.) The owner or operator may choose one of the following options to control air emission from the container:

- Must be equipped with a vapor leak-tight cover.
- Must have a design capacity less than or equal to 0.46 m^3 (approximately 119 gallons) and be equipped with a cover in compliance with HMTA requirements under 49 CFR part 178.
- Must be attached to or form part of a truck, trailer, or railcar, and must have demonstrated organic-vapor tightness within the preceding 12 months.

At closure, all hazardous wastes and residues must be removed from the storage area. All containers, liners, bases, and contaminated soil and groundwater must be removed or decontaminated (40 CFR 265.11).

Additional Requirements for Permitted Units

In addition to the above requirements, a permitted container storage unit, in accordance with 40 CFR 264.175, must have a containment system that includes the following:

- An impervious flooring that will allow the collection of any leakage.
- Flooring that is sloped to collect any leaks or spills.
- A contaminant area that has sufficient capacity to contain 10 percent of the volume of all containers or the volume of the largest container, whichever is greater.
- A system to prevent precipitation runon into the containment system unless the contaminant system has excess containment capacity.
- A collection and removal system for any released waste.

Container storage areas that store wastes and that have no *free liquids* (i.e., wastes that pass the Paint Filter Test, Method 9095) are not required to have a containment system as described in the preceding list. However, this exemption is allowed only if the storage area is sloped to drain and remove any liquid resulting from precipitation and if containers are elevated or protected from contact with any accumulation [40 CFR 264.175(c)].

The dioxin-containing wastes (F020, F021, F022, F023, F026, F027, and F028) must have a containment system that complies with the storage requirements for liquid wastes regardless of whether free liquid is present [40 CFR 264.175(d)].

Closure

At closure, all hazardous wastes and residues must be removed from the storage area. All containers, liners, bases, and contaminated soil and groundwater must be removed or decontaminated (40 CFR 264.178).

TANK SYSTEMS

The regulations contained in subpart J of 40 CFR parts 264 and 265 are applicable to *tank systems,* which include the tank, all associated ancillary equipment. and the containment system. *A tank is* a stationary device, designed to contain an accumulation of hazardous waste, that is constructed primarily of nonearthen materials (e.g., wood, steel, plastic) that provide structural support (40 CFR 260.40). *Ancillary equipment* is any device, including but not limited to piping, fittings, flanges, valves, and pumps, used to distribute, meter, or control the flow of hazardous waste from its point of generation to the storage or treatment tank(s), between hazardous waste storage and treatment tanks to a point of disposal on-site or to a point of shipment for disposal off-site (40 CFR 260.10).

The following pages contain separate discussions on these tank systems:

- Small-quantity-generator tank systems
- Large-quantity-generator tank systems
- Hazardous waste management facility tank systems
- New tank systems

Small-Quantity-Generator Tank Systems

Small-quantity generators (generators of 100 to 1,000 kg/mo) accumulating waste in a tank need to comply with the following requirements pertaining to tank systems [40 CFR 263.34(d)(3)]:

- Treatment must not generate any extreme heat, explosions, fire, flammable fumes, mists, dusts, or gases; or damage that tank's structural integrity; or threaten human health or the environment.
- Hazardous wastes or treatment reagents that can cause corrosion, erosion, or structural failure may not be placed into a tank.
- Incompatible wastes may not be placed into a tank.
- At least two feet of freeboard must be maintained in an uncovered tank unless sufficient containment capacity (i.e., capacity equals or exceeds volume of top two feet of tank) is supplied.
- Continuously fed tanks must have a waste-feed cutoff or bypass system.
- Ignitable, reactive, or incompatible wastes may not be placed into a tank unless these wastes are rendered nonignitable, nonreactive, or compatible.
- At least once each operating day, the waste-feed cutoff and bypass systems, monitoring equipment data, and waste level must be inspected.
- At least weekly, construction materials and the surrounding area of the tank system must be inspected for possible corrosion, leaks, or visible signs of erosion.
- At closure, all hazardous wastes must be removed from the tank, containment system, and discharge control system.

Large-Quantity-Generator Tank Systems

Large-quantity generators using 90-day accumulation tanks (to the degree that it is authorized by the state) in accordance with 40 CFR 262.34 must comply with most of the provisions of Subpart J of 40 CFR part 265, including the following:

- A one-time assessment of the integrity of the tank system.
- Installation standards for new tanks.
- Design standards, including an assessment of corrosion potential.
- Secondary containment phase-in provisions.

- Air emission controls.
- Closure.
- Periodic leak testing if the tank system does not have secondary containment.
- Additional response requirements regarding leaks, including reporting to the EPA Regional Administrator on the extent of any release as well as repairing or replacing leaking tanks.

However, owners or operators of 90-day accumulation tanks are not required to prepare closure or post-closure plans, prepare contingent closure or post-closure plans, maintain financial responsibility, or conduct waste analysis and trial tests [40 CFR 262.34(a)(1) and 265.201].

Non-Generator Tank Systems

Hazardous waste management facilities using tank systems to treat or store hazardous waste are subject to the following requirements:

- General operating requirements
- Secondary containment
- Air emission controls
- Inspections
- Waste analysis
- Response to leaks and spills
- Closure

General Operating Requirements. The following requirements are general operating requirements applicable to tank systems (40 CFR 264/265.19):

- Incompatible wastes or reagents must not be placed into the tank system.
- No ignitable, reactive, or incompatible wastes are to be placed into a tank unless these wastes are rendered nonignitable, nonreactive, or compatible.
- The tank must have spill prevention (e.g., check valves) and overfill prevention controls.
- The tank must maintain sufficient freeboard (distance between the top of the tank and the surface of the waste) to prevent overtopping by precipitation or wind action.

Secondary Containment. Existing tanks were required to have secondary containment, including leak-detection capability, by January 12, 1989, or the date when the tank system reaches 15 years of age, whichever is later. If the age of the tank system cannot be documented, secondary containment must have been provided by January 12, 1995. However, if a facility for which the age of the tank system cannot be documented is older than 7 years. secondary containment was to be provided by January

12, 1989, or when the facility reaches 15 years of age, whichever is later. Documentation may include a bill of sale, an installation certification, or dated engineering drawings of the system [40 CFR 264/265.193(a)].

Tank systems without secondary containment, which eventually must be retrofitted) were required to have written assessment of the tank's integrity completed and on file at the facility by 1988 (40 CFR 264/265.191). If a tank stores or treats a waste that subsequently becomes a hazardous waste (i.e., is newly listed), the assessment must be conducted within 12 months of the listing. The assessment, certified by an independent, qualified, registered professional engineer, must determine the adequacy of the tank's design and ensure that it will not rupture, collapse, or fail, based on the following considerations:

- The tank design standards.
- Hazardous characteristics of the waste.
- Existing corrosion protection.
- Documented age of the tank.
- Results of a leak test (tank tightness test), internal inspection, or other integrity test results.

If a tank handles hazardous waste that does not contain any free liquids and the tank is situated inside a building on an impermeable floor, it is not subject to the secondary containment section [40 CFR 264/265.190(a)].

All tanks without secondary containment that cannot be entered for inspection must undergo a leak test. If the assessment indicates that the tank is leaking or is unfit for use, it must be taken out of service immediately [40 CFR 264/265.191(b)(5)(i)].

The containment system must be designed to prevent any migration of wastes or liquids into the soil, groundwater. or surface water, and must be able to detect and collect any released waste [40 CFR 264/265 193(b)]. At a minimum, the containment system must be

- Constructed or lined with compatible materials that have sufficient strength and thickness to prevent failure from pressure gradients, weather, waste contact, and daily operational stress.
- Able to prevent failure from settlement, compression, or uplift.
- Provided with a leak-detection system that will detect the presence of any waste released within 24 hours of release.
- Sloped or designed to drain and remove any liquids resulting from leaks, spills, or precipitation.

In accordance with 40 CFR 264/265.193(d), each tank system must use one or more of the following containment devices:

- Liner system

- Vault system
- Double-walled tank
- Alternative design approved by EPA

Liner System. A liner system must be [40 CFR 264/265.193(e)(1)]:

- Free of gaps and cracks.
- Able to contain 100 percent of the volume of the largest tank contained in the system.
- Designed and operated to prevent runon or precipitation from a 25-year, 24-hour rainfall event.
- Designed to surround the tank completely and to cover all surrounding earth that may come into contact with any released waste.

Vault System. A vault system must be [40 CFR 264/265.193(e)(2)]:

- Designed for 100 percent containment capacity of the largest tank.
- Designed or operated to prevent precipitation runon from a 25-year, 24-hour rainfall event.
- Constructed with chemically resistant water-stops at all joints.
- Constructed with an impermeable, compatible lining or coating.
- Constructed with an outside moisture barrier to prevent migration of moisture into the vault.

Double-Walled Tanks. A double-walled tank must be [40 CFR 264/265.193(e)(3)]:

- Designed as an integral structure so that any release will be contained by the outer shell.
- Protected from corrosion if constructed of metal.
- Provided with a built-in continuous (interstitial) leak detection system capable of detecting a leak within 24 hours of a release.

Ancillary Equipment. A tank system's ancillary equipment must be provided with secondary containment (e.g., trenches, jacketing, double-walled piping), except for the following devices, provided that a daily inspection is conducted [40 CFR 264/265.193(f)]:

- Welded joints, flanges, or connections.
- Sealless or magnetic coupling pumps.
- Aboveground piping.
- Pressurized aboveground piping systems with automatic shutoff devices.

Variances A variance from secondary containment requirements may be granted if the owner or operator can demonstrate that alternative design and operating practices, together with the location characteristics, will prevent the migration of any hazardous waste or constituents into the groundwater or surface water at least as effectively as secondary containment. The considerations involved in assessing a variance are listed in 40 CFR 264/265.1 93(g).

Inspections. At least once each operating day the owner or operator must inspect the following (40 CFR 264/265.195):

- Aboveground portions of the tank to detect corrosion or releases.
- Data gathered from monitoring and leak-detection equipment.
- Construction materials and surrounding areas of the tank system for visible erosion of releases.
- Overfill and spill control equipment (part 265 and 90-day accumulation tanks only); permitted tanks must develop a schedule that will be specified in the permit for inspection of the spill control equipment.

Air Emission Controls. Tanks receiving hazardous waste with a volatile organic concentration greater than 500 ppmw must comply with the RCRA air emission controls (40 CFR 264.1084 and 265.1085). There are two levels of control requirements, depending on the tank design and capacity and the maximum organic vapor pressure of the waste in the tank. Level 1 tank controls require only a fixed roof, provided the tank meets certain criteria. Level 2 tank controls require hazardous waste to be managed in one of the following tank-types:

- Tank equipped with a fixed roof and an internal floating roof.
- Tank equipped with an external floating roof.
- Pressure tank designed to operate as a closed system with no detectable emissions at all times.
- Tank located inside an enclosure that is vented through a closed-vent system to an enclosed combustion control device.

In addition, tanks used for waste stabilization processes must use specific air emission controls. Facilities using steam strippers, distillation units, fractionation units, thin-film evaporation units, solvent extraction, or air strippers must reduce or destroy organics from all such devices at a facility to an aggregate total maximum of 3.0 lbs./hour or 3.1 tons/year or by 95 percent by weight [40 CFR 262.34(a)(1)(ii) and 264/265.1032].

Waste Analysis. If a tank is going to hold any waste that is substantially different from its previous contents, or uses a process to treat a waste that is substantially different from the previous contents or treatment action, both a waste analysis and a trial test must be conducted (40 CFR 265.200).

Permitted tank systems will have a significantly more detailed waste analysis plan and program established as a permit condition.

Response to Leaks and Spills. Any tank that has a release or is unfit for service must immediately be taken out of service. The flow of incoming waste must be stopped immediately, and the tank contents must be removed within 24 hours or as soon as possible. If the release is confined to the containment system, the contents must be removed (40 CFR 264/265.196).

Any release from a tank system into the environment that meets a Superfund reportable quantity must be reported immediately. (See Chapter 12 for further discussion.) In addition, within 30 days of the detection of a release, a report containing the following information must be submitted to EPA [40 CFR 264/265.196(d)(3)]:

- Likely route of migration.
- Characteristics of surrounding soils.
- Results of any monitoring or sampling conducted (must be sent as soon as available).
- Proximity to downgradient drinking water, surface water, and human populations.
- Description of response actions taken or planned unless the tank system is repaired before its return to service.

For all major repairs, the owner or operator must submit a certification to EPA before the tank is put into service. The certification verifying the adequacy of the tank's repairs must be performed by an independent, qualified, registered professional engineer [40 CFR 264/265.196(f)].

Closure. At closure, the owner or operator must remove or decontaminate all waste residues, contaminated containment system components, contaminated soils, and contaminated structures and equipment [40 CFR 264/265.197(a)].

If the owner or operator cannot clean close the tank system or demonstrate that decontamination or removal of all soils is practicable, the system must be closed similar to a landfill [40 CFR 264/265.197(b)].

New Tank Systems

New tank systems must be designed, installed, and operated in compliance with the new tank standards. These standards require an assessment by an independent, qualified, registered professional engineer regarding the adequacy of the tank's structural support, foundation, seams, connections, and pressure control. This assessment must ensure that the tank will not fail, collapse, or rupture (40 CFR 264.192).

The owner or operator of a tank must ensure that proper handling procedures are followed to prevent damage during the tank's installation. Before a new tank system is covered or enclosed, a qualified installation inspector must examine the system for weld breaks, punctures, scrapes in the protective coating, cracks, corrosion, or any other structural damage, inadequate construction, or improper installation procedures. New tanks, including their ancillary equipment, must be tested for tightness before being covered or enclosed. Written documentation of these inspections and assessments must be obtained and kept at the facility.

SURFACE IMPOUNDMENTS

A *surface impoundment* is a facility or part of a facility that is a natural topographic depression, man-made excavation, or diked area formed primarily of earthen materials (although it may be lined with man-made materials), which is designed to hold an accumulation of liquid wastes or wastes containing free liquids and is not an injection well. Examples of surface impoundments are holding, storage, settling, and aeration pits; ponds; and lagoons (40 CFR 260.10).

For regulatory purposes, including the land disposal restrictions, it is important to make the regulatory distinction between a tank and a surface impoundment: If all the surrounding earthen material is removed from a unit and the unit maintains its structural integrity, it is a tank; however, if the unit does not maintain its structural integrity without support from surrounding earthen material, it is a surface impoundment (EPA Regulatory Interpretation Letter No. 110).

General Operating Requirements

All surface impoundments must have a liner for all portions of the impoundment. The liner must be constructed to prevent any migration of wastes to the surrounding environment. Any new surface impoundment or expansion or replacement of a surface impoundment must have a double-liner system (including a leachate collection and removal system) that is in compliance with the minimum technological requirements of HSWA. The double-liner requirement may be waived if the unit is a monofill that handles only foundry wastes that are hazardous solely because of the characteristic of toxicity, and if the unit has at least one liner (40 CFR 264/265.221).

Ignitable or reactive waste may not be placed into the impoundment unless the waste is treated so that it no longer retains its hazardous characteristic or is managed in such a way as to prevent ignition or reaction (40 CFR 264/265.229).

To control air emissions, all surface impoundments managing hazardous waste having a volatile organic concentration greater than 500 ppmw require the installation of a cover installed (e.g., an air-supported structure or a rigid cover) that is vented through a closed-vent system to a control device (40 CFR 264.1085 and 265.1086).

Interim Status Requirements. An interim status surface impoundment must have at least 2 feet of freeboard (the space between the surface of the waste and the top of the containment device), which must be inspected daily to ensure compliance. The surface impoundment, including the dike and surrounding vegetation, must be inspected at least weekly to detect any leaks, deterioration, or failures in the impoundment (40 CFR 265.222).

All earthen dikes must have a protective cover, such as grass, shale, or rock, to minimize wind and water erosion as well as to preserve the structural integrity of the dike (40 CFR 265.233).

Permitted Surface Impoundments. The impoundment must be inspected weekly to detect any possible leakage [40 CFR 264.226(a)]. If the dike leaks or the level of liquid suddenly drops without known cause, the impoundment must be taken out of service; in which the waste inflow must be discontinued, any surface leakage must be contained, and the leak must be immediately stopped. If the leak cannot be stopped, the contents must be removed. If any leakage enters the leak detection system, EPA must be notified, in writing, within seven days. The impoundment cannot be placed back into service until all repairs are completed (40 CFR 264.227).

If the waste is removed from it, the unit is out of service. To reuse that unit, it must be retrofitted with a double liner meeting minimum technological requirements. Removing the waste and then reusing the unit is considered *replacement.*

Each permitted surface impoundment must have an approved *action leakage rate*, which is the maximum design flow rate that the leak detection system can remove without the fluid head on the bottom liner exceeding one foot (40 CFR 264.222). If the action leakage rate is ever exceeded, the response action plan must be implemented [40 CFR 264.223(a)]. In addition, EPA must be notified with 7 days, such notice to be followed by a preliminary written assessment in 14 days as to the likely cause and actions taken and planned [40 CFR 264.223(b)].

Waste Analysis

For interim status units, whenever an impoundment is to be used to chemically treat a waste different from that previously treated, or to employ a substantially different method to treat the waste, the owner or operator of the impoundment must conduct waste analyses and treatment tests (bench or pilot scale). This requirement is in addition to the waste analysis program required by the general facility standards in 40 CFR 265.13. As an alternative, the owner or operator may obtain written, documented information on similar treatment of similar waste under similar conditions to show that this treatment will not create any extreme heat, fire, explosion, violent reactions, toxic mists, fumes, gases, or dusts, or damage the structural integrity of the impoundment. The documentation or test results must be entered into the facility's operating record (40 CFR 265.225).

Permitted surface impoundments must have a detailed waste analysis plan and program that becomes a condition of the operating permit.

Closure

At closure, the owner or operator may elect to remove or decontaminate at the impoundment *all* standing liquids, waste and waste residues, the containment system (e.g., liner), and underlying and surrounding contaminated soil and groundwater, or to close the impoundment similar to a landfill (40 CFR 264/265.228). To close with waste in place, the owner or operator must

- Eliminate all free liquids.
- Stabilize remaining wastes to support a final cover.
- Provide long-term minimization of the migration of liquids.
- Promote drainage and minimize erosion or abrasion of the final cover.
- Maintain the integrity of the final cover and make repairs when necessary.
- Maintain and monitor the leak detection system.
- Maintain and monitor the groundwater monitoring system.
- Prevent runon and runoff from impacting the final cover.

If the owner or operator can demonstrate that *all* of the materials were removed or decontaminated as described, the impoundment will no longer be subject to 40 CFR parts 264 and 265. Removal of the underlying and surrounding contaminated soil must include the removal of any contaminated groundwater. This is known as a *clean closure* or closure by removal, as discussed in Chapter 5. However, if a clean closure cannot be attained, the closure plan must be modified and approved to close the surface impoundment as a landfill. A post-closure permit will be required, and when an owner or operator applies for a permit, selected RCRA corrective action provisions [§3004(u)] will be necessary, as discussed in Chapter 9.

WASTE PILES

A *waste pile* is a noncontainerized accumulation of solid, nonflowing, waste (40 CFR 260.10). All waste piles for which construction commenced after July 29, 1992, must have a double-liner and leachate collection system. Any expansion or replacement of an existing waste pile after July 29, 1992, or the creation of a new pile after this date must comply with the minimum technological requirements [40 CFR 264.251(c) and 265.254].

Requirements for Interim Status Waste Piles

A waste pile operating under interim status must comply with the following requirements:

- The pile must be protected from wind dispersion (40 CFR 265.251).
- The pile must be located on an impermeable base (40 CFR 265.253).
- No liquid may be placed on the pile (40 CFR 265.253).
- The waste pile must be separated from other potentially incompatible materials (40 CFR 265.257).

- A runon control and collection system must be installed and able to prevent a flow onto the active portion of the pile from a peak discharge from a 25-year storm, and the pile must be protected from precipitation (i.e., have a roof) and runon (40 CFR 265.253).

No ignitable or reactive wastes may be placed on the pile. However, once a waste is treated and no longer retains a hazardous waste characteristic, it may be placed onto the pile (40 CFR 265.256).

Each interim status waste pile must have an approved *action leakage rate*, which is the maximum design flow rate that the leak detection system can remove without the fluid head on the bottom liner exceeding one foot (40 CFR 265.255). If the action leakage rate is ever exceeded, the response action plan must be implemented (40 CFR 265.259). In addition, EPA must be notified with 7 days, such notice to be followed by a preliminary written assessment in 14 days as to the likely cause and actions taken and planned [40 CFR 265.259(b)].

Waste Analysis In addition to the waste analysis requirements of the general facility standards contained in 40 CFR 265.13, the owner or operator must analyze a representative sample of the waste from each incoming shipment before adding the waste to a pile, unless the only wastes that the facility receives are amenable to piling and are compatible with each other. The waste analysis must be capable of differentiating between the types of hazardous wastes that the owner or operator places on the pile to protect against inadvertent mixing of incompatible wastes (40 CFR 265.252).

Closure All contaminated soil, groundwater, liners, equipment, and structures must be removed or decontaminated in accordance with the closure plan. If not all of the materials are removed or decontaminated (i.e., clean closure does not occur), the pile must be closed as a landfill (40 CFR 265.258).

Requirements for Permitted Waste Piles

A permitted waste pile must have the following (40 CFR 265.251):

- A liner on a supporting base or foundation that prevents the migration of any waste vertically or horizontally.
- A leachate collection and removal system, situated above the liner, that is maintained and operated to remove all leachate.
- A runon control system able to collect and control the water from at least a 24-hour, 25-year storm.
- Inspections weekly and after any storm.
- Any ignitable or reactive waste must be rendered nonignitable or nonreactive before such waste is placed onto the pile.

Each permitted waste pile must have an approved *action leakage rate*, which is the maximum design flow rate that the leak detection system can remove without the fluid head on the bottom liner exceeding one foot (40 CFR 264.252). If the action leakage rate is ever exceeded, the response action plan must be implemented (40 CFR 264.253). In addition, EPA must be notified within 7 days, such notice to be followed by a preliminary written assessment in 14 days as to the likely cause and actions taken and planned [40 CFR 264.253(b)].

Closure. The owner or operator must remove or decontaminate all equipment, structures, liners, groundwater, and soils in accordance with the closure plan. If the unit cannot clean close, it must be closed as a landfill [40 CFR 264.258(b)].

LAND TREATMENT UNITS

Land treatment is the process of using the land or soil as a medium to simultaneously treat and dispose of hazardous waste. *Land treatment facility* means a facility or part of a facility at which hazardous waste is applied onto or incorporated into the soil surface; such facilities are disposal facilities if the waste will remain after closure (40 CFR 260.10). Land treatment is used primarily for petroleum wastes.

General Operating Requirements

Hazardous waste must not be placed on the land unless the waste can be made less hazardous or nonhazardous by biological degradation or chemical reactions in the soil. A treatment demonstration must be conducted before wastes are applied to verify that the hazardous constituents will be treated by the process.

Monitoring, according to a written plan, must be conducted on the soil beneath the treatment area in the unsaturated zone. The *treatment zone* is the portion of the unsaturated zone below and including the land surface where the necessary conditions will be maintained for effective degradation, transformation, or immobilization of hazardous constituents. The maximum depth of a treatment zone must be no more than 1.5 meters from the initial soil surface and more than 1 meter above the seasonal high-water table. The resulting data must be compared to data obtained on the background concentrations of constituents in untreated soils to detect any vertical migration of hazardous wastes or constituents (40 CFR 264/265.278).

Waste Analysis

In addition to the waste analysis requirements of the general standards contained in 40 CFR 264/265.13, waste analyses must be conducted (40 CFR 264.272 and 265.273) before placing wastes in or on the land to determine the concentrations of the following:

- Any substance in the waste whose concentration meets or exceeds the levels in Table I of 40 CFR 261.24 (rendering the waste hazardous by characteristic).
- Hazardous waste constituents (Appendix VIII to part 261)

- Arsenic, cadmium, lead, and mercury (if food-chain crops are grown on the land).

The growing of food-chain crops in a treated area containing arsenic, cadmium, lead, mercury, and other hazardous constituents is prohibited unless it is demonstrated that those constituents would not be transferred to the food portion of the crop, or if it is demonstrated that the constituents occur in concentrations less than those observed in identical groups grown on untreated soil in the same region (40 CFR 264/265.276).

Closure

During the closure period, the owner or operator must continue operating practices that are designed to maximize degradation, transformation, and immobilization of the wastes at the land treatment area. Operating practices designed to maximize treatment include tilling of the soil, control of soil pH and moisture content, and fertilization. These practices generally must be continued throughout the closure period. In addition, the owner or operator must continue those practices that were designed to minimize precipitation runoff from the treatment zone and to control wind dispersion (if needed) during the closure period (40 CFR 264/265.280).

Proper closure of a land treatment area includes the placement of a vegetative cover that is capable of maintaining growth without extensive maintenance. A vegetative cover is any plant material established on the treatment zone to provide protection against wind or water erosion and to aid in the treatment of hazardous constituents.

LANDFILLS

Landfilling historically has been the preferred means of disposing of hazardous waste. However, in 1984, the U.S. Congress took the position that the existing requirements for land disposal were inadequate to protect human health and the environment, and through HSWA it discouraged land disposal. This includes the land disposal ban (discussed in Chapter 8) as well as other land disposal restrictions (e.g., prohibiting liquids in landfills and waste containing free liquids).

The problems that hazardous waste landfills have presented can be divided into two broad classes, which are addressed in the interim status requirements. The first class includes fires, explosions, production of toxic fumes, and similar problems resulting from the improper management of ignitable, reactive, and incompatible wastes. Owners and operators must conduct waste analyses to provide enough information for proper management. Mixing of incompatible wastes is prohibited in landfill cells, and ignitable and reactive wastes may be landfilled only when they are rendered nonignitable or nonreactive.

The second problem is the contamination of surface water and groundwater. To prevent contamination, there are required operational controls such as prohibiting the placement in a landfill of bulk, noncontainerized liquid hazardous wastes; or nonhazardous liquid waste; or hazardous waste containing free liquids. This prohibition prevents the formation of hazardous leachate. An exemption for disposing of nonhazardous liquids may be obtained if the only reasonably available disposal method for such liquids is a landfill or unlined surface impoundment. Like surface impoundments and waste piles, a new landfill unit (including expansions or replacements) must install two or more liners and a leachate collection system (one collector above and one between the liners) in compliance with the minimum technological requirements of HSWA. Like surface impoundments and waste piles, landfills must have established action leakage rates that are used to determine if the liner system is leaking. Should the action rate be exceeded, a response plan to correct the problem must be implemented.

Other measures incorporated in the interim status regulations are diversions of precipitation runon away from the active portion of the landfill, proper closure (including a cover) and post-closure care to control erosion and the infiltration of precipitation, and crushing or shredding of most landfilled containers so that they cannot later collapse with resultant subsidence and cracking of the cover. In addition, the regulations require the collection of precipitation and other runoff from the landfill to control surface water pollution. Additionally, wastes such as acids that would mobilize, solubilize, or dissolve other wastes or waste constituents, such as heavy metals, must be segregated.

Closure and Post-Closure

A final cover must be placed over a landfill at closure. The closure plan must address the functions as well as specify the design of the final cover. The landfill must have an appropriate cover that controls the infiltration of moisture that could increase leaching and prevents erosion or escape of contaminated soil.

There are specific requirements (40 CFR 264/265.310) regarding the type, depth, permeability, and number of soil layers required for the final cover. The requirements also list a minimum set of technical factors that the owner or operator must consider in addressing the control objectives. These factors with regard to cover design characteristics, include cover materials, surface contours, porosity and permeability, thickness, slope, run length of slope, and vegetation type. The cover design should account for the number of soil compaction layers and the indigenous vegetation, avoid or make allowances for deep-rooted vegetation, and prevent water from pooling. The final cover design can be simple (such as the placement, compaction, grading, sloping, and vegetation of on-site soils), or complex (such as a combination of compacted clay or a membrane liner placed over a graded and sloped base and covered by topsoil and vegetation).

INCINERATORS

An *incinerator* is any enclosed device using controlled flame combustion that neither meets the criteria or classification as a boiler nor is listed as an industrial furnace (40 CFR 260.10).

Under RCRA, a permit must be obtained for a mobile incinerator for each site at which it is intended to operate because as it moves it meets the definition of a new facility. The owner or operator can submit trial burn data from previous operations in lieu of data from the anticipated site. When applying for a permit for a mobile incinerator, the applicant should use the model permit application and permit developed for the first permitted mobile treatment site, EPA Office of Research and Development's mobile incinerator at the Denny Farm Site in Missouri (OSWER Directive 9527-02).

Requirements for Interim Status Incinerators

The interim status requirements are primarily general operating standards with some performance standards (except for the incineration of dioxin wastes, which must meet strict performance standards).

Exemptions. Under the interim status requirements, there are specified waste streams exempted from all incinerator requirements except for the closure requirements [40 CFR 265.340(b)]. For the owner or operator to operate under an exemption, it must be documented that the exempted waste stream would not reasonably be expected to contain any Appendix VIII hazardous constituent and is classified as one of the following wastes:

- A listed hazardous waste solely because it is ignitable, corrosive, or both.
- A listed hazardous waste solely because it is reactive for reasons other than its ability to generate toxic gases, vapors, or fumes that will not be burned when other hazardous wastes are present in the combustion zone.
- A characteristic hazardous waste solely because it is ignitable, corrosive, or both.
- A characteristic hazardous waste solely because it is reactive for reasons other than its ability to generate toxic gases, vapors, or fumes that will not be burned when other hazardous wastes are present in the combustion zone.

General Operating Requirements. During the startup and shutdown phase of the incinerator, the owner or operator must not feed hazardous waste into the unit unless the incinerator is at steady-state (normal) operating conditions, including steady-state operating temperature and air flow (40 CFR 265.345).

The incinerator and all associated equipment (e.g., pumps, valves, conveyors) must be inspected at least daily for leaks, spills, fugitive emissions, and all emergency shutdown controls and alarm systems to assure proper operation. Instruments relating to combustion and emission control must be monitored at least every 15 minutes. If needed, corrections must be made immediately to maintain steady-state combustion (40 CFR 265.347).

Waste Analysis. In addition to the waste analysis required under 40 CFR 265.13 of the general facility standards, the owner or operator must sufficiently analyze any hazardous waste that has not been previously burned in the incinerator. The purpose of the additional waste analysis requirements is to establish steady-state operating conditions, which include waste and auxiliary fuel feed and airflow requirements, and to determine the type of pollutants that might be emitted (40 CFR 265.341). At a minimum, the analysis must determine the following:

- Heating value of the waste
- Halogen and sulfur content of the waste
- Concentrations of lead and mercury in the waste unless there is written documentation showing that these elements are not present

Closure. At closure, all hazardous waste and hazardous waste residues must be removed from the incinerator. Hazardous waste residues include but are not limited to ash, scrubber waters, and scrubber sludges. Because incinerators typically have regulated storage areas, appropriate cleanup of these units must be addressed as well (40 CFR 265.351).

Requirements for Permitted Incinerators

For permitted incinerators, each permit specifies the hazardous waste that is allowed to be incinerated and the operating conditions required for each waste feed. If an owner or operator wants to burn wastes for which operating conditions have not been set in a permit, a permit modification must be secured or, if the burn is to be of short duration and for the specific purposes listed in 40 CFR 122.27(b), a temporary trial burn permit must be obtained.

There are three general performance standards (40 CFR 264.343) applicable to incinerators subject to the part 264 provisions:

- A 99.99 percent destruction and removal efficiency (DRE) is required for each principal organic constituent specified in the permit.
- Incinerators burning hazardous waste containing more than 0.5 percent chlorine must remove 99 percent of the hydrogen chloride from their exhaust gas.
- Particulate matter must not exceed 180 mg per dry standard cubic meter of gas emitted through the stack.

Trial Burns. Operating restrictions and waste feed allowances are established by the owner or operator through a test period referred to as a trial burn. The trial burn, the basis on which EPA establishes a particular facility's permit standards, is a test of an incinerator's ability to meet all applicable performance standards when burning a waste under a specific set of operating conditions. During the trial burn, the applicant tests the incinerator's ability to destroy the hazardous waste or wastes to be treated at the facility. Generally, the goal in conducting the test is to identify the most efficient conditions, or range of conditions, under which the incinerator can be operated in compliance with the performance standards. The owner or operator must develop a trial burn plan and perform the test under conditions that will most likely ensure compliance with emissions standards. Once the trial burn has been completed, the facility's waste-feed and operating requirements must be those demonstrated in the trial burn.

General Operating Requirements. During startup or shutdown phases of an incinerator, hazardous wastes must not be fed into the incinerator unless it is operating within the specified conditions of operation (e.g., temperature, air feed rate) contained in the permit. The incinerator must be operated with a functioning system that can automatically cut off the waste feed to the incinerator when required to do so [40 CFR 264.345(c)].

Waste feed mixtures must be specified in the facility's permit. For each waste feed mixture identified, the permit will specify the principal organic hazardous constituents (POHCs) that must be destroyed or removed as required by the appropriate performance standard. Thus, identifying those waste-feed constituents to which the performance standard will be applied is central to the application of the regulatory program. EPA designates specific POHCs that are the most difficult to destroy, rather than all hazardous constituents of the waste. This approach ensures that less stable hazardous organic constituents also are destroyed [40 CFR 264.342(b)].

Inspections and Monitoring. The incinerator and all associated equipment (e.g., pumps, valves, conveyors) must be inspected at least daily for leaks, spills, and fugitive emissions, and all emergency waste feed cutoff controls and alarm systems must be inspected to verify proper operation [40 CFR 264 347(b)].

At a minimum, the following monitoring must be done while the unit is incinerating hazardous waste [40 CFR 264.347(a)]:

- Combustion temperature, waste-feed rate, and air-feed rate must be checked on a continuous basis.
- Carbon monoxide must be continuously monitored at a point in the incinerator downstream of the combustion zone and before release to the atmosphere.
- As specified by the permit, or upon request by EPA, sampling and analysis of the waste and exhaust emissions can be required to verify that the specified performance standards are achieved.
- Data obtained from any inspections or monitoring must be recorded and maintained in the facility's operating log.

Closure. At closure, all hazardous waste and hazardous waste residues must be removed from the incinerator. Hazardous waste residues include but are not limited to ash, scrubber waters, and scrubber sludges. Because incinerators typically have regulated storage areas, appropriate closure of these units must be addressed as well (40 CFR 264.351).

THERMAL TREATMENT UNITS

Thermal treatment is the treatment of hazardous waste in a device that uses elevated temperature, not controlled flame combustion, as the primary means to change the chemical, physical, or biological character or composition of the waste (40 CFR 265.370). By comparison, incineration uses controlled flame combustion to oxidize a waste, which is more closely associated with destruction rather than treatment.

Several methods of thermal treatment, such as molten salt pyrolysis, calcination, and wet air oxidation, are regulated under this subsection. Owners or operators who thermally treat hazardous wastes (other than incinerators) must operate the unit by following most of the requirements that are applied to an incinerator. However, the thermal treatment standards prohibit open burning of hazardous waste except for the detonation of explosives such as outdated military ordnance.

Only interim status standards are available for these units. The part 264 standards for thermal treatment units are addressed in the subpart X standards for miscellaneous units.

CHEMICAL, PHYSICAL, AND BIOLOGICAL TREATMENT

Although treatment is most frequently conducted in tanks, surface impoundments, and land treatment areas, it can also occur in other types of equipment through processes such as centrifugation, reverse osmosis, ion exchange, and filtration. Because the processes are frequently waste-specific, EPA has not developed detailed regulations for any particular type of process or equipment. Instead, general requirements have been established to ensure safe containment of hazardous wastes. In most respects, these other treatment methods are very similar to using tanks for treatment; therefore, they are essentially regulated in the same way. The requirements that must be met include avoiding equipment or process failure that could pose a hazard, restricting the use of reagents or wastes that could cause equipment or a process to fail, and installing safety systems in continuous flow operations to stop the waste inflow in case of a malfunction (40 CFR 265.401).

MISCELLANEOUS UNITS

A *miscellaneous unit* is a hazardous waste management unit in which hazardous waste is treated, stored, or disposed of, and which is not a container, tank, surface impoundment, waste pile, land treatment area, incinerator, boiler, industrial furnace, or underground injection well (40 CFR 260.10).

Previously, EPA promulgated standards for specific units, but some hazardous waste management technologies were not covered by the existing part 264 standards. Thus, owners or operators of facilities utilizing these technologies were not able to obtain the necessary RCRA operating permits for these units. To rectify this problem, EPA promulgated (52 *FR* 46946, December 10,1987) a new set of standards under the heading of Subpart X of part 264, Standards for Miscellaneous Units. These standards are applicable to owners and operators of new and existing miscellaneous units.

A miscellaneous unit must be located. designed, constructed, operated, and closed in a manner that will ensure protection of human health and the environment (40 CFR 264.601). To ensure that the requirements are met, permit conditions must incorporate the requirements under subparts I through O of part 264 that are appropriate for the unit to be permitted. In addition, the principal hazardous substance migration pathways must be protected.

Protection of human health and the environment from operations at miscellaneous units includes but is not limited to the following:

- Prevention of any releases that may have adverse effects on human heath or the environment due to migration of waste constituents in the groundwater or subsurface environment, considering the following:
  - Volume and physical and chemical characteristics of the waste in the unit, including its potential for migration through soil, liners, or other containing structures.
  - Hydrologic and geologic characteristics of the unit and the surrounding area.
  - Existing quality of groundwater, including other sources of contamination and their cumulative impact on the groundwater.
  - Quantity and direction of groundwater flow.
  - Proximity to and withdrawal rates of current and potential groundwater users.
  - Patterns of land use in the region.
  - Potential for deposition or migration of waste constituents into subsurface physical structures, and into the root zone of food-chain crops and other vegetation.
  - Potential for health risks caused by human exposure to waste constituents.
  - Potential for damage to domestic animals, wildlife, crops, vegetation, and physical structures caused by exposure to waste constituents.
- Prevention of any releases that may have adverse effects on human health or the environment due to migration of waste constituents in surface water, wetlands, or on the soil surface, considering the following:
  - Volume and physical and chemical characteristics of the waste in the unit.
  - Effectiveness and reliability of containing, confining, and collecting systems and structures in preventing migration.
  - Hydrologic characteristics of the unit and the surrounding area, including the topography of the land around the unit.
  - Patterns of precipitation in the region.
  - Quantity, quality, and direction of groundwater flow.
  - Proximity of the unit to surface waters.
  - Current and potential uses of nearby surface waters and any water quality standards established for those surface waters.

— Existing quality of surface waters and surface soils, including other sources of contamination and their cumulative impact on surface waters and surface soils.
— Patterns of land use in the region.
— Potential for health risks caused by human exposure to waste constituents.
— Potential for damage to domestic animals, wildlife, crops, vegetation. and physical structures caused by exposure to waste constituents.

- Prevention of any release that may have adverse effects on human health or the environment due to migration of waste constituents in the air, considering the following:
  — Volume and physical and chemical characteristics of the waste in the unit, including its potential for the emission and dispersal of gases, aerosols, and particulates.
  — Effectiveness and reliability of systems and structures to reduce or prevent emissions of hazardous constituents to the air.
  — Operating characteristics of the unit.
  — Atmospheric, meteorologic, and topographic characteristics of the unit and the surrounding area.
  — Existing quality of the air, including other sources of contamination and their cumulative impact on the air.
  — Potential for health risks caused by human exposure to waste constituents.
  — Potential for damage to domestic animals, wildlife, crops, vegetation, and physical structures caused by exposure to waste constituents.

CONTAINMENT BUILDINGS

A *containment building* is a completely enclosed structure (i.e., four walls, a roof, and a floor) that houses an accumulation of noncontainerized waste (primarily bulky waste, such as slag, spent potliners, contaminated debris). Containment buildings must comply with the following general requirements:

- The technical standards in 40 CFR part 265, subpart DD.
- Certification from a professional engineer that the building conforms to the design standards specified in 40 CFR 265.1101.
- Documentation that the above procedures are followed.

Owners and operators accumulating hazardous waste in containment buildings are exempt from most of the closure and financial assurance requirements in part 265, subparts G and H. However, after the useful life of the building has expired, the building must be closed in compliance with 40 CFR 265.111 and 265.114 (i.e., closure performance standard and disposal or decontamination of equipment, structures, and soils).

The technical standards for containment buildings include the following:

- It must be completely enclosed with four walls, a floor, and a roof, which must be constructed of man-made materials possessing sufficient structural strength to withstand movement of wastes, personnel, and heavy equipment within the unit.
- Dust control devices must be used as necessary to prevent fugitive dust from escaping through building exits.
- All building surfaces that come into contact with waste during treatment or storage must be chemically compatible with that waste.
- Incompatible wastes that could cause unit failure may not be placed in a containment building.
- If the building is used to manage hazardous wastes containing free liquids, the unit must be equipped with a liquid collection system, a leak detection system, and a secondary barrier.
- The floor of the unit must be free of significant cracks, corrosion, or deterioration.
- A containment building must be inspected at least once a week, with all results recorded.

If a release is discovered during an inspection, the generator must remove the affected portion of the unit from service and take all appropriate steps for repair and release containment. EPA must be notified of the discovery and of the proposed schedule for repair. Upon completion of all necessary repairs and cleanup, a qualified, registered professional engineer must verify that the plan submitted to EPA was followed.

UNDERGROUND INJECTION WELLS

Hazardous waste, primarily wastewater, has a long history of being disposed of in underground injection wells. These wells are regulated primarily by the Underground Injection Control (UIC) program established under Part C of the Safe Drinking Water Act (SDWA). SDWA has established five classes of injection wells:

Class I — Wells that are used to inject liquid wastes, including hazardous waste, below the lowermost underground source of drinking water. (*Underground sources of drinking water* are those currently used as a public drinking water supply or those that have the potential to serve as a public drinking water supply and have less than 10,000 mg/l total dissolved solids.)

Class II — Wells used in oil and gas production, primarily for the injection of produced brine, injection for enhanced recovery, and hydrocarbon storage.

Class III Wells used to inject fluids for the extraction of minerals.

Class IV Wells, which are currently banned, used to inject hazardous or radioactive waste into or above an underground source of drinking water.

Class V Wells, currently unregulated, used for purposes not identified in the first four classes of wells.

Class I UIC wells that are used to dispose of hazardous waste must have authorization under SDWA and RCRA. Although the UIC program under SDWA regulates an injection well below the wellhead, all Class I UIC wells that inject hazardous waste must also be authorized under RCRA as a hazardous waste management facility. They are eligible for interim status and must comply with 40 CFR 265.430. In addition, Class I UIC wells are eligible for a permit-by-rule under 40 CFR 270.60(b), which is explained further in Chapter 7. Class I UIC wells that apply for a permit-by-rule will then be subject to the corrective action requirements established under §3004(u) of RCRA, which is discussed in Chapter 9. In addition, Class I UIC wells are subject to the applicable requirements of the land disposal restrictions program discussed in Chapter 8.

Similar to RCRA, the SDWA UIC program allows a state to operate the UIC program with oversight from the federal government. A state that operates the program in lieu of the federal government is classified as having *primacy.* Texas and Louisiana, which contain 73 percent of all Class I hazardous waste injection wells, have primacy.

7

PERMITS AND INTERIM STATUS

Permitting is key to implementing the Subtitle C hazardous waste program. A RCRA permit is a legally binding document that establishes the authorized waste management activities a facility may engage in and under what conditions the activities must be conducted. Permits also stipulate the various administrative requirements (e.g., record keeping, reporting, waste analyses) for hazardous waste management facilities. All site-specific requirements established in a permit are known as permit conditions.

Owners or operators of hazardous waste management facilities must have a RCRA permit unless they are specifically excluded [40 CFR 270.1(c)(1)].[1] Except for some limited special permits (e.g., emergency and RD&D permits), there is only one type of permit, known as a RCRA or hazardous waste management permit. The RCRA permit application consists of two parts: Part A and Part B, and both must be completed and approved to obtain a "full" RCRA permit. However, because RCRA permits may cover different phases of operations, they may be referred to by other names (e.g., active permit, final permit, post-closure permit). Further, because it can take many years to issue a permit, Congress established an "interim" authorization allowing certain facilities to operate legally. This interim authorization is known as interim status. As depicted in Figure 7-1, interim status is available only to existing facilities under certain conditions, which are described in the following pages. Interim status is obtained by submitting Part A of the RCRA permit application. If a facility operating under interim status wants to obtain a "full" permit, Part B of the permit application must be submitted. New hazardous waste management facilities

[1] It is important to note the distinction between unit and facility. A hazardous waste management facility may have one or many units. However, a permit is issued for individual units (e.g., tank, incinerator, surface impoundment) and not for an entire facility.

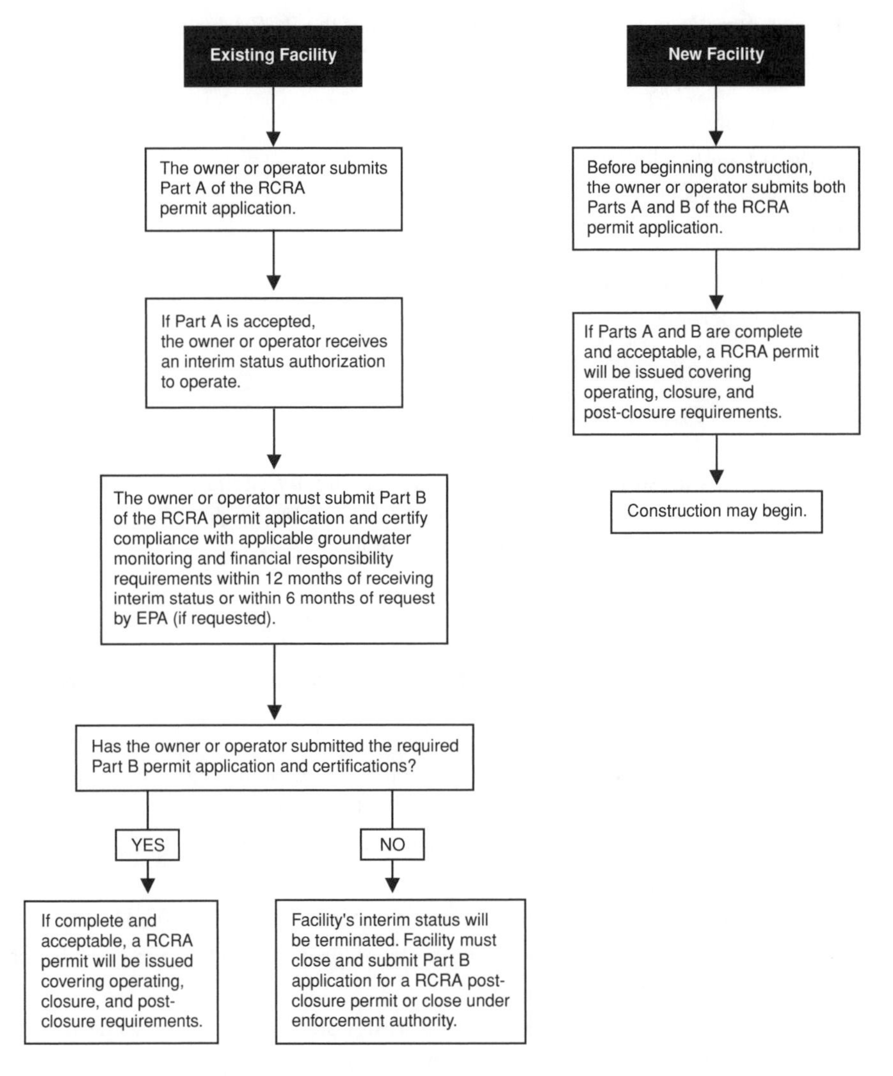

Figure 7-1. General RCRA Permit Application Process

must submit both parts of the permit application and be issued a RCRA permit before physical construction may begin. In addition, should a facility operating under interim status want to close, that facility must still submit a Part B permit application for a post-closure permit, which is a smaller version of the standard RCRA permit.[2]

[2] EPA has concluded that requiring a post-closure permit for interim status facilities is not always the best solution (62 *FR* 123456, July 30, 1997). Therefore, EPA may seek an enforcement order in lieu of a post-closure permit to ensure proper closure.

This chapter discusses the following elements of the RCRA hazardous waste permit program:

- Exclusions
- Permit application process
- Interim status
- Permits
- Post-closure permits
- Special permits

EXCLUSIONS FROM RCRA PERMITS

This section lists those facilities or activities specifically excluded from having to obtain a RCRA permit and discusses the permit-by-rule provision, which is a system to reduce duplicative federal environmental permits.

Specific Exclusions

The following are excluded from having to obtain a permit to operate under RCRA [40 CFR 270.1(c)(2)]:

- Facilities that are state approved to exclusively handle conditionally exempt small-quantity generator wastes (less than 100 kg/mo)
- Facilities that meet the definition of a totally enclosed treatment facility in 40 CFR 260.10
- A generator accumulating waste in compliance with 40 CFR 262.34
- A farmer disposing of waste pesticides from his or her own use in compliance with 40 CFR 262.51
- An elementary neutralization unit or a wastewater treatment unit as defined in 40 CFR 260.10
- A person engaged in the immediate treatment or containment of a discharge of hazardous waste
- A transporter storing manifested waste at a transfer facility in compliance with 40 CFR 262.30
- Handlers and transporters managing universal waste under 40 CFR part 273.
- The addition of absorbent material to waste or vice versa.

Permit-By-Rule

To avoid duplicative permitting, EPA has established the permit-by-rule program. A *permit-by-rule* is essentially an amendment to an existing federal environmental permit stating that a unit or activity is deemed to have a RCRA permit if it meets specified requirements, which are contained in 40 CFR 270.60.

The following are eligible for a permit-by-rule:

- Publicly owned treatment works (POTWs) that have a permit under the National Pollutant Discharge Elimination System (NPDES) under the Clean Water Act (CWA).
- Persons with ocean dumping permits under the Marine Protection, Research, and Sanctuaries Act (MPRSA).
- Permitted Underground Injection Control (UIC) facilities under the Safe Drinking Water Act (SDWA).

PERMIT APPLICATION PROCESS

The RCRA permit application process differs depending on whether the facility is new or existing. (*Existing facilities* are those facilities in existence before November 19, 1980, and facilities in existence on the effective date of the statutory or regulatory change that renders the facility subject to the requirement to obtain a RCRA permit [40 CFR 270.10(e)]. A RCRA permit application is composed of two parts: Part A and Part B.

Note: Under §3004(u) of RCRA, the submission of a permit application, which means Part B, initiates the corrective action program for solid waste management units as discussed in Chapter 9.

New Facilities

New facilities must submit both parts (A and B) of the RCRA permit application before physical construction may begin. When the "final" permit is authorized, it will cover the facility's construction, operation, closure, and post-closure.

Existing Facilities

Owners or operators of existing eligible facilities that want to manage hazardous waste may obtain interim status, described in the following section. To obtain interim status, only the Part A portion of the RCRA permit application needs to be submitted. No later than one year after receiving interim status (or within six months if requested earlier by EPA), the owner or operator must submit the Part B portion of the application in addition to a certification that the facility or unit is in compliance with applicable groundwater monitoring and financial responsibility requirements.

If the owner or operator fails to submit Part B in a timely manner, fails to certify compliance with the applicable groundwater monitoring and financial responsibility requirements, or is denied a permit, the interim status authorization is terminated and the facility must close. For any facility that closes with waste in place, a post-closure permit must be obtained. To obtain a post-closure permit, a Part B permit application and certification of compliance with the groundwater monitoring and financial responsibility requirements must be submitted. For those facilities that cannot certify compliance with the groundwater monitoring or the financial responsibility requirements, the facility may enter the post-closure period under an enforcement order.

INTERIM STATUS

Because all hazardous waste management facilities must obtain a permit (with some exceptions), that is the key to implementing the RCRA Subtitle C hazardous waste program. On the effective date of the RCRA regulations (November 19, 1980), the treatment, storage, or disposal of hazardous waste was prohibited unless the operations were either authorized by a RCRA permit or specifically excluded from having to obtain a permit. Because this provision obviously would have created a disastrous situation for the nation's hazardous waste management program as it existed then, Congress established interim status provisions. (A unit operating under interim status is considered to be operating under a permit until EPA takes final administrative action on the permit application.) *Interim status* is the statutorily conferred authorization for a hazardous waste management facility to operate pending issuance or denial of its RCRA permit. The interim status provisions, contained in 40 CFR Part 265, allow a unit to operate legally. The unit is considered to be operating under a permit until EPA takes final administrative action on that unit's permit application.

To be eligible for interim status, a waste management unit must be "in existence" on the effective date of a regulatory or statutory change. For example, a unit manages a nonhazardous waste, which is subsequently listed as a hazardous waste. The facility is then eligible for interim status (unless the facility already has a permit for which only a modification is required). To obtain interim status, which is described in more detail in the next section, the facility must properly notify EPA and submit a Part A RCRA permit application by the date specified in the rule, but no later than 90 days after promulgation of the rule [40 CFR 270.1(b)].

Application Process for Obtaining Interim Status

The Part A permit application is a request for interim status for newly regulated, existing facilities. A standardized form (EPA Form 8700-23) is used. This form requires general information about a facility and the unit, including the following:

- Name of the owner or the operator and the facility address.
- Longitudinal and latitudinal coordinates of the facility.
- The activities to be conducted that require a RCRA permit.
- The unit's design capacity.
- Description of the processes for treating, storing, or disposing of hazardous wastes.

- Description of the unit's processes generating hazardous waste.
- A scale drawing of the unit/facility, including photographs.
- A detailed description of hazardous wastes to be handled.
- A topographic map.
- A listing of all federal environmental permits obtained or applied for by the facility.

EPA may issue a Notice of Discrepancy (NOD) to any owner or operator of an existing hazardous waste management facility who submits a deficient Part A permit application. The owner or operator has 30 days after being notified to explain or amend the application before being subject to EPA enforcement. After such notification and opportunity for response, EPA will determine if that facility qualifies for interim status (OSWER Directive 9528:50-1A).

General Requirements

A unit must comply with the applicable requirements contained in 40 CFR part 265 as discussed in Chapters 5 and 6. (Chapter 5 addresses general facility requirements, whereas Chapter 6 addresses unit-specific requirements.) A unit is subject to the part 265 standards only until it receives its final permit. On the day the final permit is issued, the unit is subject to the conditions specified in its permit, which are based on the part 264 standards.

Note: The major difference between operating under interim status compared to operating under a permit is that under interim status, all requirements are self-implementing, subject to enforcement for noncompliance. Operating under a permit means compliance with the specified conditions of the permit. EPA has significantly more authority to force compliance for permitted facilities.

Facilities that have obtained interim status must submit a Part B RCRA permit application and certify compliance with all applicable groundwater monitoring and financial responsibility requirements within 12 months of obtaining interim status [40 CFR 270.73(d)]. Failure to do so is grounds for termination of interim status.

A unit that has its interim status terminated must still meet the part 265 standards, including those for closure, post-closure, and financial responsibility. A technical amendment to the interim status regulations (40 *FR* 46094, November 21, 1984), clarified that interim status standards are applicable to units that have had their interim status terminated until their closure and post-closure requirements are fulfilled. Thus, even if a unit no longer has interim status (and thus may no longer manage hazardous waste), the unit must be in compliance with all applicable regulations under interim status until it is satisfactorily closed.

Specific Operational Requirements under Interim Status

There are specific operational constraints (beyond the operating standards of interim status of part 265) established for interim status facilities. These operational constraints, contained in 40 CFR 270.71 and 270.72, are as follows:

- A unit may not treat, store, or dispose of any hazardous waste not specified in the unit's Part A permit applications.
- A unit may not employ processes not specified in the Part A permit application.
- A unit may not exceed the design capacities that were specified in the Part A permit application.
- No changes may be made to the facility during interim status that constitute reconstruction of that facility. (*Reconstruction* is a capital investment exceeding 50 percent of the capital cost of a comparable new hazardous waste management facility.) However, changes to treat, or store in containers or tanks, hazardous wastes subject to the land disposal prohibitions imposed by part 268 are allowed, provided that such changes are made solely for the purpose of complying with part 268.

Exceptions to the above constraints (except reconstruction) are allowed under certain conditions. For example, new wastes may be handled simply by submitting a revised Part A permit application. The process and design capacity changes require a revised Part A permit application, written justification for a change, and state or federal approval (40 CFR 270.72).

A change in ownership or operational control of a facility is allowed if the facility submits a revised Part A permit application at least 90 days before the change. The previous owner or operator must comply with the financial requirements until the new owner or operator can demonstrate compliance with the financial requirements to EPA. The new owner or operator must comply with the financial requirements within six months of the change of ownership or operational control.

HAZARDOUS WASTE MANAGEMENT PERMITS

A RCRA hazardous waste management permit (the "RCRA permit") is a legal document that describes how a treatment, storage, or disposal facility (TSDF) will be designed, constructed, operated, and maintained. The permit also describes how releases to the environment (e.g., groundwater, soil) will be prevented, how emergencies will be handled, what financial assurance mechanisms will be used, whether and how corrective action will be instituted for past releases, and how the facility will be closed.

The actual permit consists simply of written approval of the completed permit application, which can be quite large and can encompass volumes. The permit requires the applicant to adhere to all statements made in the application and may include specific conditions (e.g., corrective action) with which the applicant must comply.

As discussed within the "Permit Application Process" section of this chapter, new facilities generally must complete and submit Parts A and B of the RCRA permit application, while existing facilities may have to submit only Part A, initially.

Part B Permit Application Process

Owners and operators of new hazardous waste management facilities and those operating under interim status must submit a Part B RCRA permit application that addresses the design, operation, maintenance, and closure of each hazardous waste management unit as outlined in 40 CFR Part 270.[3]

Note: The Part B permit application must be submitted within one year for all land disposal units that were granted interim status [40 CFR 270.73(d)]. New facilities must submit the Part B and receive a permit before physical construction of the facility may occur. Otherwise, EPA will request a facility's Part B, but will give at least six months' notice [40 CFR 270.10(e)(4)].

The general steps in the Part B permit application process, depicted in Figure 7-2, are as follows:

1. The owner or the operator of a hazardous waste management unit completes and submits the Part B portion of a RCRA permit application.
2. EPA conducts an administrative and technical review of the permit application for completeness. EPA may issue a notice of deficiency and request additional information.[4]
3. If required, the applicant prepares and submits additional information.
4. EPA again reviews the original and any additional submittals and notifies the applicant, in writing, that the application is complete.
5. EPA prepares a draft permit or issues a notice of intent to deny the application.
6. EPA sends copies of the document prepared in Step 5 to the applicant and simultaneously notifies the public. The notice is to allow for public and applicant comment.
7. If a public hearing is requested, one will be scheduled and announced at least 30 days before the hearing date.
8. EPA prepares and issues a final permit decision.

Contents of the Application. There is no standard or required form for the Part B permit application. It is a document prepared by the permit applicant that addresses each point of concern specified in 40 CFR 270.22, and it may be presented in sev-

[3] If the operator of a unit submits the permit application, the owner of the unit also must sign it; if the owner fails to sign the application, then a RCRA permit will not be issued (OSWER Directive 9523.0). Both the owner and operator are the "permitees" on the permit; however, it is common for the operator to assume responsibility for meeting permit conditions. Both the owner and operator are liable during the facility's operating life and during closure and post-closure of the unit, unless the closure and post-closure plans specify that the owner of the unit is becoming the operator as well as the owner. This action would be accompanied by a permit modification and relieve the original operator from liability (under RCRA) during the closure and post-closure period.

[4] At *any time* during the process, a Notice of Deficiency (NOD) may be issued to the permittee for deficient information. The permit applicant then must either submit the requested information, withdraw the application, or face possible denial of the permit.

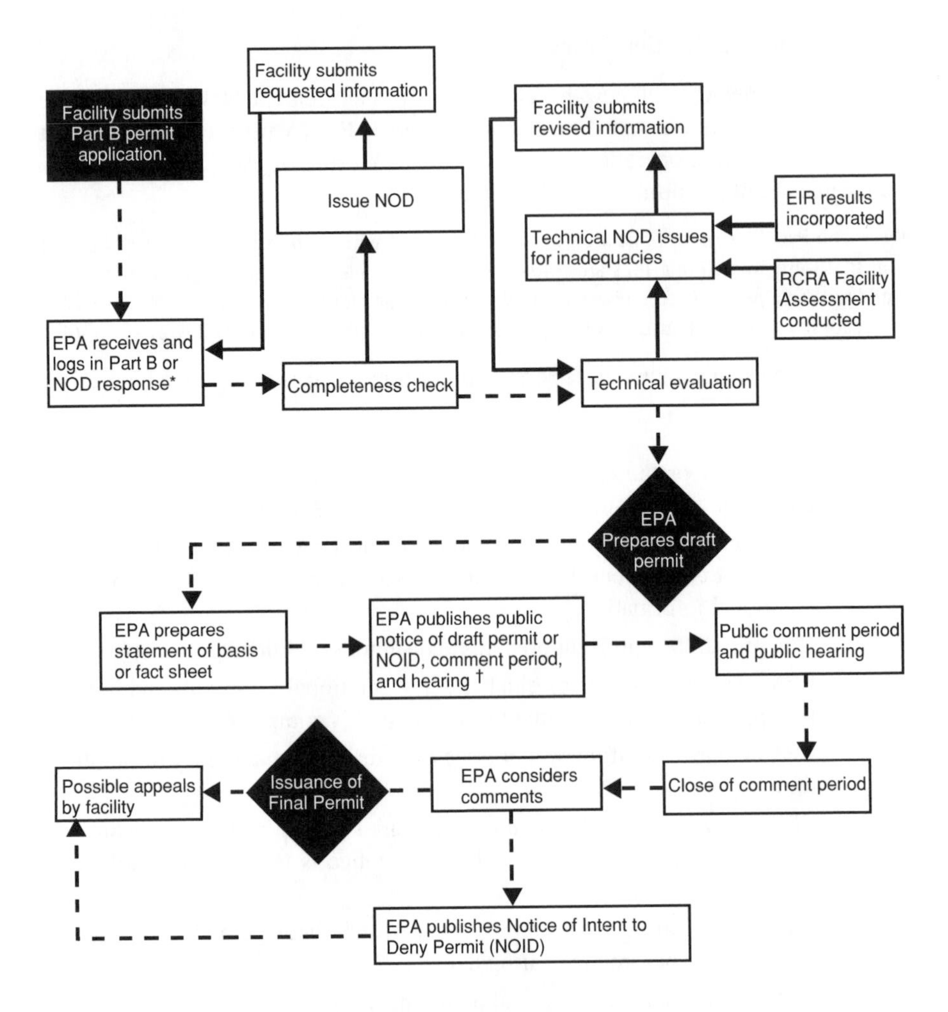

Figure 7-2. Part B Permit Application Process

eral ways. The actual application format is left to the discretion of the applicant. However, EPA suggests, in its *Permit Applicant's Guidance Manual for Land Disposal, Treatment, and Storage Facilities,* a general format for the submission of a Part B permit application, as follows:

Part I General Information Requirements

- A copy of the Part A application (required of both new and existing facilities).
- General description of the unit/facility.

- The process codes (from Part A) that identify the type of units for which the permit is required.
- Chemical and physical analysis of hazardous waste to be handled and a copy of the waste analysis plan.
- Security description for the active portion of the facility.
- General inspection schedule and description of the procedures, including the specific requirements for particular units.
- Preparedness and prevention documentation.
- Contingency plan documentation.
- Documentation of preventive procedures, structures, and equipment for control of unloading hazards, waste runoff, water supply contamination, effects of equipment failure and power outages, and undue personnel exposure to wastes.
- Documentation of prevention of accidental ignition or reaction, including specific requirements for particular unit types.
- Facility traffic documentation.
- Facility location documentation.
- Personnel training program documentation.
- Closure plan documentation, including specific requirements for particular unit types.
- Post-closure plan, when applicable, including specific requirements for particular unit types.
- Documentation for deed notice.
- Closure cost estimate and documentation of financial assurance mechanism.
- Post-closure cost estimate and documentation of financial assurance mechanism.
- Topographical map showing contours with 0.5- to 2.0-meter intervals; map scale; map date; 100-year floodplain area; surface waters including intermittent streams; surrounding land uses; wind rose; north orientation; legal boundaries of facility; access control; injection and withdrawal wells; buildings and other structures; utility areas; barriers for drainage or flood control; and location of operating units, including equipment cleaning areas. Each hazardous waste management unit should be shown on the map with a unique identification code and the associated process code from the Part A.

Part II Specific Information Requirements

The information contained in this section depends on the particular type of unit for which the permit is required. Refer to subparts I, J, K, L, M, N, O, W, X, AA, BB, CC, and DD of 40 CFR part 264 and subpart B of part 270 for the appropriate unit-specific information requirements.

Part III Additional Information Requirements

- Summary of groundwater monitoring data obtained during the interim status period.
- Identification of aquifers beneath the facility.
- Delineation of waste management area and point of compliance for groundwater monitoring on topographic map.
- Description of any existing plume of contamination in the groundwater.
- Detection monitoring program description, if applicable.
- Compliance monitoring program description, if applicable.

Part IV Information Requirements for Solid Waste Management Units

- Location of SWMUs on topographic map.
- Type of SWMU (e.g., pile, landfill).
- Dates of operation.
- An inventory of all wastes managed at the SWMU.
- Information on releases of hazardous waste or constituents.
- Dimensions and structure descriptions.

Part V Exposure Information Report

All permit applications for surface impoundments and landfills must be accompanied by an exposure information report [40 CFR 270.10(j)]. The major goals of this report are to identify human exposures to past releases from subject units and potential exposures from future releases so that there can be an attempt to mitigate the exposure potential with specific permit conditions.

At a minimum, the report needs to address these three issues:

- Reasonably foreseeable potential releases from both normal operations and accidents at the unit, including releases associated with transportation to or from the unit.
- The potential pathways of human exposure to hazardous wastes or constituents resulting from such releases.
- The potential magnitude and nature of the human exposure resulting from such releases.

Confidentiality of Permit Applications. An applicant may claim information in the permit application as confidential in compliance with 40 CFR part 2, which sets forth the general requirements and procedures for EPA's handling of confidential information. The permit applicant is required to attach a cover sheet, use a stamp, type a notice of confidentiality on each page of the information, or otherwise identify its confidential portion(s) (40 CFR 270.12). Whenever possible, the applicant should separate the information contained in the application into confidential and nonconfidential sections and submit them under separate cover letters.

If there is a public request for information under the Freedom of Information Act, EPA will give the applicant an opportunity to substantiate the claim before releasing any information for which a claim of confidentiality has been made, and will then determine whether the information warrants confidential treatment. If it is considered confidential by EPA, such information will not be released.

Administrative and Technical Review

When EPA receives a Part B permit application, the application is reviewed for administrative and technical completeness. The completeness review generally takes 60 days for existing facilities and 30 days for new ones.

If the application is incomplete, EPA will request the missing information through a Notice of Deficiency (NOD) letter, which details the information needed to complete the application and specifies the date for submission of the data. EPA may issue a warning letter to accompany the NOD, requiring submission of the necessary information within a specified additional period of time. If an applicant fails or refuses to correct the deficiencies in the application, the permit may be denied, and appropriate enforcement actions may be taken under statutory provisions including RCRA §3008 (OSWER Directive 9521.01).

EPA will determine when the Part B application is complete. This determination, however, is not necessarily a determination that the application is free of deficiencies. During detailed review of the application and drafting of the permit conditions, it may become necessary for the applicant to clarify, modify, or supplement provisions previously submitted before a draft permit is granted or a decision to deny is made.

Technical Review. The purpose of the technical review is to determine whether a specific unit should be granted a permit; in other words, to determine if the unit has satisfied all the siting, design, and operation criteria, as well as closure, post-closure, and financial requirements.

The steps of the technical review are as follows:

1. Preliminary review
2. Site visit
3. Verification of accuracy
4. Compliance assessment

The *preliminary review* involves primarily a secondary completeness check to ensure that the applicant has responded adequately to any previous notices of deficiency.

The *site visit* enables the permit reviewer/writer to inspect the facility and its hazardous waste management units to ensure that the permit application accurately reflects the stated conditions.

The *verification of accuracy* involves a check on the reasonableness of data and the accuracy of computations. It also requires a professional assessment of information, assumptions, and methodology presented to identify weaknesses that may require a response from the permit applicant.

The *compliance assessment,* conducted concurrently with the accuracy verification step, is a process intended to ensure that the permitted operations will be in compliance with part 264.

Upon completion of the technical review, EPA tentatively decides whether to issue or deny a RCRA permit. If the tentative decision is to issue the permit, EPA prepares a draft permit. EPA must also prepare a fact sheet or statement of basis for the public, which explains in simple language each condition included in the draft permit and the reasons for each condition. The draft permit, prepared primarily for public review, contains tentative conditions.

Public Involvement During the Permitting Process

In 1995, EPA promulgated the Expanded Public Involvement Rule, which increased the opportunity for public involvement in all facets of the RCRA permit process, especially the initial stages of a permit application. Before the Expanded Public Involvement Rule, such involvement was limited to the determination of whether a permit was to be granted or denied and certain cases where a permittee was requesting a modification to a permit. Specifically, the rule requires the following:

- Before the submission of a Part B permit application, the applicant must hold at least one meeting with the public to solicit questions from members of the community and inform them of proposed hazardous waste management activities [40 CFR 124.31(a)].[5]
- The applicant must also allow attendees to provide their names and addresses. As part of the Part B permit application, a summary of the meeting, a list of attendees, and copies of written comments must be submitted [40 CFR 124.31(c)].
- The permitting agency must announce the submission of a permit application by sending a notice to everyone on the facility mailing list.[6] The announcement must also notify the public of the physical location where the application may be examined.

[5] The applicant must provide public notice at least 30 days prior to the meeting. Public notice must include a newspaper advertisement, a notification sign near the facility, a broadcast media announcement (radio or television), and a notice to EPA.

[6] EPA is required under 40 CFR 124.10(c)(viii) to compile and maintain a mailing list for each RCRA permitted facility. The list must include all persons who have asked in writing to be on the list (e.g., in response to public solicitations from EPA). Also, it generally includes both local residents in the vicinity of the facility and statewide organizations that have expressed interest in receiving such information on permit modifications.

- The permitting agency may require a facility to establish an information repository at any point during the permitting process.
- The permitting agency must notify the public before a trial (or test) burn at a combustion unit by sending a notice to everyone on the facility mailing list.

In addition to the above public participation requirements, the public must also be notified, in accordance with 40 CFR 124.10(a), whenever

- a permit application has been tentatively denied,
- a draft permit has been prepared,
- an informal public hearing has been scheduled, and
- an appeal has been granted by EPA regarding a final permit decision.

EPA must issue a public notice identifying the applicant and the facility and must state where copies of the draft permit and other related information may be obtained (e.g., statement of basis or fact sheet). The notice must be circulated in local newspapers for major permits and mailed to various agencies and parties expressing interest.

Public notice provides interested persons a minimum of 45 days to comment on the draft permit. If written opposition to EPA's intent to issue a permit and a request for a hearing are received during the comment period, a public hearing may be held. Notification of the hearing is issued at least 30 days before the scheduled date, and the public comment period is extended until the close of the public hearing.

Statement of Basis and Fact Sheet. The statement of basis and the fact sheet, which must accompany the public notice of the draft permit, briefly set forth the principal facts and the significant factual, legal, methodological, and policy questions considered in preparing the draft permit.

The fact sheet must be prepared for all *major facilities* (i.e., land disposal facilities and incinerators) and for those facilities that EPA finds are "subject to widespread public interest or raise major issues" (40 CFR 124.8).

Both documents must include the following information:

- A description of the activity that requires the draft permit.
- The type and quantity of wastes that are proposed to be or are being treated, stored, or disposed of.
- Reasons why any requested variances or alternatives to required standards do or do not appear justified.
- A description of the procedures used in reaching a final decision on the draft permit, including
  - —The extent of the public comment period,
  - —Procedures for requesting a hearing,
  - —Any other procedures in which the public may participate

- Name and telephone number of a contact.

After the close of the public comment period (which includes the public hearing period), EPA must either grant or deny the permit application. In either case, the applicant, those persons submitting questions, and those requesting notification must be notified of the decision and must be given information regarding appeal procedures.

Standards Applicable to All Permits

Under 40 CFR 270.30, the following conditions apply to all RCRA permits and are incorporated into the permits either expressly or by reference:

- Compliance with all conditions of the permit.
- Proper operation and maintenance, including the operation of backup or auxiliary units when necessary to achieve compliance.
- Halting production when necessary to ensure compliance and taking all reasonable steps to minimize releases to the environment.
- Providing all relevant information requests to EPA regarding facility operation, including the reporting of any noncompliance that may endanger health or the environment within 24 hours of the time when the owner or the operator becomes aware of such conditions.
- Allowing authorized representatives to inspect the facility upon presentation of credentials.
- Maintaining and reporting all records and monitoring information necessary to document and verify compliance, including data such as continuous-strip chart recordings from monitoring instruments.
- Reporting all planned changes and anticipated periods of noncompliance.

EPA's Omnibus Permit Authority. RCRA's §3005(c)(3), which gives the permit-issuing agency (state or federal) broad authority concerning permit conditions, reads: "Each permit issued under this Section [3005] shall contain such terms and conditions as the Administrator [or state] determines necessary to protect human health and the environment."

This provision allows the issuing agency to impose permit conditions beyond those contained in the Subtitle C regulations as may be necessary to protect human health and the environment from risks posed by a particular unit. For example, this authority could be used if a unit may adversely affect vulnerable groundwater in part because of a lack of adequate location standards in the regulations.

Permit-as-a-Shield Provision. The regulations provide that compliance with a RCRA permit during its term constitutes compliance with the RCRA statute. The purpose of the shield provision is to protect permittees who are in full compliance with their permit but who may not be in full compliance with other provisions under

RCRA that may have been recently promulgated. The permit shield provision protects the permittee from enforcement action brought by EPA and states as well as civil actions from citizen groups. However, this provision does not provide a defense to an imminent hazard action under §7003 of RCRA (45 *FR* 33428, May 19, 1980).

Permit Compliance Schedules. Permit compliance schedules may be used to allow facilities to construct or install equipment that is not mandated under part 265 but is required under part 264. However, a permit compliance schedule cannot be used to satisfy the information requirements under part 270. A permit compliance schedule must be specific and enforceable, allow for public notice and comment, and provide the applicant additional time only when it is legitimately needed (OSWER Directive 9524.01).

Draft Permit

A draft permit functions only as a tentative decision on the issuance, modification, reissuance, or termination of a permit. It is only a proposal, and is subject to change based on comments received during the public comment period and hearing. The draft permit must be accompanied by a statement of basis or a fact sheet (described within the "Public Involvement" section), and must be based on the administrative record. When public notice of the draft permit is given, as required by 40 CFR 124.10(a), the comment period on the permit application begins. It is during this time that interested persons must raise all issues. Failure to do so may limit the ability to raise an issue on appeal.

Permit Issuance

Final RCRA permits become effective 30 days after the date of the public notice of decision to issue a final permit unless a later date is specified in the permit, or when the conditions of 40 CFR 124.15(b) are met. At the time the final RCRA permit is issued, EPA also issues a response to any significant public comments received and indicates any provisions of the draft permit that have been changed and the reasons for the changes. The responses to comments become part of the administrative record.

Permit Duration. A permit may be issued for any length of time for up to 10 years. All land disposal permits must be reviewed every 5 years and modified if necessary. However, a permit may be terminated at any time if noncompliance with any condition of the permit occurs, if any false information was included in the application, or if the unit's activity endangers human health or the environment.

Permit Denial

A permit may be issued or denied for one or more units at a facility without simultaneous issuance or denial of a permit to all of the units at the facility [40 CFR

270.1(c)(4)]. (The interim status of any unit for which a permit has not been issued or denied does not affect the issuance or denial of a permit for any other unit at a facility).

If EPA tentatively decides to deny a RCRA permit, a notice of intent to deny a permit is prepared. This notice is considered a type of draft permit and follows the same procedures as any other draft permit. These procedures include preparation of a statement of basis or a fact sheet containing reasons supporting the tentative decision to deny the permit, public notices of the denial, acceptance of comments, a possible hearing, preparation of a final decision, and possible receipt of a request for appeal.

A permit may be denied for the following reasons:

- The unit may not meet the requirements set forth in 40 CFR part 264.
- Activities at the unit would endanger human health or the environment.
- An applicant is believed not to have fully disclosed all relevant facts in the application or during the RCRA permit issuance process.
- An applicant has misrepresented relevant facts.
- The application did not fully meet the requirements of part 270 (e.g., did not have the signature(s) of both the owner and operator).

The denial of a permit for the active life of a hazardous waste management unit does not affect the requirement to obtain a post-closure permit for land-based units [40 CFR 270.1(c)].

Permit Appeals

An owner or an operator wanting to appeal a permit denial must follow the procedures in 40 CFR 124.19, which addresses recourse for permit denial.

Persons who submitted comments on the draft RCRA permit or participated in any public hearing are allowed 30 days after the final permit decision to file a notice of appeal and a petition for review with the EPA Administrator in Washington, DC, who will review and then grant or deny the petition within a reasonable time. If the Administrator decides to conduct a review, the parties are given the opportunity to file briefs in support of their positions. Within the 30-day period, the Administrator may, on his or her own motion, decide to review the decision to grant or deny a hearing, which is usually handled by an Administrative Law Judge. The Administrator then notifies the parties and schedules a hearing. On review, the Administrator has several options regarding the final decision. It may be summarily affirmed without opinion, modified, set aside, or remanded for further proceedings. This petition for review is a prerequisite for judicial review of the Administrator's final decision.

Permit Reapplication

If a permittee wishes to continue an activity regulated by a permit after the permit's expiration date, the permittee must apply and obtain a new permit. A new application must be submitted at least 180 days before the permit's scheduled expiration date unless other arrangements have been made with the permit-issuing authority. However, the date may not be extended past the permit's expiration date [40 CFR 270.10(h)].

Permit Modifications

The RCRA permit establishes a unit's opening conditions for hazardous waste management. However, over time, the unit may need to be modified because of such considerations as improved equipment, new management techniques, corrective action, or changes in response to new standards. Recognizing this, EPA established procedures for modifying permits.

The permit modification program establishes procedures that apply to changes owners and operators may want to make at their facilities. EPA has categorized selected permit modifications into three classes and established administrative procedures for approving modifications in each of these classes.

Appendix I to 40 CFR 270.42 (see Appendix D of this book) contains a list of specific modifications and assigns to each one of three designations, which may be described as follows:

- *Class 1 permit modifications* address routine and administrative changes.
- *Class 2 permit modifications* primarily address improvements in technology and management techniques.
- *Class 3 permit modifications* deal with major changes to a unit and its operations.

EPA has established the following procedures [40 CFR 270.42(d)] for a permittee wishing to make a permit modification covering an activity that does not have a predesignated class: submit a Class 3 modification request, or, alternatively, ask EPA for a determination that Class 1 or 2 modification procedures should apply. In making this determination, EPA will consider the similarity of the requested modification to listed modifications and also will apply the general definitions of Class 1, 2, and 3 modifications.

Class 1 Modifications. Class 1 modifications are those that do not substantially alter the permit conditions or significantly affect the overall operation of the facility. They cover changes necessary to correct minor errors in the permit, upgrade plans and records maintained by the facility, or make routine changes to the facility or its operation. Generally, these modifications include the correction of typographical errors; necessary updating of names, addresses, or phone numbers identified in the permit or its supporting documents; updating of sampling and analytical methods to conform with revised EPA guidance or regulations; updating of certain types of schedules identified in the permit; replacement of equipment with functionally equivalent equipment; and replacement of damaged groundwater monitoring wells.

The approval procedures for Class 1 modifications are specified in 40 CFR 270.42(a). There are two categories of Class 1 modifications: those that do not require EPA approval and those that do.

Under the Class 1 modification procedures, the permittee may, at any time, put into effect any Class 1 modification that does not require EPA approval. The permittee must notify EPA, by certified mail or by any other means that establishes proof of delivery, within seven calendar days of making the change. The notice must specify the change being made to the permit conditions or documents referenced in the permit and explain briefly why it was necessary.

The permittee must also notify, by mail, persons on the *facility mailing list* within 90 days of making the modification.

The approval procedure is analogous to the former minor modification procedure; that is, a Class 1 permit modification requiring prior EPA approval (see Appendix D) may be made only with the prior written approval of EPA. In addition, upon approval of such a request, the permittee must notify persons on the facility mailing list of the decision within 90 calendar days after it is made.

There are no time-frame requirements for EPA action concerning a decision on a Class 1 modification. However, 40 CFR 270.42(a)(3) allows a permittee to elect to follow the Class 2 process (instead of the Class 1 process). The Class 2 process assures that a decision will be made on the modification request within established time frames (usually 90 to 120 days).

Although the permittee may make most Class 1 modifications without EPA approval or prior public notice, under 40 CFR 270.42(a)(iii) the public may ask the permitting agency to review any Class 1 modification.

In the event that such a review is conducted, and EPA denies a Class 1 modification request, EPA must notify the permittee in writing of its ruling, and the permittee is required to comply with the original permit conditions.

Class 2 Modifications. Class 2 modifications cover changes that are necessary to enable a permittee to respond, in a timely manner (see Figure 7-3 for a timeline for Class 2 permit modifications), to

- Common variations in the types and quantities of the wastes managed under the facility permit.
- Technological advancements.
- Regulatory changes, where such changes may be implemented without substantially altering the design specifications or management practices prescribed by the permit.

Class 2 modifications include increases of up to 25 percent in a facility's non-land-based treatment or storage capacity, authorizations to treat or store new wastes that do not require different unit design or management practices, and modifications to improve the design of hazardous waste management units or improve management practices.

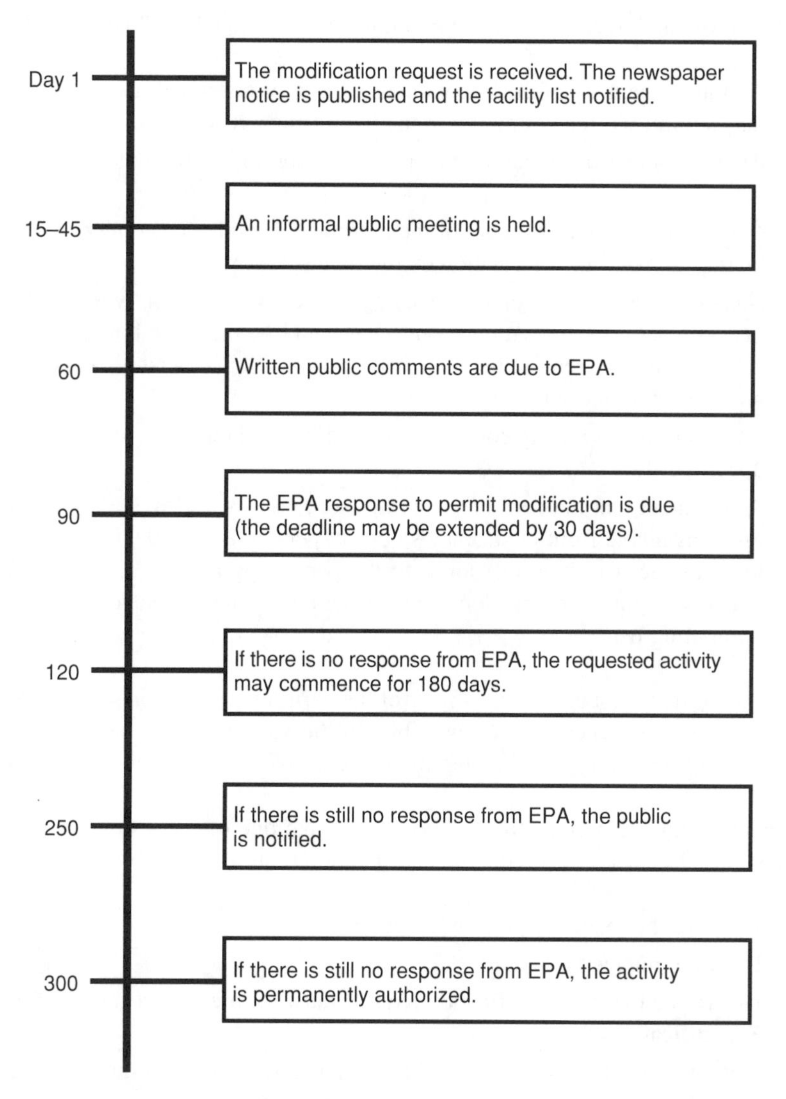

Figure 7-3. Timeline for Class 2 Permit Modifications

Under 40 CFR 270.42(b)(1), a permittee who wishes to make a Class 2 modification must submit to EPA a modification request describing the exact change to be made to the permit conditions. The permittee must also submit supporting documents that identify the modification as a Class 2 modification required by 40 CFR 270.13 through 270.21, 270.62, and 270.63.

Under 40 CFR 270.42(b)(2), the permittee must notify persons on the facility mailing list and appropriate units of state and local government and publish a notice in a local newspaper regarding the modification requests.

The following information is required in the notice [40 CFR 270.42(b)(2)]:

- Announcement of a 60-day comment period during which interested persons may submit written comments to the permitting agency.
- Announcement of the date, time, and place for an informational public meeting.
- Name and telephone number of the permittee's contact person whom the public may contact for information on the request.
- Name and telephone number of an agency contact person, whom the public may contact for information about the permit, the modification request, applicable regulatory requirements, permit modification procedures, and the permittee's compliance history.
- Information on viewing copies of the modification request and any supporting documents.
- A statement that the permittee's compliance history during the life of the permit is available from the agency's contact person. Also, 40 CFR 270.42(b)(2) requires the permittee to submit to the permitting agency evidence that this notice was published in a local newspaper and mailed to persons on the facility mailing list.

The permittee must also make a copy of the permit modification request and supporting documents accessible to the public in the vicinity of the permitted facility (e.g., at a public library, local government agency, or location under control of the owner).

Under 40 CFR 270.42(b)(6)(i), EPA must make one of the following five decisions within 90 days of receiving the modification request:

- Approve the request with or without changes
- Deny the request
- Determine that the modification request must follow the procedures for Class 3 modifications
- Approve the request, with or without changes, as a temporary authorization having a term of up to 180 days
- Notify the permittee that it will make a decision on the request within 30 days

If the permitting agency notifies the permittee of a 30-day extension for a decision (or if it fails to make any of the decisions), it must, by the 120th day after receiving the modification request, make one of the following decisions:

- Approve the request, with or without changes,
- Deny the request,
- Determine that the modification request must follow the procedures for Class three modifications, or

- Approve the request as a temporary authorization for up to 180 days.

In accordance with 40 CFR 270.42(b)(6)(vii), EPA is allowed to extend the deadlines for action on a Class 2 request with the written consent of the permittee. This option may be useful if EPA requests additional information from the permittee, or the permittee wishes to conduct additional public meetings.

If, however, EPA fails to make one of the four decisions listed above by the 120th day, the activities described in the modification request, as submitted, are authorized for a period of 180 days as an "automatic authorization" without EPA action. However, at any time during the term of the automatic authorization, EPA may approve or deny the permit modification request. If EPA does so, this action will terminate the automatic authorization. If EPA has not acted on the modification request within 250 days of receipt of the modification request (i.e., 50 days before the end of the automatic authorization), under 40 CFR 270.42(b)(6)(iv) the permittee must notify persons on the facility mailing list within 7 days, and make a reasonable effort to notify other persons who submitted written comments, that the automatic authorization will become permanent unless EPA acts to approve or deny it.

If EPA fails to approve or deny the modification request during the term of the automatic authorization, the activities described in the modification request become permanently authorized without EPA action on the day after the end of the term of the automatic authorization. However, if the owner or operator fails to notify the public when EPA has not acted on an automatic authorization 50 days before its termination date, the clock on the automatic authorization will be suspended. The permanent authorization will not go into effect until 50 days after the public is notified. Until the permanent authorization becomes effective, EPA may approve or deny the modification request at any time. In addition, the owner or the operator will be subject to potential enforcement action. This permanent authorization lasts for the life of the permit unless modified later by the permittee (40 CFR 270.42) or EPA (40 CFR 270.41). This procedure for automatic authorization is commonly referred to as the "default" provision.

Class 3 Modifications. Class 3 modifications cover changes that substantially alter the facility or its operations. Generally, they include the following:

- Increases in the facility's land-based treatment, storage, or disposal capacity.
- Increases of more than 25 percent in the facility's non-land-based treatment or storage capacity.
- Authorization to treat, store, or dispose of wastes not listed in the permit that require changes in unit design or management practices.
- Substantial changes to landfill, surface impoundment, and waste pile liner and leachate collection/detection systems.
- Substantial changes to the groundwater monitoring systems or incinerator operating conditions.

The first steps in the application procedures for Class 3 modifications are similar to the procedures for Class 2. Under 40 CFR 270.42(c)(1), the permittee must submit a modification request to EPA indicating the change to be made to the permit, identifying the change as a Class 3 modification, explaining why the modification is needed, and providing applicable information required by 40 CFR 270.13 through 270.21, 270.62, and 270.63. As with Class 2 modifications, the permittee is encouraged to consult with EPA before submitting the modification request.

The permittee must notify persons on the facility mailing list and local and state agencies concerning the modification request. This notice must occur not more than 7 days before and not more than 7 days after the date of submission. The notice must contain the same information as the Class 2 notification, including an announcement of a public informational meeting. The meeting must be held no fewer than 15 days after and no fewer than 15 days before the end of the comment period [40 CFR 270.42(c)(2)].

After the conclusion of the 60-day comment period, the permitting agency will then initiate the permit issuance procedures of 40 CFR part 124 for the Class 3 modifications, publish a notice, allow a 45-day public comment period on the draft permit modification, hold a public hearing on the modification if requested, and issue or deny the permit modification.

Temporary Authorizations. EPA may grant a temporary authorization, without prior public notice and comment, for a permittee to conduct activities necessary to respond promptly to changing conditions [40 CFR 270.42(e)]. An EPA-issued temporary authorization may be obtained for activities that are necessary to

- Facilitate timely implementation of closure or corrective action activities.
- Allow treatment or storage in tanks or containers of restricted wastes in accordance with part 268.
- Avoid disrupting ongoing waste management activities at the permittee's facility.
- Enable the permittee to respond to changes in the types or quantities of wastes being managed under the facility's permit.
- Carry out other changes to protect human health and the environment.

Temporary authorizations may be granted for any Class 2 modification that meets these criteria, or for a Class 1 modification that is necessary to

- Implement corrective action or closure activities.
- Allow treatment or storage in tanks or containers of restricted waste.
- Provide improved management or treatment of a waste already listed in the permit when necessary to avoid disruption of ongoing waste management, to allow the permittee to respond to changes in waste quantities, or to facilitate other changes to protect human health and the environment.

A temporary authorization will be valid for a period of up to 180 days. The term of the temporary authorization will begin at the time of its approval by EPA or at some specified effective date shortly after the time of approval. The authorized activities must be completed at the end of the authorization.

RCRA POST-CLOSURE PERMITS

Hazardous waste management units must have a permit during their active life, which also includes the closure period. If a facility closes with no hazardous waste or residues in place (clean closure), there is no further responsibility under Subtitle C of RCRA. However, if hazardous waste or residues will remain when a unit is closed, the unit must enter a post-closure care period. Permitted facilities either have already addressed the issue of post-closure care (e.g., landfills), or may modify their RCRA permit accordingly. (Thus, if a unit currently has a permit, there is no separate post-closure permit required because it is incorporated as a condition of the "operating" permit.) However, facilities with only interim status that close with hazardous waste or residues in place must obtain a post-closure permit.[7] Separate post-closure permits are intended for interim status units that close without an operating permit. Thus, if a surface impoundment loses its interim status or voluntarily closes under interim status, EPA may request ("call in") the facility's Part B permit application for a post-closure permit. Failure to submit a Part B on time or failure to submit complete information is grounds for termination of a unit's interim status.

The information required for post-closure permits is generally less than the standard operating permit. At a minimum, an owner or an operator must submit the following information:

- The post-closure plan.
- A copy of the post-closure inspection schedule.
- Location information, including a delineation of the floodplain area.
- Documentation of the notice in the deed.
- Cost estimate for post-closure.
- A copy of the financial mechanism to be used.
- Exposure information.

[7] The denial of a permit for the active life of a hazardous waste management unit does not affect the requirement to obtain a post-closure permit. However, EPA reinterpreted this requirement to mean that EPA *may* require a post-closure permit (62 *FR* 55778, December 8, 1997). This is because in some cases, there is no incentive for an owner or operator to obtain a post-closure permit and can delay the process for many years. Furthermore, any facility that is forced to close because it could not comply with the 40 CFR part 265 requirements for groundwater monitoring and financial assurance would not be expected to obtain a post-closure permit because the issuance of such a permit also depends on certification of these very standards as required by §3005(c) of RCRA. Therefore, EPA has stated that it may require the issuance of a post-closure permit or may use other enforcement mechanisms to achieve corrective action at the site [e.g., §3008(h) or §7003 enforcement orders as described in Chapter 10].

- Groundwater data and a demonstration of compliance with the part 264, subpart F.
- Information pertaining to the RCRA §3004(u) corrective action provision (e.g., location and type of SWMUs).
- Demonstration of financial responsibility for corrective action, if required.

SPECIAL PERMITS

There are two types of special permits addressed in this section: first, research, demonstration, and development permits; and then, emergency permits.

Research, Demonstration, and Development Permits

The Hazardous and Sold Waste Amendments of 1984 (HSWA) provides EPA with authority to issue permits for research, development, and demonstration (RD&D) treatment activities. EPA has authority to issue permits independent of existing regulations relating to hazardous waste management processes. EPA is directed to include certain provisions in each permit as well as any other requirements deemed necessary to protect human health and the environment (40 CFR 270.65).

The EPA Administrator is authorized to issue RD&D permits for innovative and experimental treatment technologies or processes for which permit standards have not been established under part 264. The Administrator may establish permit terms and conditions for the RD&D activities as necessary to protect human health and the environment. The statute allows the Administrator to select the appropriate technical standards for each RD&D activity to be permitted. EPA is required to address construction (if appropriate), limit operation for not longer than one year, and place limitations on the waste that may be received to those types and quantities of wastes deemed necessary to conduct RD&D activities. The permit must include the financial responsibility requirements currently in EPA's regulations and other such requirements as necessary to protect human health and the environment. Other possible requirements include but are not limited to provisions regarding monitoring, operation, closure, remedial action, testing, and information reporting. EPA may decide not to permit an RD&D project if it determines that the project, even with restrictive permit terms and conditions, may threaten human health or the environment.

Emergency Permits

If EPA finds that an imminent and substantial endangerment to human health or the environment exists, a temporary emergency permit may be issued (40 CFR 270.61). This permit may be issued to a nonpermitted unit to allow treatment, storage, or disposal of hazardous waste, or to a permitted unit to allow treatment, storage, or disposal of wastes not covered by an effective permit.

The permit may be oral or written. If it is oral, a written emergency permit is to be sent to the facility within five days. The conditions of an emergency permit (40 CFR 270.61) are that it

- Shall not exceed 90 days.
- Shall specify the types of wastes to be received and their methods of handling (e.g., disposal or treatment).
- Can be terminated at any time by EPA.
- Shall be accompanied by a public notice.
- Shall incorporate, to the greatest extent possible, all applicable standards in parts 264 and 266.

8

LAND DISPOSAL RESTRICTIONS

The Hazardous and Solid Waste Amendments of 1984 (HSWA) forced a major shift in how the U.S. manages its hazardous waste. Before 1984, land disposal, especially landfills and injection wells, was the primary method of managing hazardous waste, mainly because of low cost. In the early 1980s, however, Congress became acutely aware that Superfund, as originally enacted, was dramatically underfunded and that the hazardous waste land disposal problem was far more pervasive than originally thought. Thus, if Congress was to continue funding Superfund, reliance on land disposal had to be reduced to prevent the creation of future Superfund sites. As a result, Congress, in 1984, enacted HSWA, which contained the land disposal restrictions as a centerpiece. The land disposal restrictions are contained in §1002(b)(7) of RCRA.

The land disposal restrictions program, codified in 40 CFR part 268, requires hazardous waste to be treated in a specified manner or treated to meet specified constituent levels before land disposal may occur. This program has effectively eliminated the land disposal of untreated hazardous waste. Wastes that are not treated as required or that do not meet their respective constituent levels may not be disposed of on or in land unless EPA has granted a specific variation.

APPLICABILITY

The land disposal restrictions are applicable only to wastes identified as hazardous under RCRA. (Figure 8-1 depicts the applicability of the land disposal restrictions.) Consequently, if the waste is not defined as a solid and hazardous waste under RCRA (see Chapter 2), the regulations of 40 CFR part 268 do not apply. Hazardous

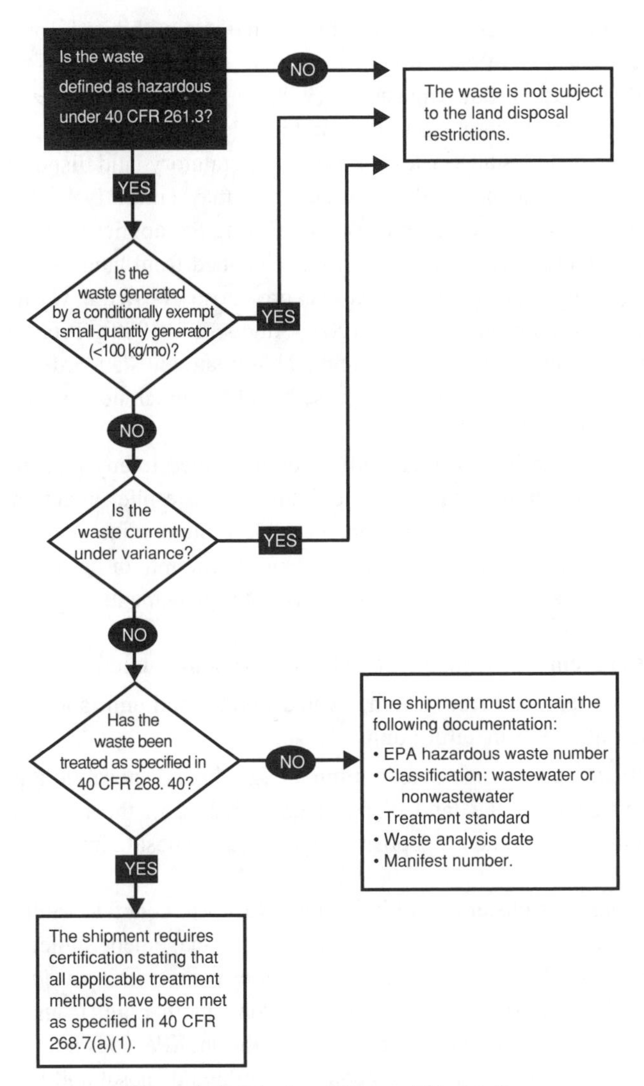

Figure 8-1. The Land Disposal Restrictions

wastes generated by conditionally exempt small-quantity generators also are not subject to the land disposal restrictions [40 CFR 268.1(e)].

Additionally, the land disposal restrictions apply only prospectively to wastes that are/were land disposed *after* the effective date of the respective restrictions (i.e., the land disposal restrictions do not require that wastes land-disposed before the effective date of the restrictions be removed and treated). However, if restricted wastes land-disposed before the applicable effective date were to be removed from the disposal unit, subsequent placement of such wastes in or on the land would be subject to the applicable prohibitions and treatment provisions.

Understanding the distinction between restricted and prohibited wastes is important. *Restricted wastes* are hazardous wastes that are subject to the restrictions on land disposal (53 *FR* 31208, August 17, 1988). A waste becomes restricted under the land disposal restrictions on its statutory deadline. Thus, the universe of restricted wastes includes all wastes for which the statutory land disposal restriction deadline has passed, including those wastes that may currently be land-disposed because of an extension or a variance or by meeting the applicable treatment standard, as well as those wastes that are currently banned from land disposal. *Prohibited wastes*, a subset of restricted wastes, are those restricted wastes currently banned from land disposal (53 *FR* 31208, August 17, 1988). Prohibited wastes include all wastes that do not meet the applicable treatment standards for which the applicable disposal restriction date has passed, and for which no variances or extensions have been granted.

Land disposal is defined to include but not be limited to any placement of hazardous wastes in a landfill, surface impoundment, waste pile, injection well, land treatment facility, salt dome formation, or underground mine or cave (40 CFR 268.2). *Placement* means movement of hazardous waste into or onto a land disposal unit. The following are examples of what constitutes placement:

- Wastes are removed from a unit and put into another unit.
- Wastes are removed from a unit, treated in a second unit, and redeposited in the original unit or in a third unit.
- A treatment unit is constructed within a larger land disposal unit, and material is excavated from the land disposal unit, treated in the newly constructed treatment unit, and then redeposited in the land disposal unit.

EPA also considers placement of hazardous wastes in concrete vaults or bunkers intended for disposal purposes as a type of waste management subject to the land disposal restrictions. However, waste consolidation within a unit, *in situ* waste treatment (provided the wastes were not moved from another unit), and capping of wastes in place do not constitute placement. In addition, EPA does not consider open burning and detonation to be methods constituting land disposal and has concluded that the land disposal restrictions program is not applicable in these instances (50 *FR* 40580, November 7, 1986). HSWA required EPA to establish treatment standards for each of seven groups of hazardous wastes by specific dates referred to as the statutory deadlines. Since these dates have all passed, the land disposal of nearly all hazardous wastes is now restricted (except for some wastes under variances).

Any waste listed as hazardous after 1984 becomes restricted and thus subject to the land disposal restrictions within six months of the listing. However, the waste is not affected should EPA miss the six-month deadline.

RCRA does not impose an absolute ban on the land disposal of hazardous wastes. A waste may be excluded from the land disposal restrictions under the following circumstances [40 CFR 268.1(c)]:

- The waste meets the prescribed treatment standards found in 40 CFR 268.40.
- A nationwide extension to the effective date is granted because of a lack of available treatment capacity.
- EPA grants a site-specific variance that demonstrates there will be no migration of hazardous constituents from the disposal unit for as long as the waste remains hazardous.
- An individual extension to an effective date is granted based on the characteristics of a specific waste.
- Untreated waste may be treated in a surface impoundment that complies with the minimum technological requirements (i.e., double liner with leachate collection and removal system) if the treatment residues that are hazardous are removed within a year of placement in the impoundment.

REQUIREMENTS

The land disposal restrictions dictate specific requirements for owners and operators of hazardous waste management facilities and generators.

Hazardous Waste Management Facilities

Owners and operators of hazardous waste management facilities using land disposal have the ultimate responsibility for verifying that only wastes meeting the treatment standards are land-disposed. The owner or operator of the land disposal facility must maintain documentation to demonstrate that wastes are in compliance with the applicable treatment standards.

The storage of hazardous wastes restricted from land disposal is prohibited except where storage is needed to accumulate sufficient quantities to allow for proper recovery, treatment, or disposal [40 CFR 268.50(a)]. Owners and operators of hazardous waste management facilities may store restricted wastes as needed to accumulate sufficient quantities to allow for proper recovery, treatment, or disposal. However, when storage lasts beyond one year, the owner or operator bears the burden of proving, in the event of an enforcement action, that such storage is used solely for the purpose of accumulating sufficient quantities to allow for proper recovery, treatment, or disposal. For periods less than or equal to one year, the burden of demonstrating whether or not a facility is in compliance with the storage provisions lies with EPA [40 CFR 268.50(b)]. These facilities may store restricted wastes in containers, tanks, and containment buildings; each storage unit must be clearly marked to identify its content and the date when the hazardous waste entered storage [40 CFR 268.50(a)(2)].

Because of the derived-from rule [40 CFR 261.3(c)(2)(i)], all residuals resulting from the treatment of the original listed waste are likewise considered to be listed wastes, and thus must meet the treatment standards if these residuals are to be disposed of in or on the land (53 *FR* 31142, August 17, 1988).

Documentation. Documentation that must be maintained by an off-site disposal facility includes the following:

- Waste analysis data obtained through testing of the waste.
- A copy of the notice and certification required by the owner or operator of a treatment facility under 40 CFR 268.7(b)(1) and (2).
- A copy of the notice and certification required by the generator in cases where the wastes meet the treatment standard and can be land-disposed without further treatment [40 CFR 268.7(a)(2)].
- Records of the quantities (and date of placement) for each shipment of hazardous waste placed in the unit under an extension of the effective date (a case-by-case extension or a two-year extension of the effective date) or a no-migration petition and a copy of the notice under 40 CFR 268.7(a)(3).

If a treatment facility ships a restricted hazardous waste off-site, documentation and notification requirements similar to those for generators (discussed in the following section) also must be followed.

Generators

Generators of hazardous waste must determine if a hazardous waste is a restricted waste (i.e., subject to the land disposal restrictions) at the point of generation by testing and analysis or applying their knowledge [40 CFR 268.7(a)]. If the waste is a restricted waste, the generator must determine if the waste meets either the applicable treatment standards or constituent levels and thus is allowed to be land disposed, or if the waste does not meet the treatment standards and may not be land disposed.

Waste Analysis Plan. If generators are treating hazardous waste on-site in compliance with 40 CFR 262.34 to meet the treatment standards of the land disposal restrictions, they must also prepare a waste analysis plan [40 CFR 268.7(a)(4)]. The waste analysis plan must be based on a detailed physical and chemical analysis of the waste and contain all information necessary to treat the waste in accordance with the treatment standards. This plan must be filed with EPA or the state at least 30 days before the treatment activity.

Documentation. Each shipment of hazardous waste must be accompanied with documentation identifying the land disposal restriction status of that particular waste.

If a generator determines that the waste does not meet the specified treatment levels (and is thus prohibited from land disposal), the documentation must contain the following [40 CFR 268.7(a)(1)]:

- EPA hazardous waste number
- Whether the waste is wastewater or nonwastewater

- Corresponding treatment standard
- Waste analysis data (where applicable)
- Manifest number associated with the shipment of waste

If the generator determines that a restricted waste may be land-disposed without further treatment, each shipment of that waste to a designated facility must have a notice and certification stating that the waste meets all of the applicable treatments [40 CFR 268.7(a)(1)]. The signed certification must state the following:

> I certify under penalty of law that I have personally examined the waste through analysis and testing and through knowledge of the waste to support this certification that the waste complies with the treatment standards specified in 40 CFR 268.32 or RCRA §3004(d). I believe that the information submitted is true, accurate, and complete. I am aware that there are significant penalties for submitting a false certification, including the possibility of a fine or imprisonment.

If a generator's waste qualifies for an exemption from a treatment standard, such as a capacity variance, case-by-case extension, or no-migration exemption, the generator must submit the above notice with a statement identifying the date the waste will become subject to the land disposal restrictions.

Reclamation of Restricted Wastes

Restricted wastes may be recovered or reclaimed (it is the land disposal that is restricted). However, while the reclamation operation itself is exempt from regulation under RCRA, storage of restricted wastes before reclamation is still subject to the provisions specified in 40 CFR 268.50. Still bottoms and other residues from reclamation of restricted wastes remain subject to the land disposal restrictions if they meet the definition of hazardous waste.

Schedule of Restrictions

Congress set forth a schedule of land disposal restrictions in HSWA for all hazardous wastes. The statute automatically prohibited the land disposal of hazardous wastes if EPA failed to set a treatment standard by the statutory deadline. This type of self-implementing provision, regardless of EPA action, is known as a hammer provision.

The statute also required EPA to make determinations on prohibiting land disposal within indicated time frames for the following:

- At least one-third of all ranked and listed hazardous wastes by August 8, 1988 (the "First Third").
- At least two-thirds of all ranked and listed hazardous wastes by June 8, 1989 (the "Second Third").

- All remaining ranked and listed hazardous wastes and all hazardous wastes identified by a characteristic by May 8, 1990 (the "Third Third").

EPA promulgated the entire schedule on May 28, 1986 (51 *FR* 19300), under 40 CFR part 268, subpart B. The schedule was based on a ranking of the listed hazardous waste by intrinsic hazard and volume generated. High-volume hazardous wastes with a high intrinsic hazard were scheduled first, and low-volume wastes with a lower intrinsic hazard were scheduled last.

For any waste listed after November 8, 1984, EPA must make a determination on land disposal within six months of the listing. However, there is no automatic prohibition (hammer) on land disposal of such wastes if EPA fails to meet this deadline.

TREATMENT STANDARDS

HSWA requires EPA to promulgate treatment standards for the applicable hazardous wastes (restricted wastes) by each statutory deadline. Wastes that meet these treatment standards may be directly land-disposed. Wastes that do not meet these standards must be treated before they are placed in a land disposal unit. Previously, EPA promulgated the various treatment standards in a series of tables. Since then, EPA has consolidated the various treatment standards into one table, found in 40 CFR 268.40.

EPA set one of three types of treatment standards for restricted wastes:

- Constituent concentrations in milligrams per kilogram (mg/kg) of waste, which must be met before land disposal.
- Constituent concentrations in an extract of the waste in milligrams per liter (mg/l), which must be met before land disposal.
- Treatment standards expressed as specified technologies and represented by a five-letter code contained in 40 CFR 268.42.

All three types of treatment standards are established based on the best demonstrated available technology (BDAT) identified for that waste. Dilution, however, is not normally considered an allowable treatment method. As stated in 40 CFR 268.3, "No generator, transporter, handler, or owner or operator of a treatment, storage, or disposal facility shall in any way dilute a restricted waste or the residual from treatment of a restricted waste as a substitute for adequate treatment to achieve compliance with Subpart D of this part." However, treatment that necessarily involves some degree of dilution (such as biological treatment or steam stripping) is acceptable. Also, mixing wastes together before treatment is not considered dilution. Dilution is prohibited if it is conducted in lieu of adequate treatment for purposes of attaining the applicable treatment standards.

If a new technology is shown to be more effective in reducing the concentration of hazardous constituents in the waste (or the waste extract) than the existing technology upon which the treatment standard has been based, EPA may revise the treatment standard.

Wherever possible, EPA establishes treatment standards as performance standards rather than requiring a specific treatment method. In such cases, any method (other than inappropriate solidification practices that would be considered dilution to avoid adequate treatment) that may meet the treatment standard is acceptable. Solidification may, nonetheless, be a necessary prerequisite to land disposal to comply with the prohibition against free liquids in landfills (40 CFR 265/265.314). When EPA has specified a technology as the treatment standard, the applicable wastes must be treated by using the specified technology.

Universal Treatment Standards

During the early phases of the land disposal restrictions program, different treatment requirements and levels for the same constituent in different wastes were unintentionally set. As a result, EPA established universal treatment standards to ensure consistency for the same constituent. The universal treatment standards and the list of common constituents are found in 40 CFR 268.48.

VARIANCES FROM THE LAND DISPOSAL RESTRICTIONS

Under certain, limited conditions, a waste not meeting the specified treatment standards may be land disposed. These conditions, known as variances, include the following:

- National capacity variances
- No-migration petitions
- Case-by-case extensions
- Treatment variances

National Capacity Variances

If the capacity of alternative treatment, recovery, or disposal facilities is insufficient for a particular waste or group of wastes nationwide, EPA may grant a nationwide extension to the effective date of the restriction and has done so in the past. The purpose of the extension is to allow time for the development of this capacity. This extension may not exceed a period of two years beyond the applicable statutory deadline for the waste.

No-Migration Petitions

A facility owner or operator may petition EPA to allow continued land disposal of a specific waste at a specific site. The applicant must demonstrate that the waste may be contained safely in a particular type of disposal unit so that no migration of any hazardous constituents occurs from the unit for as long as the waste remains hazardous. If EPA grants the petition, the waste is no longer prohibited from land disposal in that specific unit at that site [40 CFR 268.6(a)]. However, a successful demonstration of "no-migration" is difficult. Owners and operators of land treatment areas treating petroleum wastes and wood preservative wastes are most apt to be granted a no-migration petition.

The statutory standard for the evaluation of no-migration petitions requires that the petitioner demonstrate, to a reasonable degree of certainty, that there will be no migration of hazardous constituents from the disposal unit or injection zone for as long as the wastes remain hazardous [RCRA §3004(d)].

EPA requires applicants to submit no-migration petitions directly to the EPA Administrator. When possible, EPA intends to process Part B permit applications and no-migration petitions concurrently. However, if the review of Part B permit applications or no-migration petitions may be unduly delayed by concurrent reviews, EPA typically will process such applications separately. Applications for no-migration petitions are reviewed at EPA headquarters, whereas EPA regional offices or authorized states are responsible for issuing RCRA permits. EPA headquarters coordinates reviews with appropriate regional and state staff responsible for reviewing Part B applications for the same facility [40 CFR 268.6(a)].

Case-By-Case Extensions

Any person who generates or manages a restricted hazardous waste may submit an application to the EPA Administrator for a case-by-case extension of the applicable effective date. The applicant must demonstrate the following:

- He or she has made a good-faith effort to contract with a treatment, storage, or disposal facility.
- A binding contract has been entered into to construct or otherwise provide alternative capacity.
- This alternative capacity cannot reasonably be made available by the applicable effective date because of circumstances beyond the applicant's control.

There is no deadline for submitting these applications to EPA. However, case-by-case extensions cannot extend beyond 48 months from the date of the statutory land disposal restriction [40 CFR 268.5(e)].

Treatment Variances

Generators or owners and operators of hazardous waste management facilities may petition EPA for a variance from a treatment standard. Wastes may be granted a variance because of physical and chemical characteristics that are significantly different from those of the wastes evaluated by EPA in setting the treatment standards. For example, in some cases, it may not be possible to treat a restricted waste to the applicable treatment standards [40 CFR 268.44(a)].

In accordance with 40 CFR 268.44(g), during the petition review process the applicant must comply with all restrictions on land disposal under part 268 once the effective date for the waste has been reached.

9

RCRA CORRECTIVE ACTION

The main objective of the RCRA corrective action program is to clean up releases of hazardous waste and/or hazardous constituents from hazardous waste treatment, storage, and disposal facilities (TSDFs). The primary method for implementing various components of the corrective action program is the RCRA permitting process.

Before the Hazardous and Solid Waste Amendments (HSWA) of 1984, EPA had limited authority in requiring corrective action at TSDFs.[1] The authority was limited to §7003 of RCRA, which required EPA to make a finding of "imminent and substantial endangerment" (discussed in Chapter 10). If the facility had a RCRA permit, the permit did require corrective action at land disposal units when leakage exceeded specified levels (i.e., beyond the groundwater protection standard, which is discussed in Chapter 5). However, few facilities had RCRA permits in 1984 because the RCRA program was still relatively new. Thus, EPA had to pursue lengthy legal actions or attempt to issue the facility a permit with stricter requirements. Both avenues were time consuming. In response, HSWA amended RCRA by establishing three new corrective action authorities that gave EPA explicit authority to compel corrective action at hazardous waste management facilities. As summarized in Table 9-1, the three additional RCRA corrective action authorities are the following:

[1] Corrective action at TSDFs before HSWA was nearly nonexistent. The only corrective action provision was for permitted units that leaked hazardous constituents into the groundwater above specified levels. However, in 1984, very few facilities had permits or had sufficient groundwater monitoring systems; so even the knowledge of whether the underlying groundwater violated acceptable limits and/or required corrective action was typically nonexistent.

Table 9-1. Summary of RCRA Corrective Action Authorities

Authority	Principle Focus	Comments
§3004(u)	Prior releases at facility seeking permit	A facility that applies for a RCRA permit must address releases of hazardous waste and/or constituents from solid waste management units regardless of when the waste was placed in the unit.
§3008(h)	Interim status corrective action orders	EPA is granted explicit authority to issue a corrective action order at a facility operating under interim status without having to issue a permit.
§3004(v)	Corrective action beyond facility boundaries	Hazardous waste management facilities must address corrective action of releases that have migrated beyond the facility boundary.
§7003	Imminent and substantial endangerment	EPA may seek judical relief or issue an administrative order compelling corrective action to any facility that may cause an imminent and substantial endangerment to human health or the environment.
40 CFR 264.100	Groundwater protection at permitted facilities	If leakage from a land-based unit required to monitor groundwater exceeds the predetermined groundwater protection standard, corrective action is automatically triggered.

- §3004(u), which addresses *prior releases* at a facility seeking a permit
- §3008(h), which addresses *interim status corrective action orders*
- §3004(v), which addresses corrective action *beyond facility boundaries*

The remainder of this chapter describes the above corrective action provisions in more detail, followed by a discussion of the actual RCRA corrective action process.

CORRECTIVE ACTION FOR PRIOR RELEASES

The prior release program, also known as the §3004(u) program, requires corrective action (i.e., to air, soil, groundwater, and surface water) for releases of hazardous waste or hazardous constituents from any solid waste management unit (SWMU) or hazardous waste unit (a subset of solid waste management units) at a RCRA hazardous waste management facility (40 CFR 264.101). The prior release program is initiated through the permit application, which includes operating permits, post-closure permits, and permits-by-rule (July 15, 1985, 50 *FR* 28712).

Definitions

A *facility* is the entire facility as defined in 40 CFR 260.10. Thus, even if a permit is being sought for a unit only, the owner or operator must address possible releases from all waste management units at the facility under this program. Included in the EPA definition of a facility are different parcels of land under common ownership that are separated by a road or public right-of-way. Such parcels are considered to be a single facility for purposes of corrective action (61 *FR* 19442, May 1, 1996).

The term *release* refers to the Superfund definition contained in §10(22):

> *Release* means any spilling, leaking, pumping, pouring, emitting, emptying, discharging, injecting, escaping, leaching, dumping, or disposing into the environment, but excludes (A) any release which results in exposure to persons solely within a workplace, with respect to a claim which such persons may assert against the employer of such persons, (B) emissions from the engine exhaust of a motor vehicle, rolling stock, aircraft, vessel, or pipeline pumping station engine, (C) release of source, byproduct, or special nuclear material from a nuclear incident, as those terms are defined in the Atomic Energy Act of 1954, if such release is subject to requirements with respect to financial protection established by the Nuclear Regulatory Commission under §170 of such Act, or, for purposes of §104 of this title or any other response action, any release of source byproduct, or special nuclear material from any processing site designated under §120(a)(1) or 302(a) of the Uranium Mill Tailings Radiation Control Act of 1978, and (D) the normal application of fertilizer.

The terms *hazardous waste* and *hazardous constituent* mean any waste meeting the statutory definition of hazardous (as opposed to the regulatory definition in 40 CFR 261.3) and any 40 CFR part 261, Appendix VIII hazardous constituent (61 *FR* 19443, May 1, 1996). (See Appendix B of this book for a list of hazardous waste constituents.) Thus, a nonhazardous waste could have released a hazardous constituent and still trigger corrective action for that unit.[2]

A *solid waste management unit* (SWMU) is any discernible waste management unit at a RCRA facility from which hazardous waste or hazardous constituents might migrate, irrespective of whether the unit was intended for the management of solid and/or hazardous waste. Thus, if a facility is seeking a RCRA permit and has nonhazardous waste management units (or units that accepted waste before November 19, 1980) on-site, the owner or operator must address those units for possible corrective action. Even though some units are currently exempt from permit stan-

[2] The Appendix VIII constituents, found in Appendix VIII of 40 CFR Part 264, are known as the *hazardous constituents*. The Appendix VIII list is composed of priority pollutants under the Clean Water Act, the Department of Transportation's hazardous substances, EPA's Carcinogen Assessment Group (CAG) list, and substances on the National Institute of Occupational Safety and Health (NIOSH) Registry of Toxic Effects of Chemical Substances (RTECS) that are potential carcinogens or that have a low LD_{50}. These constituents are used as a trigger for corrective action.

dards (i.e., wastewater treatment units, elementary neutralization units), they are considered SWMUs under this provision.

According to OSWER Directive 9502.00-6c, the solid waste management unit definition includes the following:

- Containers, tanks, surface impoundments, containers, storage areas, waste piles, land treatment units, landfills, incinerators, underground injection wells, and other physical, chemical, and biological units, including units defined as *regulated units.*
- Recycling units, wastewater treatment units, and other units that EPA has generally exempted from standards applicable to waste management units.
- Areas associated with production processes at facilities that have become contaminated by routine, systematic, and deliberate discharges of waste or constituents.

One-time spills of hazardous waste or constituents are subject to §3004(u) only if the spill occurred from a solid waste management unit. A spill that cannot be linked to a discernible solid waste management units is not itself a SWMU. Likewise, leakage from product storage and other types of releases associated with production processes would not be considered a SWMU, unless those releases were routine, systematic, and deliberate (50 *FR* 28712, July 15, 1985). Routine and systematic releases constitute, in effect, management of wastes; the area at which this activity has taken place can thus reasonably be considered an SWMU. *Deliberate* does not require a showing that the owner or operator knowingly caused a release of hazardous waste or constituents. Rather, the term *deliberate* was included to indicate EPA's intention not to exercise its §3004(u) authority against one-time accidental spills that cannot be linked to a SWMU. An example of this type of release would be an accidental spill from a truck at a RCRA facility (OSWER Directive 9502.00-6c).

It should be noted that EPA has stated that it believes Congress did not intend to limit corrective action exclusively to releases from SWMUs, but entire RCRA facilities (61 *FR* 19443, May 1, 1996). Although §3004(u) is limited to SWMUs, EPA's omnibus permit authority under 40 CFR 270.32(b)(2) allows EPA to impose such permit conditions necessary to protect human health and the environment. Furthermore, EPA also may use §3008(h) and §7003 to address releases of hazardous waste from non-SWMUs.

Implementation

The RCRA §3004(u) provision is initiated when an owner or an operator applies for a permit, which includes operating permits, post-closure permits, and permits-by-rule. It is important to note that it is the application of a permit that triggers corrective action, and not the issuance of a permit. Thus, even if a permit is denied, the §3004(u) provisions are applicable. The RCRA Part B permit application requires specified information concerning solid waste management units to be included [40

CFR 270.14(c)]. (In the "corrective measures implementation" section later in this chapter, the mechanisms to obtain this more detailed information are addressed.) New facilities seeking a permit that were not previously engaged in hazardous waste operations also are subject to the §3004(u) provisions. Section 3004(u) states that corrective action is required "for all releases of hazardous waste or constituents from any solid waste unit at a treatment, storage, or disposal facility seeking a permit . . . under Subtitle C of RCRA . . . regardless of the time at which waste was placed in such unit. . . ." Therefore, any solid waste management unit located at a facility is seeking a permit is subject to corrective action even if there has never been any previous authorization for hazardous waste activity at the site.

Section 3004(u) states that corrective action for a facility shall be required as a condition of each permit issued after November 8, 1984. Thus, a previously issued facility permit being reviewed for reissuance also is subject to the 3004(u) corrective action provisions.

A facility that is not required to obtain a permit under §3005(c) of RCRA will not have to comply with §3004(u) (i.e., a surface impoundment that clean closes). However, if EPA found evidence of a release of hazardous waste or hazardous constituents from hazardous or solid waste, it could order corrective action under the interim status corrective action order authority in §3008(h). Section 3008(h) orders may be issued both before, during, and after closure.

The owner or operator must submit the following information in the Part B permit application concerning SWMUs [40 CFR 270.14(c)]:

- The location of the unit(s) on the required topographical map
- Designation of type of unit (e.g., landfill, impoundment)
- General dimensions and structural description (supplying any available drawings)
- The dates of operation
- Specification of all wastes that have been managed at each unit

As discussed later in this chapter, if corrective action is required under §3004(u), the corrective action process will be implemented as conditions of the facility's permit. The permit must include schedules of compliance and financial assurance that corrective action will be completed at the facility.

INTERIM STATUS CORRECTIVE ACTION ORDERS

HSWA also gave EPA authority to order facilities to undertake corrective action outside the permitting process for interim status facilities. As authorized by §3008(h), EPA may require corrective actions or other measures necessary to protect human health and the environment whenever there is or has been a release of hazardous wastes or hazardous constituents from an interim status facility. Such orders may also revoke or suspend a facility's interim status and/or assess penalties of up to $25,000 per day for noncompliance with previous corrective action orders.

Section 3008(h) authorizes the EPA Administrator to issue corrective action orders to address releases of hazardous wastes into the environment from facilities *authorized* to operate under interim status [§3005(e) of RCRA]. This authority extends to include those facilities that should have had interim status but failed to notify EPA under §3010 of RCRA or failed to submit a Part A application (OSWER Directive 9901.1). For example, facilities with closed units may remain in interim status indefinitely, and thus are potentially subject to enforcement action under §3008(h).

The authority under §3008(h) is not confined to addressing releases from solid waste management units, as is §3004(u). Thus, one-time spills and other types of contamination at facilities may be addressed under §3008(h). In addition, §3004(u) is initiated only when an owner or an operator is applying for a permit, whereas §3008(h) has no such limitations for interim status facilities.

CORRECTIVE ACTION BEYOND FACILITY BOUNDARIES

Owners and operators of permitted and interim status hazardous waste management units are required to institute corrective action beyond the facility's boundary when necessary to protect human health and the environment. However, the provision does not apply in cases where the owner or operator is denied access to adjacent property, provided the owner or operator uses best efforts to gain access. In determining *best efforts,* EPA considers case-by-case circumstances; however, at a minimum, the effort should include a certified letter to the adjacent property owner (52 FR 45790, December 1, 1987).

Even if permission from the adjacent landowner to institute corrective action beyond the facility's boundary is denied (despite the best efforts of the owner or the operation), the facility is not necessarily relieved of its responsibility to undertake corrective measures to address releases that have migrated beyond the facility boundary. EPA may request that the facility implement on-site corrective measures in an attempt to clean up releases beyond its boundary. Any corrective measures will be based on their feasibility and appropriateness, according to case-by-case circumstances, considering hydrogeologic conditions and other relevant factors (52 *FR* 45790, December 1, 1988).

THE CORRECTIVE ACTION PROCESS

Although the RCRA corrective action authorities were enacted in 1984 (by HSWA), few regulations have been promulgated that define the requirements to implement the RCRA corrective action program. There have been attempts to propose regulations, but the resulting controversy has derailed the effort. However, despite the lack of a formal regulatory program, numerous policies and guidance documents

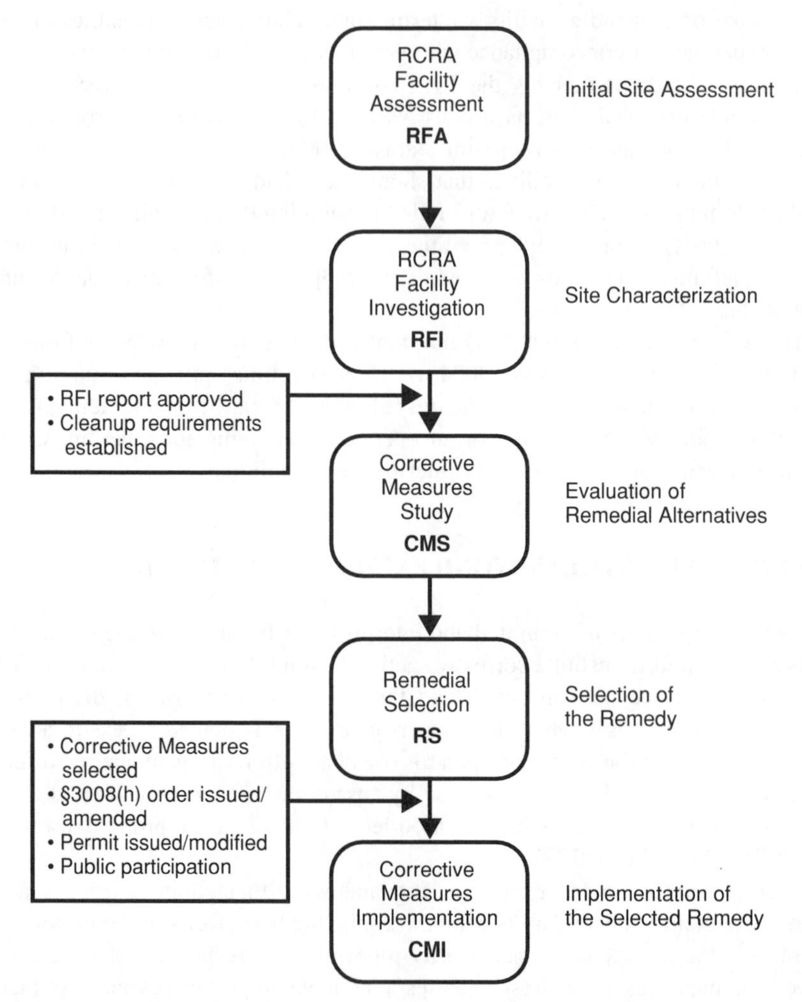

Figure 9-1. Corrective Action Process for RCRA Sites

released since 1984 have defined the requirements and will be used until final regulations are finalized.[3]

The RCRA corrective action process, consisting of five phases, is used to evaluate the nature and extent of contamination, identify appropriate corrective measures, and implement corrective measures. The five phases of the RCRA corrective action process, as outlined in Figure 9-1, are used to identify, evaluate, and implement corrective measures. The phases are as follows:

[3] See the RCRA Corrective Action Plan (OSWER Directive 9902.3-2A, May 1994) for guidance on directing corrective action activities.

1. *RCRA Facility Assessment (RFA).* The RFA is an investigation conducted by EPA to identify possible hazardous waste or hazardous constituent releases.
2. *RCRA Facility Investigation (RFI).* If the RFA determines that there is a suspected release, the owner or operator must conduct an RFI, which characterizes the nature and extent of the release.
3. *Corrective Measures Study (CMS).* If the RFI determines that a corrective action is necessary, a CMS is performed to determine the most effective cleanup alternative.
4. *Remedy Selection.* After the CMS, EPA then selects a suitable remedy and either incorporates it into the permit or issues a corrective action order specifying the remedy and the time when it must be implemented.
5. *Corrective Measures Implementation (CMI).* After the remedy has been selected, the corrective measures are implemented.

Note: Throughout the corrective action process, EPA has the authority to initiate Interim Measures (IMs); primarily when there is an immediate threat to human health or the environment. For example, IMs may include requiring the immediate cleanup of a unit, constructing a fence, or segregating wastes.

The RCRA corrective action program has been modeled after the Superfund remedial action program. A comparison of these two programs is outlined in Figure 9-2.

RCRA Facility Assessment

A RCRA facility assessment (RFA) is conducted for every facility seeking a RCRA permit (OSWER Directive 9502.00-6c). The purpose of the RFA is to identify actual or potential releases of hazardous wastes or hazardous constituents from solid waste management units.

EPA is responsible for conducting RFAs. Typically, EPA uses contractors to assist them in conducting these investigations, but EPA retains overall responsibility. However, in some cases, the facility owner or operator is requested to conduct certain sampling activities (OSWER Directive 9502.00-5).

The RFA identifies potential releases from solid waste management units and hazardous waste management units that need to be cleaned up. EPA uses the information gathered in the RFA to determine if further investigation is necessary, to assess potential risk to human health and the environment, and to require an immediate cleanup if warranted.

The RFA consists of three basic stages:

- Preliminary review
- Visual site inspection
- Sampling visit

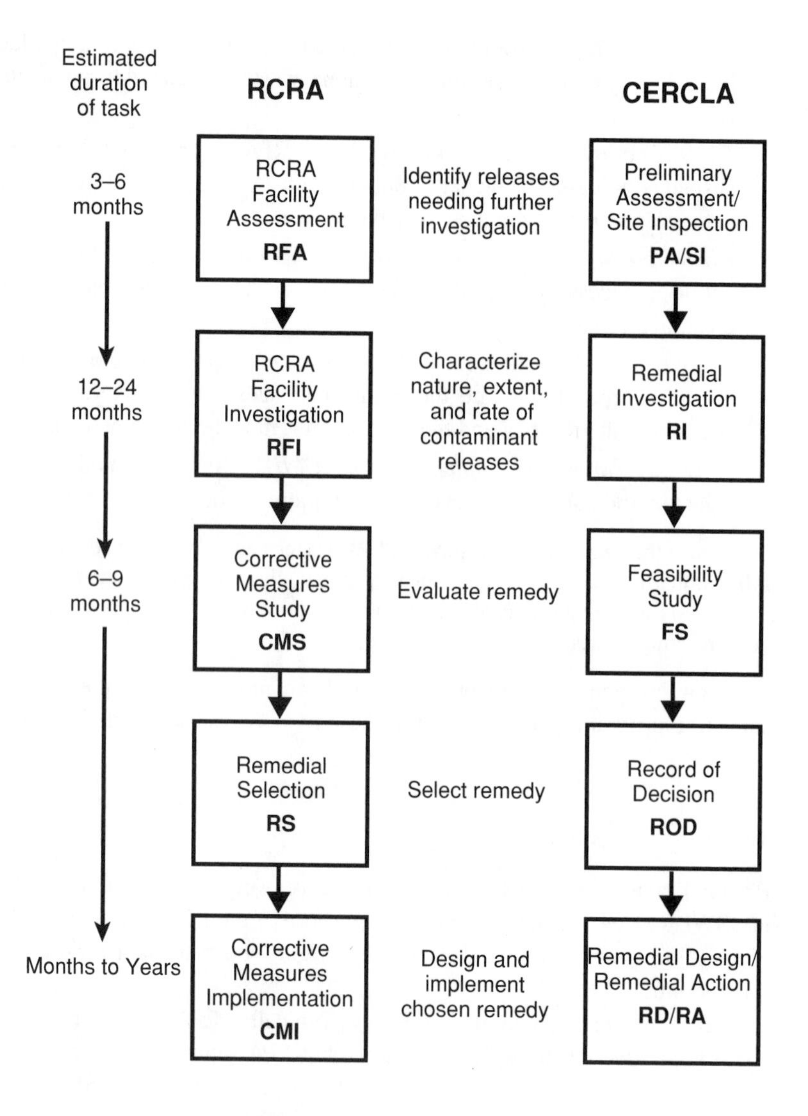

Figure 9-2. Comparison of RCRA Corrective Action and CERCLA Remedial Processes

The RFA begins with a preliminary review of pertinent information concerning the facility. During the preliminary review, EPA collects and evaluates information, such as state and facility files, inspection reports, monitoring reports, and information collected through interviews with people familiar with the facility. In the visual site inspection, an EPA representative tours the facility and looks for evidence of solid waste management units and releases, fills in data gaps identified in the preliminary review, and determines the need for a sampling visit, an interim measure, or an RFI.

During the sampling visit, which is the optional part of the RFA, EPA conducts sampling to fill data gaps found during the preliminary review and visual site inspection. The data gathered during this stage of the corrective action process are limited compared to the data gathered during the RFI.

One of the primary goals of the RFA is to identify releases or potential releases from solid waste management units. When determining the potential for releases from a solid waste management unit, EPA considers such factors as unit characteristics, waste characteristics (i.e., some wastes migrate more quickly than others), pollution-migration pathways (the paths a pollutant can follow when it is released from a unit), evidence of a release, and exposure potential. These factors are evaluated separately to detect releases to each medium, including groundwater, surface water, air, subsurface gas (such as methane generated in landfills), and soil.

The findings of the RFA will have one or more results, as follows:

- No further action will be required because there is no evidence of a release.
- A RCRA facility investigation by the owner or operator will be required if the information collected indicates that a release exists and further information is needed.
- Interim measures by the owner or operator will be required if EPA determines that expedited action is necessary to protect human health or the environment.

National Corrective Action Prioritization System. The National Corrective Action Prioritization System (NCAPS) is used by EPA to set priorities for corrective action, which are generally based on information gathered during the RFA. The NCAPS is necessary because the number of facilities and limitations on EPA's resources make prioritization essential. NCAPS is based on risk, although not a traditional site-specific risk assessment. NCAPS considers the environmental setting of a facility, potential receptors, actual and potential releases of hazardous waste or constituents, and the toxicity of the waste and or constituents. Based on the information, a facility is then placed into a high-, medium-, or low-priority group.

RCRA Facility Investigation

The RCRA facility investigation (RFI), which is analogous to the remedial investigation in the Superfund program, is the second part of the RCRA corrective action process. If the RFA indicates that a release has occurred, or if there is a significant potential for a release, EPA requires the owner or operator of a facility to conduct an RFI, with EPA overseeing the entire process. Generally, permit applicants are required to set out a schedule for the RFI as part of the permit application, which, when issued, will make the RFI a condition of the permit.

The RFI characterizes solid waste management units, the nature and extent of releases of hazardous wastes or constituents from these units, the environmental setting, and potentially exposed populations and affected environments. The RFI is also used to determine the need for interim measures or a corrective measures study and whether wastes have migrated beyond the facility boundary. The general requirements include the identification of each solid waste management unit, the types of investigation required for each medium, and the level of detail required. The schedule of compliance may also include constituent concentration levels that, if exceeded, would require the owner or operator to proceed to the next step of the corrective action process, the corrective measures study, which is used to determine the most suitable cleanup option.

The RFI uses a phased approach, whereby data are collected and analyzed and then used to revise future sampling criteria for more specific sampling if necessary. This approach ensures that all data are incorporated into future portions of the investigation; it is especially important when little is known about the site and the actual location of the unit, or when a release needs to be verified.

The RFI is divided into media-specific investigations, one each for groundwater, surface water, soil, air, and subsurface gas. The unit characteristics, waste characteristics, and release characteristics, and how they affect the potential for and the extent of exposure, are assessed. In addition, each of the media-specific investigations takes into account the unique characteristics of each medium (i.e., the groundwater investigation measures the rate, direction, and flow of groundwater and the depth of the water table, whereas the soil investigation studies surface features and the erosion potential of the soil).

Corrective Measures Study

If the RFI determines the need for corrective measures, the owner or operator must conduct a corrective measures study (CMS), which will most likely require a permit modification. The purpose of the CMS is to identify and recommend specific corrective measures that will adequately correct the release. (The CMS process is analogous to a feasibility study under the Superfund program.)

The CMS can range from a highly focused study of a single treatment technology to a complex study of several cleanup options. The feasibility and the effectiveness of each cleanup technology are examined. The selected corrective action must protect human health and the environment, meet cleanup levels specified by EPA, and minimize any further releases.

A corrective measures study typically includes the following:

- Evaluation of performance, reliability, ease of implementation, and potential impacts of the remedy, including safety impacts, cross-media impacts, and control of exposure to any residual contamination.
- Assessment of the effectiveness of potential remedies in achieving adequate control of sources and cleanup of hazardous waste and constituents released from SWMUs.

- Assessment of the time required to begin and complete the remedy.
- Estimation of the costs of remedy implementation.
- Assessment of the institutional requirements, such as state or local permit requirements, or other environmental or public health requirements that may substantially affect implementation of the remedy.

The CMS also evaluates any possible impacts that a corrective action might cause, such as cross-media contamination (i.e., when groundwater containing a volatile organic is run through an air stripper treatment system, volatile organics are typically transferred from the groundwater to the air).

Because sites differ, EPA has not established national uniform standards for cleanup per se. *EPA-recommended limits or factors* (discussed in Chapter 5 in the section on closure) generally are used as cleanup levels. These factors are health-based levels (i.e., maximum contaminant levels, reference doses, and carcinogen slope factors). As in most other EPA programs, carcinogens must be cleaned up to the risk range of 10^{-4} to 10^{-6}, with 10^{-6} as the point of departure. Site-specific factors also are used to determine cleanup levels. These factors include groundwater use (i.e., drinking water), depth of contamination, mobility and persistence of contaminants, and potential receptors.

Upon completion and receipt of the corrective measures study, EPA evaluates its adequacy. If the plan is deficient, EPA either modifies the plan or requires the owner or operator to make the appropriate modifications before approval of the remedy selection.

Remedy Selection

The corrective measures study presents corrective action alternatives with a preferred alternative. The remedy selection process requires the evaluation of a number of factors before a remedy is selected and the corrective measures implementation phase begins.

The selected remedy must

- Be protective of human health and the environment.
- Attain the cleanup standards.
- Control the sources of releases to reduce or eliminate threats to human health and the environment.
- Comply with the applicable hazardous waste management regulations.

A remedy that meets the above standards also must meet the following technical standards:

- Long-term reliability and effectiveness
- Reduction of toxicity, mobility, and/or volume
- Short-term effectiveness

- Implementability

Furthermore, in selecting the remedy, EPA has the following expectations (61 *FR* 19448, May 1, 1996):

- Treatment is to be used to address the principal threats posed by a site whenever practicable and cost-effective. Contamination that represents principal threats for which treatment is most likely to be appropriate includes contamination that is highly toxic, or highly mobile, or that cannot be reliably contained.
- Engineering controls, such as containment, are to be used for wastes and contaminated media that can be reliably contained or pose relatively low long-term threats, or for which treatment is impracticable.
- Institutional controls, such as land-use and water restrictions, are to be used primarily to supplement engineering controls (but not be the sole control) as appropriate for short- and long-term management, to prevent or limit exposure to hazardous wastes and constituents.
- Innovative technology is to be used when such technology offers the potential for comparable or superior treatment performance or implementability, less adverse impact, or lower costs for acceptable levels of performance when compared to more conventional technologies.
- Usable groundwaters are to be returned to their maximum beneficial uses wherever practicable, within a time frame that is reasonable given the site conditions. When restoration is not practicable, the migration of the plume must be prevented and exposure to the contaminated groundwater must be reduced.
- Soils need to be remediated as necessary to prevent or limit direct exposure to human health and the environment or limit the transfer via other transportation processes (e.g., leaching, runoff, air).

Corrective Measures Implementation

The purpose of the corrective measures implementation (CMI) phase is to design, construct, implement, maintain, and monitor the performance of selected corrective measures.

The implementation of corrective measures is the actual cleanup process. The selected corrective measure is implemented by the owner or operator, with close supervision by EPA, and the cleanup is conducted until concentration levels set by EPA are attained. The facility is monitored during and after the CM process to ensure that all wastes or constituents have been cleaned up and will not be released over time.

The corrective measures may be implemented through compliance schedules contained in the facility's permit, which will likely require a permit modification and subsequent public involvement. On June 22, 1987 (52 *FR* 2344), EPA promulgated amendments to allow the information related to detailed corrective action planning, required by 40 CFR 270.14(c)(7), (c)(8)(iii), and (c)(8)(iv) to be developed at EPA's discretion, after the issuance of a permit through the use of compliance schedules.

The owner or operator must obtain advanced written authorization from EPA waiving submittal of the corrective action information if the requirements are issued in compliance schedules. This waiver applies only to the design of the corrective action program and not to information required to assess the need for the extent of the corrective action.

Corrective Action Management Units. On February 16, 1993 (58 *FR* 8658), EPA promulgated regulations for corrective action management units (CAMUs). This rule provides regulatory relief from specific RCRA standards such as permitting and the land disposal restrictions by creating a new type of RCRA unit. The CAMU designation may be made during the corrective action process to manage RCRA remediation wastes specifically. The design, operating, closure, and post-closure requirements for CAMUs are determined by EPA based on site-specific conditions. The regulations governing CAMUs are contained in 40 CFR 264.552. A similar type of unit, the Temporary Unit (TU), is addressed in 40 CFR 264.553.

Financial Assurance. EPA has the authority to require financial assurance for corrective action under §3004(a)(6). Although EPA has made attempts to promulgate final regulations for this requirement, none have been successful. As a result, EPA may require the demonstration of financial assurance during the permit application process and require financial assurance for corrective action as a permit condition. Instruments accepted by EPA to demonstrate such assurance for corrective action include insurance, surety bonds, trust funds, letters of credit, corporate guarantee, and a financial test of strength.

Public Participation

Public participation under RCRA is generally limited to the permit application and modification processes. Because the corrective action program is implemented through permits, it is through the permit process that the public has the opportunity to participate.

When a permit application is submitted, a mailing list must be prepared by EPA for the community in which the facility is located [40 CFR 124.10(c)(1)(viii)]. The list serves as a communication tool to enable EPA to reach interested members of the public with announcements of meetings, hearings, events, reports, and documents.

After developing a draft permit, EPA must provide public notice that a draft permit has been prepared and is available for public review (40 CFR 124.6). A 45-day public comment period on the draft permit must follow the public notice. The comment period for the draft permit provides the public an opportunity to comment on corrective action conditions contained in the permit. In most cases, requirements for the RFI will be included in the schedule of compliance in the draft permit.

A more detailed discussion of public participation during the permit process is discussed in Chapter 7.

10

ENFORCEMENT AND STATE AUTHORIZATION

It is RCRA's explicit intent to establish a viable state-federal partnership to carry out the provisions of RCRA and ultimately help states to operate the Nation's hazardous waste management program in lieu of the federal government [§1003(a)(7)]. However, before a state can operate such a program, it must meet the following primary requirements: the program must be *equivalent* to the federal program, must be *consistent* with the federal program and other state programs, and must provide for adequate enforcement.

The first part of this chapter discusses one of the most important components of state authorization—enforcement. The remainder of this chapter discusses the requirements and conditions that a state must meet to obtain and maintain RCRA authorization.

RCRA'S ENFORCEMENT PROGRAM

The purpose of any enforcement program, including RCRA's, is to promote and compel compliance with a set of regulations. The RCRA enforcement program has many tools, both legal and administrative, to accomplish this goal. The effectiveness of the hazardous waste management regulatory program under RCRA depends on whether or not the regulated community complies with the requirements. Monitoring a facility's compliance and the initiation of legal action when noncompliance exists are the fundamental components of an effective enforcement program. Compliance monitoring also allows EPA to determine which industries tend to have chronic compliance problems and why, and to evaluate the effectiveness of enforcement initiatives and programs.

Compliance Evaluation Inspections

The primary means of determining a facility's compliance status is the facility compliance evaluation inspection, which is a formal visit to a facility to review records, obtain samples, and determine the facility's compliance with the requirements by observing facility operations. RCRA provides the authority for EPA, an authorized state, or designated representatives (e.g., contractors) to enter any facility that has handled hazardous waste to inspect and examine the facility's records and to obtain samples of waste (§7003).[1]

It should be noted that with increased use of computers and improved recordkeeping capabilities at the federal and state level, compliance monitoring also occurs outside of the traditional inspection. Thus, some compliance evaluation inspections can be targeted as a result of information culled or "flagged" from records or reports required to be submitted to EPA. These "flags," however, are not solely a result of RCRA records and reports. Other records and reports, such as NPDES compliance monitoring reports, EPCRA Form Rs, U.S. Securities and Exchange Commission financial reports, and other reports submitted under environmental and non-environmental statutes also can identify inconsistencies or anomalies that can "flag" a facility for an inspection.

At reasonable times, an inspector may enter any facility where *hazardous wastes* are or have been generated, or stored, or disposed of.[2] This authorization includes allowance to inspect the facility, obtain samples, and obtain and copy records and information related to hazardous wastes (OSWER Directive 9938.0). The specific objective of the inspection does not have to be written, but the inspection must deal specifically with the generation, management, or transportation of hazardous waste (OSWER Directive 9948.0).

The primary purpose of any inspection is to determine whether the facility is managing all hazardous wastes appropriately, operating in accordance with the terms of its permit, complying with interim status standards, following the necessary generator requirements, or complying with other applicable regulations.

A inspection typically consists of the following steps:

1. Before visiting the facility, the inspector reviews the facility's permit or other records to identify any problems that may be encountered, or to "flag" areas of probable noncompliance.

[1] The Federal Facilities Act of 1992 explicitly affirmed that federal facilities are subject to the same requirements and enforcement actions under RCRA as commercial entities.

[2] It is important to note that the statutory provision authorizing inspections is not limited to the regulatory definition of hazardous waste in 40 CFR 261.3, but the much broader statutory definition of hazardous waste found in §1004(5) of RCRA.

2. The inspector enters the facility, produces the necessary credentials, and describes the nature of the inspection. An opening conference is held with the owner or operator to describe the information and samples to be gathered.
3. The facility is inspected. The inspection includes examination of facility records, possible collection of samples, and observation of the facility, including any hazardous waste management operations. The inspector may also observe all associated activities, such as unloading of wastes, lab work, and safety procedures. The inspector may use field notebooks and checklists and, if appropriate, take photographs for documentation.
4. The inspector holds a closing conference with the owner or operator to respond to questions about the inspection and provide additional information.
5. The inspector prepares a report summarizing the results of the inspection, including the results of sampling. Violations are documented in the report.

Any records, reports, or other information obtained from an inspection are available to the public, unless a claim of confidentiality is asserted under EPA's business confidentiality regulations (contained in 40 CFR part 2).

Before the Hazardous and Solid Waste Amendments of 1984 (HSWA), RCRA did not mandate inspections of facilities. Because critics claimed there were too few inspections to effectively monitor compliance with the regulations (as a result of EPA's self-policing initiative), HSWA requires that all federal- or state-operated facilities be inspected annually. Furthermore, EPA must inspect all privately owned facilities at least once every two years [§3007(e)]. Facilities may also be inspected at any time there is reason to suspect that a violation has occurred or is occurring.

Enforcement Options

If a facility is found to be out of compliance with applicable RCRA Subtitle C regulations, an enforcement action may be initiated.

EPA considers both owners and operators of a facility to be responsible for regulatory compliance. For this reason, EPA may initiate an enforcement action either against the owner, operator, or both. Normally, a compliance order is issued to the person responsible for the daily operations at the facility because that person is most likely to be in a position to rectify the problems. If the operator is unable or unwilling to rectify the problems, then EPA may issue a separate compliance order to the owner.

There are several types of enforcement actions available under RCRA, as summarized in Table 10-1, and including the following:

- Administrative actions
- Civil actions
- Criminal actions

Table 10-1. Available RCRA Enforcement Tools

Enforcement Tool	Coverage	Statutory Authority
Administrative Actions	**Nonjudicial enforcement action**	
Informal Actions	Used to attain compliance with minor violations or as the first step	
Compliance Orders	Used to force compliance with a RCRA provision	§3008(a)
Corrective Action Orders	Used to institute corrective action of a release	§3008(h)
Monitoring and Analysis Orders	Used to force the investigation of contamination or a release	§3013(a)
Imminent Endangerment Orders	Used to force action where an imminent threat to health or the environment exists	§7003
Civil Actions	**Formal lawsuit**	
Compliance Actions	Used to force compliance with a RCRA provision	§3008(a)
Corrective Action Orders	Used to institute corrective action of a release	§3008(h)
Monitoring and Analysis Orders	Used to force the investigation of contamination or a release	§3013(a)
Imminent Endangerment Orders	Used to force action where an imminent threat to health or the environment exists	§7003
Criminal Actions	Authorized for the "knowing" violation of prescribed provisions	§3008(d)

Administrative Actions. An *administrative action* is a nonjudicial enforcement action taken by EPA or a state under its own authority. These actions require less preparation than a lawsuit, which is a civil action. There are two types of administrative actions: informal actions and formal actions (i.e., administrative orders). Both constitute an enforcement response outside of the court system.

Informal Administrative Actions. An *informal administrative action* is any communication from EPA or a state that notifies a facility that it is not in compliance with a specified provision of the regulations. EPA typically uses Notices of Violations (NOVs), also known as warning letters, as informal administrative actions. If the owner or operator does not take steps to comply within a certain time period after receiving a communication, a more formal action can be taken. An NOV sets out specified actions to be taken to bring the facility back into compliance and sets out the enforcement actions that will follow if the facility fails to take the required steps.

An NOV generally contains the following information:

- Identification, citation, and explanation of the violation.
- A deadline for achieving full compliance with the appropriate regulatory or statutory requirements.
- A statement indicating that continued noncompliance beyond a particular date would generally result in the issuance of a §3008 compliance order or other enforcement action, including the assessment of civil penalties of up to $25,000 per day per violation.
- The name and telephone number of a contact person.

Administrative Orders. When a violation is detected that is more severe than those requiring an informal administrative action, EPA or the state may issue an administrative order[3] directly under the authority of RCRA. Such an order imposes enforceable legal requirements. Orders may be used to force a facility to comply with specific regulations; to initiate corrective action; to conduct monitoring, testing, and analysis; or to address a threat to human health or the environment. Four types of administrative orders may be issued under RCRA:

- Compliance orders
- Corrective action orders
- Monitoring and analysis orders
- Imminent endangerment orders

Compliance Orders. Section 3008(a) of RCRA authorizes the use of an order requiring any person who is not complying with a requirement thereunder to take steps to come into compliance. A compliance order may require immediate compliance or may set out a timetable to be followed to move toward compliance. The order may include a penalty of up to $25,000 for each day of noncompliance and may suspend or revoke the facility's permit or interim status. When an agency issues a compliance order, the person who receives the order may request a hearing on any of its factual provisions. If no hearing is requested, the order becomes final 30 days after being issued.

Corrective Action Orders. Section 3008(h) of HSWA authorizes the use of an order requiring corrective action at a facility when there has been a release of a hazardous waste or constituents into the environment, as discussed in Chapter 9. Such orders may be issued to require corrective action regardless of when waste was placed in the unit. EPA may also use §§3004(u), 3004(v), and 3008(a) to compel corrective action. These orders may impose penalties of up to $25,000 for each day of noncompliance.

[3] Refer to 40 CFR 270.4(a) and Chapter 9 of this book for a discussion of permit-as-a-shield, which limits enforcement actions that may be brought against a permitted hazardous waste management facility.

Monitoring and Analysis Order. If EPA or the state finds that a substantial hazard to human health or the environment exists at a facility, a monitoring and analysis order under §3013(a) may be issued. This order is used to evaluate the extent of the *alleged* problem through monitoring, analysis, and testing. It may be issued to either the current owner of the facility or to a past owner if the facility is not currently in operation.

A §3013(a) order requires the person to whom the order was issued to submit to EPA, within 30 days of the issuance of the order, a proposal for carrying out the required monitoring, testing, analysis, and reporting. After providing the person an opportunity to confer with EPA, the agency may require the *execution of* the proposal, as well as make any modifications in the proposal that the EPA deems reasonable to ascertain the nature and extent of the hazard. EPA may commence a civil judicial action against any person who fails or refuses to comply with a §3013(a) order. Such an action is brought in the U.S. district court in which the defendant is located, resides, or is doing business. Furthermore, the government may perform the sampling and analysis and subsequently recover the costs from the responsible party.

Imminent Endangerment Orders. Under §7003 of RCRA, in any situation in which an imminent and substantial endangerment to human health or the environment has been caused or is caused by the handling of hazardous wastes, the responsible agency may order persons contributing to the problem to take steps to rectify the situation. EPA may bring actions against past or present generators, transporters, or owners or operators of a site. Violation of a §7003 order may result in fines of up to $5,000 per day. Evidence possessed to support the issuance of a §7003 order must show that the "handling, storage, treatment, transportation, or disposal of any solid or hazardous waste *may present* an imminent and substantial endangerment to health or the environment." The words "may present" indicate that Congress established a standard of proof that does not require a certainty. The evidence is not required to demonstrate that an imminent and substantial endangerment to pubic health or the environment definitely exists. Instead, an order may be issued if there is a sound reason to believe that such an endangerment may exist.

Civil Actions. A civil action is a formal lawsuit, filed in federal district court, against an individual or a facility that either has failed to comply with a regulatory requirement or administrative order or who has contributed to a threat to human health or the environment. Civil actions are generally reserved for situations that present repeated or significant violations or serious threats to the environment. (The U.S. Department of Justice represents EPA in any civil or criminal prosecution or defense.)

RCRA provides authority for filing four types of civil actions:

- Compliance action
- Corrective action
- Monitoring and analysis action

- Imminent hazard action

Frequently, several of the civil action authorities are used together in the same lawsuit. This is likely when a facility has been issued an administrative order for violating a regulatory requirement, has ignored that order, and is in continued noncompliance. In this circumstance, a lawsuit may be filed that seeks penalties for violating the original requirement, penalties for violating the order, and a court order requiring future compliance with the requirement and the administrative order.

Compliance Action. EPA or an authorized state may file a compliance action in the form of a civil suit to force a person to comply with applicable RCRA regulations. The court also may impose a penalty of up to $25,000 per day for noncompliance.

Corrective Action. In a situation where there has been a release of hazardous waste from a facility, EPA or a state may sue to have the court order the facility to correct the problem and take any necessary response measures. The court also may suspend or revoke a facility's interim status as a part of its order.

To exercise the corrective action authority, EPA must first have information that there is or has been a release at the facility. Additional sources that may provide information on releases include the following: inspection reports, RCRA facility assessments, RCRA Part A and Part B permit applications, responses to §3007 information requests, information obtained through §3013 orders, notifications required by §103 of CERCLA, information-gathering activities conducted under §104 of CERCLA, or citizens' complaints corroborated by supporting information.

Monitoring and Analysis Action. If EPA or the state finds that a facility poses a substantial hazard to human health or the environment, a monitoring and analysis order under §3013(a) may be issued. This order is used to evaluate the extent of the *alleged* problem through monitoring, analysis, and testing. If EPA or a state has issued a monitoring and analysis order under §3013 of RCRA and the facility to which the order was issued fails to comply, EPA or a state may sue to have a court require compliance with the order. In this type of case, the court may levy a penalty of up to $5,000 for each day of noncompliance with the order.

Imminent Hazard Action. As with a §7003 administrative order, when any facility or person has contributed or is contributing to an imminent hazard to human health or the environment, EPA or a state may sue the person or facility and request the court to require that person or facility to take action to remove the hazard or remedy any problem. If an agency first issued an administrative order, the court may also impose a penalty of up to $5,000 for each day of noncompliance with the order.

Criminal Actions. A *criminal action* is an action that may result in the imposition of fines or imprisonment. There are seven instances identified in §3008(d) of RCRA that carry criminal penalties. Each of these are based on whether the accused "knowingly" violated the law. Six of the seven criminal acts carry a penalty of up to

$50,000 per day or up to five years in jail. Briefly, these "knowing" violations include the following:

- Transportation of hazardous waste to a nonpermitted facility.
- Treatment, storage, or disposal of hazardous waste without a permit or in violation of a condition of a permit.
- Omission of required information from a label or manifest.
- Generation, storage, treatment, or disposal of hazardous waste without compliance with RCRA's record-keeping and reporting requirements.
- Transportation of hazardous waste without a manifest.
- Export of hazardous waste without the consent of the receiving country.

The seventh criminal act is the knowing transportation, treatment, storage, disposal, or export of any hazardous waste in such a way that another person is placed in imminent danger of death or serious bodily injury. This act carries a possible penalty of up to $250,000 or 15 years in prison for an individual or a $1 million fine for a corporation.

Imposing Fines. EPA uses the *RCRA Civil Penalty Policy* for assessing administrative penalties under RCRA. The penalty policy is a matrix table used to derive potential fines for a violation after several factors are considered. These factors include the class of violation and violator.

Violation Classes. The RCRA enforcement program defines different classes of violations, which are used to define priorities for enforcement response and to assess penalties (OSWER Directive 9900.0-1A). The program classifies individual violations into one of two classes:

Class I Violation: the most serious violation, including a release or threat of release of hazardous waste to the environment. In addition, Class I violations include failure to assure that groundwater will be protected from hazardous waste, that proper closure and post-closure activities will be undertaken, or that hazardous waste will be destined for or delivered to nonpermitted facilities.

Class II Violation: any RCRA violation that does not meet the criteria listed for a Class I violation.

Violators are further classified according to the nature of their violation(s) and their compliance history. Three categories of violators have been established: high-, medium-, and low-priority violators.

High-Priority Violator: is a handler who has either

- Substantially deviated from RCRA statutory or regulatory requirements;

- Deviated from conditions of a permit, order, or decree by not meeting the requirements in a timely manner or failing to perform the required actions;
- Caused a substantial likelihood of exposure to hazardous waste or has caused actual exposure; or
- Is a chronic or recalcitrant violator.

High-priority violators are the violators who merit the most stringent and immediate enforcement response. The goal of any enforcement action against a high-priority violator is to impose sanctions that will compel a rapid return to compliance, penalize the violator, recover economic gains the violator may have accrued, and deter other members of the regulated community from violating the law. Once a violation discovery is made, it is expected that within 90 days either a formal administrative action will be taken or a referral will be made for judicial action.

Medium-Priority Violator: a handler with one or more Class I violations who does not meet the criteria for a high-priority violator. Handlers with only Class II violations may also be medium-priority violators when the compliance official believes that an administrative order is the appropriate response. The appropriate response is either the issuance of an administrative order or a less formal response that results in compliance within 90 days of violation discovery. The issuance of an administrative order with penalties is the preferred response.

Low-Priority Violator: a handler who has only Class II violations and is not a medium- or high-priority violator. A low-priority violator normally will receive a notice of violation or a warning letter as the initial response within 60 days of the discovery of the violation.

Self Policing. EPA issued a self-policing compliance program on December 22, 1995 (60 *FR* 66706). Basically, the policy authorizes reduced fines and penalties for voluntary disclosure. If a facility voluntarily evaluates, discloses, and corrects a violation, the facility may be eligible for substantially reduced or eliminated penalties. There may also be a deferral of criminal enforcement for violations disclosed and corrected in compliance with the policy.

Supplemental Environmental Projects. Over the past few years, EPA has increasingly relied on Supplemental Environmental Projects (SEPs) to gain significant environmental benefits in conjunction with the settlement of enforcement cases. Instead of relying on monetary penalties that are paid to the U.S. Treasury's General Fund, SEPs are alternative penalties that simultaneously punish and benefit the violator and benefit the environment. SEPs have included the installation of pollution control equipment, conducting studies to identify pollution prevention opportunities, substitution of highly toxic chemicals with less toxic chemicals, and the initiation of training programs. In exchange for the SEP project, the violator is typically relieved from some or all of the normally applied fine.

Noncompliance by Major Facilities. On the last working day of May, August, November, and February, each State Director must submit to the appropriate EPA Regional Administrator a quarterly report containing information on all *major* facilities that are in noncompliance with their RCRA permit. The EPA Regional Administrator must prepare a similar report to be submitted to EPA headquarters in Washington, DC. All of these reports must be made available to the public for inspections and/or copying by the final calendar day of the reporting month [40 CFR 370.5(c)].

EPA Enforcement Action in Authorized States

States with authorized programs have the primary responsibility for ensuring compliance with the RCRA program requirements.[4] Nevertheless, §3008 of RCRA specifically provides EPA with the authority to take enforcement action in authorized states under certain conditions. EPA has taken and will take enforcement actions in an authorized state when the state asks EPA to do so or the state fails to take timely and appropriate action.

If the state has failed to issue an order or complete a referral within 90 days after the discovery of a high-priority violator (or 60 days after deciding to issue an order to a Class I violator), the EPA Regional Office may take action. Before this action, the EPA Regional Office will notify the state of its intention. The EPA Regional Office may also choose to assess a penalty against a high-priority violator if the state's action failed to include one. The Memorandum of Agreement (MOA) or Grant Agreement (GA) between EPA and each state (discussed in a separate section in this chapter) should establish the mechanism by which notice will be provided. The EPA Regional Office may need to conduct its own case development inspection and prepare additional documentation before initiating an action. If the state has made reasonable progress in returning the facility to compliance or in processing an enforcement action, EPA may delay a response.

Citizen Suits

Section 7002 of RCRA authorizes any citizen to commence a civil action (suit) against the following:

- Any person or governmental agency alleged to be in violation of any permit, standard, regulation, condition, requirement, or order that has become effective under RCRA.
- Any past or present generator, transporter, or owner or operator of a facility who has contributed to or is contributing to a condition that may present an imminent and substantial endangerment to human health or the environment.

[4] Each authorized state that is approved to issue permits must prepare and submit quarterly noncompliance report on permittees (40 CFR 270.5).

- The EPA Administrator, where there is alleged failure of the Administrator to perform any act or duty under RCRA that is not a discretionary action.

The right of citizens to bring suits under §7002, however, is limited, as the following list indicates:

- No suit may be brought if the EPA or state has commenced, and is "diligently prosecuting," an enforcement action against the alleged violator.
- Suits may not be used to impede the issuance of a permit or the citing of a facility (except by state and local governments).
- Transporters are protected from citizen suits in response to incidents arising after delivery of waste.
- A facility actively engaged in a removal action under CERCLA may not be sued under RCRA.

Before filing a suit under §7002, a citizen must give a 60-day notice to EPA, to the state in which the alleged violation occurred, and to the alleged violator. However, if a citizen is suing a past or present generator presenting a substantial endangerment under §7002(a)(1)(B), a 90-day notice must be given to EPA, to the state in which the alleged violation occurs, and to the alleged violator.

There is no prescribed format for the presuit notification. A simple, concise letter (certified) stating the alleged violation(s), the alleged violator, the location of the alleged violation, and the intent to file suit is sufficient.

The notice to EPA should be sent to the following address:

The Administrator
U.S. Environmental Protection Agency
401 M Street, SW
Washington, DC 20460

The suit must be brought in the district court for the district in which the alleged violation or endangerment occurred. If a person is bringing suit against the EPA Administrator, the suit may be brought in either the district court in which the alleged violation occurred or the District Court of the District of Columbia. In addition, whenever action is brought against a past or present contributing generator, the plaintiff must serve a copy of the suit to the Attorney General of the United States and to the EPA Administrator.

The court, in issuing any final order in any action brought under §7002, may award costs of litigation (including reasonable attorney and expert witness fees) to the prevailing or substantially prevailing party, whenever the court determines such an award is appropriate.

STATE AUTHORIZATION

It is RCRA's explicit intent (§3006) to have the bulk of the Nation's hazardous waste management program administered by qualified states with only minimal oversight from the federal government. (Authorized states are also provided with financial resources to assist in this partnership.) Before a state can be authorized to be responsible for the Subtitle C program, it must develop a state hazardous waste program and the EPA must approve it, as outlined in 40 CFR part 271. Because EPA's hazardous waste regulations were developed in stages, the states have been given the opportunity to implement a phased approach to the Subtitle C program. Depending on the requirement, even if a state does not have the authority to implement or enforce a particular provision, the federal government may implement and enforce the provision until the state obtains the authority and necessary approval.

Program Elements

Any state that seeks authorization for its hazardous waste program must submit to the EPA Administrator an application consisting of the following elements (40 CFR 271.5):

- A letter from the state's governor requesting program approval.
- Copies of all applicable state statutes and regulations, including those governing state administrative procedures.
- A description of the program.
- Statement by the state's attorney general.
- A Memorandum of Agreement (MOA).

Program Description

As the name implies, the required program description submitted to EPA must detail the state's hazardous waste program (40 CFR 271.6) and must include descriptions of the following:

- The scope, structure, coverage, and processes of the state program.
- The state agency or agencies that will have responsibility for running the program.
- The state-level staff who will carry out the program.
- The state's compliance tracking and enforcement program.
- The state's manifest system.
- An estimate of the number of generators, transporters, and TSDFs, and an estimate of the annual quantities of hazardous waste generated and managed within the state.
- The estimated costs involved in running the program and an itemization of the sources and amounts of funding available to support the program's operation.

If a state chooses to develop a program that is more stringent and/or extensive than the one required by federal law, the description should address those parts of the program that go beyond what is required under RCRA Subtitle C.

Attorney General's Statement

Any state that wants to assume the responsibility for RCRA Subtitle C must demonstrate to EPA that the laws of the state provide adequate authority to carry out all aspects of the state program. This demonstration must be in the form of a statement written by the state's attorney general that includes references to the statutes, regulations, and judicial decisions that the state will rely on in administering its program (40 CFR 271.1).

Memorandum of Agreement

Although a state with an authorized program assumes primary responsibility for administering RCRA Subtitle C, EPA still retains some responsibility and oversight powers in relation to the state's execution of its program. The Memorandum of Agreement (MOA) between the State Director and the EPA Regional Administrator outlines the nature of these responsibilities and oversight powers and the level of coordination between the state and EPA in operating the program [40 CFR 271.8(b)]. The MOA includes provisions for the following:

- Specification of the frequency and contents of reports that the state must submit to EPA
- Coordination of compliance monitoring activities between the state and EPA
- Joint processing of permits for those facilities that require a permit from both the state and EPA under different programs
- Specification of which permit applications will be sent to the EPA Regional Administrator for review and comment

Approval Process

Before approval, EPA will review the state's application to determine if it meets all the requirements and that it is *consistent* with and as stringent as the federal program. For the purposes of this approval, *consistency*, as is defined in 40 CFR 271.4, includes the following:

- State program must not unreasonably restrict, impede, or operate as a ban on the free movement of hazardous waste across state boundaries for treatment, storage, or disposal.
- Any component of the state's program that acts as a prohibition on the treatment, storage, or disposal of hazardous waste must be based on human health or environmental protection.
- The state's manifest system must comply with the federal manifest program.

EPA has 90 days after receiving the state's application to make a tentative determination, which is published in the *Federal Register* in conjunction with the opening of the 30-day public comment period. Ninety days after the close of the public comment period, EPA must publish its final decision in the *Federal Register.*

States with Different Programs

A state with final or interim authorization may have a program more stringent or broader in scope than EPA's. A requirement is *broader in scope* if it increases the size or scope of the regulated community; for example, a state regulates a nonhazardous waste such as asbestos as a RCRA hazardous waste. A requirement is *more stringent* if it is stricter than its federal equivalent; for example, a state may require annual reporting by a generator instead of the biennial reporting required by the federal program. Table 10-2 lists all of the states and key differences between each state's program and the federal RCRA program for selected provisions.

The distinction between broader in scope and more stringent is significant because EPA may enforce a more stringent state requirement, but not state requirements that are broader in scope. RCRA §3008(a)(2) allows EPA to enforce any provision of an authorized state's approved program. State requirements that are more stringent fall into this category. State provisions that are broader in scope and not part of the federally approved RCRA program, according to 40 CFR 271.1(i), are not enforceable by EPA (Program Implementation Guidance 84-1 and 82-3).

Revising State Programs

As federal and state statutory or regulatory authority relating to RCRA is modified or supplemented, it is often necessary to revise the state program accordingly. Such revisions may be initiated by the state or required because of changes in the federal program (40 CFR 271.21). The state must submit documentation to EPA similar to the materials discussed above in the Program Elements section. Although the actual documentation will vary depending on the scope of the state modifications and previously submitted application materials, all revision applications must include a certification from the state attorney general and a copy of the state's pertinent legal authorities.

Before the enactment of HSWA, a state with final authorization administered its hazardous waste program in lieu of the federal program. This meant the federal requirements no longer applied. When a new, more stringent federal requirement was enacted, the state was obligated to enact equivalent authorities within specified time frames, but the new federal requirements were not effective in the authorized state until the state adopted the new requirements as state law and received authorization for them. However, HSWA amended RCRA by adding §3006(g), which states that new requirements and prohibitions imposed by HSWA take effect in authorized states at the same time they take effect in unauthorized states. These regulations are known as HSWA provisions. EPA implements HSWA requirements in authorized states until the state is granted explicit authorization to implement the new regulation(s). While states must still adopt HSWA provisions as state law to

Table 10-2. Summary of State Hazardous Waste Management Programs in Relation to the Federal RCRA Programs

State	CESQG	SQG	On-Site Treatment by Generators	Hazardous Waste Fees	Generator or Transporter Permits	Reporting	Additional Hazardous Wastes
Alabama	Yes	Yes	No	Landfills	Transporters	Biennial	No
Alaska	Yes	Yes	Yes	No	No	Biennial	No
Arizona	Yes, but with restrictions	Yes, but with restrictions	Yes, but with restrictions	No	Yes	Annual	No
Arkansas	Yes, but with restrictions	Yes	Yes	Yes	Yes	Annual	PCBs
California	No	Yes, but with restrictions	No	Yes	Transporters	Annual	No
Colorado	Yes	Yes	No	Yes	No	Annual for TSDFs, biennial for generators	No
Connecticut	Yes, but with restrictions	Yes, but with restrictions	Yes	Yes	Transporters	Annual	No
Delaware	Yes	Yes, but with restrictions	Yes, but with restrictions	No	No	Annual	No
District of Columbia	Yes	No	Yes	No	No	Biennial	No
Florida	Yes	Yes, but only up to 180 days	Yes	No	No	Biennial	Mercury-containing lamps
Georgia	Yes	Yes	Yes	Yes	No	Biennial	Mercury-containing lamps (universal waste)

Note: An explanation of codes and entries appears at the end of this table.

continued

Table 10-2. Continued

State	CESQG	SQG	On-Site Treatment by Generators	Hazardous Waste Fees	Generator or Transporter Permits	Reporting	Additional Hazardous Wastes
Hawaii	Yes	Yes	Yes	No	No	Biennial	Oil, gas, and geothermal exploration, development, and production wastes
Idaho	Yes	Yes	Yes	Only for disposal	No	Annual	No
Illinois	Yes	Yes	Yes	Yes	Transporters	Annual	There are designate "special" wastes. Fluorescent lamps, if they exhibit a characteristic
Indiana	Yes	Yes	Yes	Yes	No	Biennial	Mercury-containing lamps (universal waste)
Iowa	Yes	Yes	Yes	No	No	Biennial	No
Kansas	No	No, LQG is >25 kg	Yes	Yes	No	Biennial	No
Kentucky	Yes	Yes	Yes	Yes	No	Annual	Nerve and blistering agents

Louisiana	Yes	No, LQG is >100 kg	Yes	Yes	No	Annual	No
Maine	No	Yes	Yes, but with restrictions	Yes	Yes	Annual	PCBs <50 ppm
Maryland	Yes	No, LQG is >100 kg	No	No	TSDFs and transporters	Biennial	PCBs
Massachusetts	Yes, but with restrictions	Yes, but with restrictions	Yes, but with restrictions	Yes	Transporters	Annual	Waste oil, PCBs, pain-related wastes
Michigan	Yes	Yes	Yes	Yes	Transporters	Monthly	
Minnesota	Yes, but with restrictions	Yes, but with restrictions	Yes, but with restrictions	Yes	No	Annual	PCBs
Mississippi	Yes	Yes	Yes, in tanks	Yes	No	Biennial	No
Missouri	Yes	Yes	Yes, but restricted	Yes	Transporters	SQGs, quarterly or annual; LQGs, quarterly or annual, and biennial; TSDFs, quarterly and biennial.	PCBs and used oil, not recycled
Montana	Yes	Yes	Yes	Yes	No	Annual	No
Nebraska	Yes	Yes	Yes	TSDFs	No	Biennial	No
Nevada	Yes	Yes	Yes	TSDFs	No	Biennial	No
New Hampshire	No, all generators <100 kg are considered SQGs	No, all generators >100 kg are regulated as LQGs	Yes	Generators and TSDFs	No	Quarterly reporting by TSDFs and LQGs; annual reporting also by LQGs	Used oil, strontium sulfide, solid corrosives

continued

Table 10-2. Continued

State	CESQG	SQG	On-Site Treatment by Generators	Hazardous Waste Fees	Generator or Transporter Permits	Reporting	Additional Hazardous Wastes
New Jersey	Yes	Yes	Yes	Generators, TSDFs, and transporters	Transporters need a DEP license	Biennial reporting for generators and TSDFs	
New Mexico	Yes	Yes	Yes	Yes	No	Biennial	Mercury-containing lamps that exhibit a hazardous characteristic
New York	Yes, but with restrictions	No	Yes	Yes	No	Annual	PCBs
North Carolina	Yes	Yes	Yes, but in the original accumulation unit	Yes	No	Annual and biennial	No
North Dakota	Yes	Yes	Yes	No	Transporters	Biennial	No
Ohio	Yes	Yes	No	Transporters	TSDFs	Annual	No
Oklahoma	Yes	Yes	Yes	Yes	Transporters	Quarterly reports for generators, monthly reports for TSDFs	Drum cleaning waste
Oregon	Yes	Yes	Yes, but with restrictions	Yes	No	Annual	Pesticide residues, nerve agents

Pennsylvania	Yes, but with restrictions	Yes, but with restrictions	Yes, but with restrictions	Yes	Transporters	Biennial	No exclusion for residues from “empty” containers
Rhode Island	No	No	No	Transporters and TSDFs	Transporters	Biennial	Solid corrosives, ignitable waste with flash point <200°F, PCBs, used oil
South Carolina	Yes	Yes, but only up to 180 days	Yes	TSDFs	Transporters	Quarterly	No
South Dakota	Yes	Yes	Yes	No	No	Biennial	No
Tennessee	Yes	Yes	Yes	Yes	Transporters	Annual	No
Texas	Yes	Yes, but with restrictions	No	Yes	Transporters	Annual	Some used oil
Utah	Yes	Yes	Yes	Commercial TSDFs	No	Biennial	No
Vermont	Yes, but with restrictions	Yes, but only up to 180 days	Yes, but with restrictions	No	Transporters	Annual	PCBs, petroleum distillates, pesticides, infectious waste, paint-related waste, waste ethylene-glycol-based coolants, metal grinding wastes

continued

Table 10-2. Continued

State	CESQG	SQG	On-Site Treatment by Generators	Hazardous Waste Fees	Generator or Transporter Permits	Reporting	Additional Hazardous Wastes
Virginia	Yes	Yes, but only up to 180 days	Yes, but with restrictions	No	Transporters	Annual	Fluorescent lamps that exhibit a characteristic
Washington	Yes	Yes	Yes	Yes	No	Annual	No
West Virginia	Yes	Yes	Yes	Yes	No	Biennial	No
Wisconsin	No	Yes, but with restrictions	Yes, but with restrictions	Yes	Transporters	Annual	F500-wastes containing halogenated compounds
Wyoming	Yes	Yes	Yes	TSDFs	No	Biennial	No

Source: Information is based on the results of a written survey conducted by the author in late 1997.

Explanation of Codes

Yes = The state has adopted the federal provision as written.

CESQG = Conditionally exempt small-quantity generator (<100 kg/mo)

SQG = Small quantity generator (100 to 1,000 kg/mo)+

On-site Treatment by Generators = This refers to onsite treatment without a permit in compliance with 40 CFR 262.34.

Hazardous Waste Management Fee = This refers to fees charged by the state for generators, transporters, or waste management facilities.

Generator or Transporter Permits = This refers to the requirement for generators or transporters to obtain a permit or license. AU hazardous waste management facilities must have a permit.

Reporting = The federal program requires biennial reporting (LQG = large quantity generator).

Additional Hazardous Wastes = This refers to the classification, identification, or regulation of any hazardous waste beyond the federal program.

maintain authorization, HSWA requirements are implemented by EPA in authorized states until such adoption.

For purposes of authorization, EPA has segmented the elements of the federal program into individual rules that are grouped into annual or multiyear groups called *clusters*. As a result, states receive authorization for clusters of requirements instead of having to respond to every new or modified federal regulation. The purpose of the cluster system is to facilitate the revision process for a state. However, a state is not required to request authorization for an entire cluster. A state may request authorization for parts of a cluster, provided the state eventually adopts all of the required provisions. Each cluster has due dates by which authorized states must submit applications to revise their program to include that group of rules [40 CFR 271.21(e)]. Normally, a state need only amend its regulations and is given one year from the closing date of the cluster to apply for revisions. Sometimes, however, the state lacks the statutory authority to implement certain programs. If a statutory change is required before regulations may be issued, a state is authorized for two years [40 CFR 271.21(e)(2)(v)]. If a state is late as established by the cluster rule, EPA may grant the state an extension or establish a compliance schedule [40 CFR 271.21(g)]. The exception to this requirement is for changes in the federal program that are deemed less stringent, which are optional for states.

Program Reversion

Authorized state programs are continually subject to review. If EPA finds that a state's program no longer complies with the appropriate regulatory requirements, EPA may withdraw program approval and revert the program back to federal control. (However, to date, this has not occurred.) Such circumstances include a failure to

- Issue permits that conform to the regulatory requirements
- Inspect and monitor activities subject to regulation
- Comply with terms of the MOA
- Take appropriate enforcement action

In some cases (e.g., when there is a lack of sufficient resources), states with approved hazardous waste management programs may voluntarily transfer the programs back to EPA (e.g., Iowa). In such a case, the appropriate EPA Regional Office will administer the RCRA program in the state. If an authorized state determines that it can no longer comply with RCRA, the state may voluntarily transfer program responsibilities to EPA. In doing so, the state must provide EPA with at least 180 day's notice of the proposed transfer. At least 30 days before the approved transfer occurs, EPA must publish notices of the transfer in the *Federal Register.*

PART II

SUPERFUND

The Comprehensive Environmental Response, Compensation, and Liability Act (CERCLA), more commonly known as Superfund,[1] provides the federal government with broad authority to respond to emergencies involving uncontrolled releases of hazardous substances, to develop long-term solutions for the most serious sites containing hazardous substances, and to arrange for the restoration of damaged natural resources.[2] Superfund also establishes liabilities for responsible parties involving the release of hazardous substances and outlines a claims procedure for parties who have cleaned up sites.

RCRA (the Resource Conservation and Recovery Act), a system of cradle-to-grave regulation of hazardous waste enacted in 1976, was the first federal law giving EPA the explicit authority to regulate the management of hazardous waste. However, it became apparent that the government's authority under RCRA was inadequate to clean up abandoned hazardous waste sites.[3]

At New York's Love Canal in 1978, President Jimmy Carter declared a state of emergency in a neighborhood where long-buried chemical wastes were seeping into homes and reports of adverse health effects were growing. This emergency declaration was necessary in part to allow the government to respond and to secure the nec-

[1] The Comprehensive Environmental Response, Compensation, and Liability Act and the Superfund Amendments and Reauthorization Act of 1986 are collectively referred to as Superfund in this book.

[2] The term *hazardous substances* includes a number of specifically listed chemicals and compounds including RCRA hazardous wastes, as discussed in detail in Chapter 12.

[3] This aspect of RCRA has since been modified. Section 3004(u) of RCRA, discussed in Chapter 9, provides authority for the federal government to respond to some abandoned hazardous waste sites.

essary funds. The declaration, coupled with the inability of the government to respond in a coordinated and energetic manner, provoked intense national attention. This attention subsequently triggered the discovery of thousands of dumpsites throughout the nation, which led to the fear of a potential Love Canal in every city. The government itself lacked the authority and the organization to respond, and there was no explicit authority to force other parties, those responsible, to respond to releases of hazardous substances. This series of events and intense public pressure mobilized the administration and Congress; CERCLA was hastily written and passed by Congress in late 1980 during the final days of President Carter's term. The bill incorporated under one statute those federal authorities whose duty is was to respond to releases of hazardous substances, whether they be intentional or accidental, continuous releases or a one-time spill.

A principal thrust of Superfund is to arrange for or compel potentially responsible parties (PRPs) to clean up sites where a hazardous substance has been released into the environment. Superfund provides EPA with the explicit authority and funding to initiate cleanup activities or to require others to undertake immediate cleanup without first having to determine who is liable. If the responsible party cannot be found or is bankrupt, money from the Hazardous Substance Response Trust Fund (the Superfund) can be used. If a responsible party refuses to clean a site, EPA may do so with federal monies and sue the responsible party for treble damages.

To finance these actions, the original act established the $1.6 billion Hazardous Substance Response Trust Fund. The monies for this fund were generated from a tax on specified feedstock chemicals; however, Superfund's taxing authority ceased on October 1, 1985. Although the statute was still in effect, without money, CERCLA would have been short-lived. But, on October 17, 1986, President Ronald Reagan signed the Superfund Amendments and Reauthorization Act of 1986 (SARA). The basic principles of Superfund, including requiring PRPs to finance cleanups, did not change. However, this new reauthorization changed the cleanup approach and standards and allowed for more public involvement throughout the cleanup process. SARA also provided an additional $8.5 billion to finance additional cleanups. Since that time, Congress has periodically appropriated additional monies for Superfund cleanups. Although constantly threatening to do so, Congress has not made any substantial changes to Superfund since 1986.

Under the original CERCLA, relatively few regulations were promulgated. Instead of mandating EPA to promulgate regulations, Congress simply stated the requirements, with no action necessarily required by EPA. Thus, because of a lack of legal clarification normally obtained through the rulemaking procedure, the courts supplied many of the legal interpretations through case law. EPA also used the original National Contingency Plan (NCP) under the Clean Water Act and periodically published policy directives clarifying the applicability or requirements under the NCP for hazardous substances.

In writing and enacting the reauthorization of Superfund (SARA), Congress basically incorporated directly into statutory language many of the policies and procedures established by EPA based on case law and the NCP.

SUPERFUND'S RELATIONSHIP TO OTHER MAJOR LAWS

Because the intent of Superfund was to incorporate into one law the authorities for responding to various releases of hazardous substances, many environmental laws have connections to Superfund. However, the primary relationship discussed here is with RCRA because of the subject of this book—hazardous waste. (Additional relationships are discussed in Chapter 12.) In addition, the relationship between Superfund and the National Environmental Policy Act (NEPA) is discussed because of the potential NEPA-compliance implications for governmental organizations.

Resource Conservation and Recovery Act

No clear line distinguishes when Superfund or RCRA is solely applicable in a given situation because many provisions of the statutes overlap. It is generally perceived that Superfund addresses past activities and RCRA addresses current activities; however, this generalized distinction can be altered by such factors as financial viability, type of waste managed, waste management practices, and dates of actions. In general, if a facility comes under the purview of RCRA, EPA will not pursue Superfund actions for that facility (51 *FR* 21057, June 10, 1986). For example, RCRA generally cannot be applied to a facility that disposed of hazardous wastes and ceased this practice before November 19, 1980 (the effective date of Phase I of the RCRA regulations); therefore, that facility would be subject to Superfund. However, if the facility managed waste after November 19, 1980, it would be subject to RCRA. But if a RCRA corrective action order cannot be issued, or the facility cannot financially pursue corrective action measures, Superfund may be applied.

National Environmental Policy Act

Section 102(2)(c) of the National Environmental Policy Act (NEPA) requires that environmental impact statements (EISs) be prepared for "all major actions significantly affecting the quality of the human environment." This requirement exists unless Congress has specifically exempted a federal action from NEPA, or the entity follows procedures that serve as a "functional equivalent" of an impact statement.

Superfund does not contain a NEPA exemption; however, EPA has asserted that the Superfund remedial action is functionally equivalent to the EIS process under NEPA. This opinion, dated September 1, 1982, from Robert M. Perry, Associate Administrator, Office of General Counsel, to Rita Lavelle, Assistant Administrator for Office of Solid Waste and Emergency Response, states that under the functional equivalent exception, an agency with expertise in environmental matters is not obligated to comply with the formal EIS process prior to taking a particular action if two criteria are met. First, the agency's authorizing statute must provide "substantive and procedural standards that ensure full and adequate consideration of environmental issues" [*Environmental Defense Fund, Inc. v. EPA,* 489 F.2d 1247, 1257 (D.C. Cir. 1973)]. Second, the agency must afford an opportunity for public participation in the evaluation of environmental factors prior to arriving at a final decision (*Portland Cement Association v. Ruckelshaus, supra,* 486 F.2d at 386). The opinion further states that "it must be stressed that remedial actions do not automatically qualify for the functional equivalent exception to §102(2)(C) of NEPA. Rather, the availability of the exception is contingent upon structuring remedial actions to satisfy the requirements for environmental assessment and public participation underlying the exception. If EPA complies with the procedures for environmental evaluation contained in the NCP and provides for public comment during the decision-making process, a strong argument can be made that the exception is applicable. However, if these precautions are not taken, there is a considerable risk that a court will find remedial actions to be subject to the EIS requirements."

To date, no organization has successfully challenged an EPA-led Superfund cleanup on the basis of insufficient compliance with NEPA. However, it is important to note that federal agencies other than EPA would probably not qualify under the functional equivalence exemption and would, therefore, be required to comply with NEPA.[4]

[4] For further discussion on this issue, see Wagner, Travis and Leigh Benson, "Compliance with NEPA at Federal Superfund Sites," *The Environmental Professional*, vol. 14 (July 1992): 109–116.

11

KEY DEFINITIONS UNDER SUPERFUND

This chapter contains the major regulatory definitions of Superfund terms used throughout Part II. These definitions are from §101 of CERCLA and 40 CFR 300.6 and are marked accordingly. Definitions appear here verbatim as they appear in their respective federal statutes.

Act of God means an unanticipated grave natural disaster or other natural phenomenon of an exceptional, inevitable, and irresistible character, the effects of which could not have been prevented or avoided by the exercise of due care or foresight [CERCLA §101(1)].

Administrator means the Administrator of the United States Environmental Protection Agency [CERCLA §101(2)].

Alternative Water Supplies includes, but is not limited to, drinking water and household water supplies [CERCLA §101(34)].

Applicable Requirements means those Federal requirements that would be legally applicable, whether directly, or as incorporated by a Federally authorized State program, if the response actions were not undertaken pursuant to CERCLA section 104 or 106 (40 CFR 300.6).

Barrel means forty-two United States gallons at sixty degrees Fahrenheit [CERCLA §101(5)].

Claim means a demand in writing for a sum certain [CERCLA §101(4)].

Claimant means any person who presents a claim for compensation under this Act [CERCLA §101(5)].

Coastal Waters, for the purposes of classifying the size of discharges, means the waters of the coastal zone except for the Great Lakes and specified ports and harbors on inland rivers (40 CFR 300.6).

Contractual Relationship means:

(A) for the purpose of Section 107(b)(3), includes, but is not limited to, land contracts, deeds, or other instruments transferring title or possession, unless the real property on which the facility concerned is located was acquired by the defendant after the disposal or placement of the hazardous substances on, in, or at the facility, and one or more of the circumstances described in clause (i), (ii), or (iii) is also established by the defendant by a preponderance of the evidence.

- (i) At the time the defendant acquired the facility the defendant did not know and had no reason to know that any hazardous substance which is the subject of the release or threatened release was disposed of on, in, or at the facility.
- (ii) The defendant is a government entity which acquired the facility by escheat, or through any other involuntary transfer or acquisition, or through the exercise of eminent domain authority by purchase or condemnation.
- (iii) The defendant acquired the facility by inheritance or bequest. In addition to establishing the foregoing, the defendant must establish that he or she has satisfied the requirements of section 107(b)(3)(a) and (b).

(B) To establish that the defendant had no reason to know, as provided in clause (i) of subparagraph (A) of this paragraph, the defendant must have undertaken at the time of acquisition all appropriate inquiry into the previous ownership and uses of the property consistent with good commercial and customary practice in an effort to minimize liability. For purpose of the preceding sentence the court shall take into account any specialized knowledge or experience on the part of the defendant, the relationship of the purchase price to the value of the property if uncontaminated, commonly known or reasonably ascertainable information about the property, the obviousness of the presence or likely presence of contamination at the property, and the ability to detect such contamination by appropriate inspection.

(C) Nothing in this paragraph or in Section 107(b)(3) shall diminish the liability of any previous owner or operator of such facility who would otherwise be liable under this Act. Notwithstanding this paragraph, if the defendant obtained actual knowledge of the release or threatened release of a hazardous substance at such facility when the defendant owned the real property and then subsequently transferred ownership of the property to another person without disclosing such knowledge, such defendant shall be liable under section 107(a)(1) and no defense under section 107(b)(3) shall be available to such defendant. (D) Nothing in this paragraph shall affect the liability under this Act of a defendant who, by any act or omission, caused or contributed to the release or threatened release of a hazardous substance which is the subject of the action relating to the facility [CERCLA §101(35)].

Damages means damages for injury or loss of natural resources as set forth in Section 107(a) or 111(b) of this Act [CERCLA §101(6)].

Drinking Water Supply means any raw or finished water source that is or may be used by a public water system (as defined in the Safe Drinking Water Act) or as drinking water by one or more individuals [CERCLA §101(7)].

Environment means (A) the navigable waters, the waters of the contiguous zone, and the ocean waters of which the natural resources are under the exclusive management authority of the United States under the Fishery Conservation and Management Act of 1976, and (B) any other surface water, ground water, drinking water supply, land surface or subsurface strata, or ambient air within the United States or under the jurisdiction of the United States [CERCLA §101(8)].

Facility means (A) any building, structure, installation, equipment, pipe or pipeline (including any pipe into a sewer or publicly owned treatment works), well, pit, pond, lagoon, impoundment, ditch, landfill, storage container, motor vehicle, rolling stock, or aircraft, or (B) any site or area where a hazardous substance has been deposited, stored, disposed of, or placed, or otherwise come to be located; but does not include any consumer products in consumer use or any vessel [CERCLA §101(9)].

Feasibility Study is a process undertaken by the lead agency (or responsible party if the responsible party will be developing a cleanup proposal) for developing, evaluating, and selecting remedial actions which emphasizes remedial data analysis. The feasibility study is generally performed concurrently and in an interdependent fashion with the remedial investigation. In certain situations, the lead agency may require potentially responsible parties to conclude initial phases of the remedial investigation prior to initiation of the feasibility study. The feasibility study process uses data gathered during the remedial investigation. These data are used to define the objectives of the response action and to broadly develop remedial action alternatives. Next, an initial screening of these alternatives is required to reduce the number of alternatives to a workable number. Finally, the feasibility study involves a detailed analysis of a limited number of alternatives which remain after the initial screening stage. The factors that are considered in screening and analyzing the alternatives are public health, economics, engineering practicality, environmental impacts, and institutional issues (40 CFR 300.6).

Federally Permitted Release means:

(A) discharges in compliance with a permit under section 402 of the Federal Water Pollution Control Act.

(B) discharges resulting from circumstances identified and reviewed and made part of the public record with respect to a permit issued or modified under section 402 of the Federal Water Pollution Control Act and subject to a condition of such permit.

(C) continuous or anticipated intermittent discharges from a point source, identified in a permit or permit application under section 402 of the Federal Water Pollution Control Act, which are caused by events occurring within the scope of relevant operating or treatment systems.

(D) discharges in compliance with a legally enforceable permit under section 404 of the Federal Water Pollution Control Act.

(E) releases in compliance with a legally enforceable final permit issued pursuant to section 3005 (a) through (d) of the Solid Waste Disposal Act from a hazardous waste treatment, storage, or disposal facility when such permit specifically identifies the hazardous substances and makes such substances subject to a standard of practice, control procedure or bioassay limitation or condition, or other control on the hazardous substances in such releases.

(F) any release in compliance with a legally enforceable permit issued under section 102 or section 103 of the Marine Protection, Research, and Sanctuaries Act of 1972.

(G) any injection of fluids authorized under Federal underground injection control programs or State programs submitted for Federal approval (and not disapproved by the Administrator of the Environmental Protection Agency) pursuant to part C of the Safe Drinking Water Act.

(H) any emission into the air subject to a permit or control regulation under section 111, section 112, title I part C, title I part D, or State implementation plans submitted in accordance with section 110 of the Clean Air Act (and not disapproved by the Administrator of the Environmental Protection Agency), including any schedule or waiver granted, promulgated, or approved under these sections.

(I) any injection of fluids or other materials authorized under applicable State law (i) for the purpose of stimulating or treating wells for the production of crude oil, natural gas, or water, (ii) for the purpose of secondary, tertiary, or other enhanced recovery of crude oil or natural gas, or (iii) which are brought to the surface in conjunction with the production of crude oil or natural gas and which are reinjected.

(J) the introduction of any pollutant into a publicly owned treatment works when such pollutant is specified in and in compliance with applicable pretreatment standards of section 307(b) or (c) of the Clean Water Act and enforceable requirements in a pretreatment program submitted by a State or municipality for Federal approval under section 402 of such Act, and

(K) any release of source, special nuclear, or by-product material, as those terms are defined in the Atomic Energy Act of 1954, in compliance with a legally enforceable license, permit, regulation, or order issued pursuant to the Atomic Energy Act of 1954 [CERCLA §101(10)].

Fund or *Trust Fund* means the Hazardous Substance Response Trust Fund established by section 221 of CERCLA (40 CFR 300.6).

Ground Water means water in a saturated zone or stratum beneath the surface of land or water [CERCLA §101(11)].

Guarantor means any person, other than the owner or operator, who provides evidence of financial responsibility for an owner or operator under this Act [CERCLA §101(13)].

Hazardous Substance means:

(A) any substance designated pursuant to section 311(b)(2)(A) of the Federal Water Pollution Control Act.

(B) any element, compound, mixture, solution, or substance designated pursuant to section 102 of this Act.

(C) any hazardous waste having the characteristics identified under or listed pursuant to section 3001 of the Solid Waste Disposal Act (but not including any waste the regulation of which under the Solid Waste Disposal Act has been suspended by Act of Congress).

(D) any toxic pollutant listed under section 308(a) of the Federal Water Pollution Control Act.

(E) any hazardous air pollutant listed under section 112 of the Clean Air Act, and

(F) any imminently hazardous chemical substance or mixture with respect to which the Administrator has taken action pursuant to section 7 of the Toxic Substances Control Act.

The term does not include petroleum, including crude oil or any fraction thereof which is not otherwise specifically listed or designated as a hazardous substance under subparagraphs (A) through (F) of this paragraph, and the term does not include natural gas, natural gas liquids, liquefied natural gas, or synthetic gas usable for fuel (or mixtures of natural gas and such synthetic gas) [CERCLA §101(14)].

Indian Tribe means any Indian tribe, band, nation, or other organized group of community, including any Alaska Native village but not including any Alaska Native regional or village corporation, which is recognized as eligible for the special programs and services provided by the United States to Indians because of their status as Indians [CERCLA §101(36)].

Lead Agency means the Federal agency (or State agency operating pursuant to a contract or cooperative agreement executed pursuant to section 104(d)(1) of CERCLA) that has primary responsibility for coordinating response actions under this Plan. A Federal lead agency is the agency that provides OSC or RPM as specified elsewhere in this Plan. In the case of a State as lead agency, the State shall carry out the same responsibilities delineated for OSCs/RPMs in this Plan (except coordinating and directing Federal agency response actions) (40 CFR 300.6).

Liable or *Liability* under this title shall be construed to be the standard of liability which obtains under section 311 of the Federal Water Pollution Control Act [CERCLA §101(32)].

Management of Migration means actions that are taken to minimize and mitigate the migration of hazardous substances or pollutants or contaminants and the effects if such migration. Management of migration actions may be appropriate where the hazardous substances or pollutants or contaminants are no longer at or near the area where they were originally located or situations where a source cannot be adequately identified or characterized. Measures may include, but are not limited to, provision of alternative water supplies, management of a plume of contamination, or treatment of a drinking water aquifer (40 CFR 300.6).

National Contingency Plan means the national contingency plan published under section 311(c) of the Federal Water Pollution Control Act or revised pursuant to section 105 of this Act [CERCLA §101(31)].

Natural Resources means land, fish, wildlife, biota, air, water, ground water, drinking water supplies, and other such resources belonging to, managed by, held in trust by, appertaining to, or otherwise controlled by the United States (including the resources of the fishery conservation zone established by the Fishery Conservation and Management Act of 1976), any State or local government, or any foreign government [CERCLA §101(16)].

Navigable Waters or Navigable Waters of the United States means the waters of the United States, including the territorial seas [CERCLA §101(15)].

Offshore Facility means any facility of any kind located in, on, or under any of the navigable waters of the United States, and any facility of any kind which is subject to the jurisdiction of the United States and is located in, on, or under any other waters, other than a vessel or a public vessel [CERCLA §101(17)].

On-Scene Coordinator (OSC) means the federal official predesignated by the EPA or USCG to coordinate and direct Federal responses under Subpart E and removals under Subpart F of this Plan; or the DOD official designated to coordinate and direct the removal actions from releases of hazardous substances, pollutants, or contaminants from DOD vessels and facilities (40 CFR 300.6).

Onshore Facility means any facility (including, but not limited to, motor vehicles and rolling stock) of any kind located in, on, or under, any land or nonnavigable waters within the United States [CERCLA §101(18)].

Operable Unit is a discrete part of the entire response action that decreases a release, threat of release, or pathway of exposure (40 CFR 300.6).

Otherwise Subject to the Jurisdiction of the United States means subject to the jurisdiction of the United States by virtue of United States citizenship. United States vessel documentation or numbering, or as provided by international agreement to which the United States is a party [CERCLA §101(19)].

Owner or Operator means:

(A) (i) In the case of a vessel, any person owning, operating, or chartering by demise, such vessel,

(ii) in the case of an onshore facility or an offshore facility, any person owning or operating such facility, and

(iii) in the case of any facility, title, or control of which was conveyed due to bankruptcy, foreclosure, tax delinquency, abandonment, or similar means to a unit of State or local government, any person who owned, operated, or otherwise controlled activities at such facility immediately beforehand.

(B) In the case of a hazardous substance which has been accepted for transportation by a common or contract carrier and except as provided in section 107(a)(3) or (4) of this Act, (i) the term owner or operator shall mean such common carrier or other bona fide for hire carrier acting as an independent contractor during such transportation, (ii) the shipper of such hazardous substances shall not be considered to have caused or contributed to any release during such transportation which resulted solely from circumstances or conditions beyond his control.

(C) In the case of a hazardous substance which has been delivered by a common or contract carrier to a disposal or treatment facility and except as provided in section 107(a)(3) or (4)(i) the term owner or operator shall not include such common or contract carrier, and (ii) such common or contract carrier shall not be considered to have caused or contributed to any release at such disposal or treatment facility resulting from circumstances or conditions beyond his control.

(D) The term owner or operator does not include a unit of State of local government which acquired ownership or control involuntarily through bankruptcy, tax delinquency, abandonment, or other circumstances in which the government involuntarily acquires title by virtue of its function as sovereign. The exclusion provided under this paragraph shall not apply to any State or local government which has caused or contributed to the release or threatened release of a hazardous substance from the facility, and such a State or local government shall be subject to the provisions of this Act in the same manner and to the same extent, both procedurally and substantively, as any nongovernmental entity, including liability under section 107 [CERCLA §101(20)].

Person means an individual, firm, corporation, association, partnership, consortium, joint venture, commercial entity, United States Government, State, municipality, commission, political subdivision of a State, or any interstate body [CERCLA §101(21)].

Plan means the National Oil and Hazardous Substances Pollution Contingency Plan published under section 311(c) of the CWA and revised pursuant to section 105 of CERCLA (40 CFR 300.6).

Pollutant or Contaminant, as defined by section 104(a)(2) of CERCLA shall include, but not be limited to, any element, substance, compound, or mixture, including disease causing agents, which after release into the environment and upon exposure, ingestion, inhalation, or assimilation into any organism, either directly from the environment or indirectly by ingesting through food chains, will or may reasonably be anticipated to cause death, disease, behavioral abnormalities, cancer, genetic mutation, physiological malfunctions (including malfunctions in reproduction), or physical deformation in such organisms or their offspring. The term does not include petroleum, including crude oil, and any fraction thereof which is not otherwise specifically listed or designated as a hazardous substance under section 101(4) (A) through (F) of CERCLA, nor does it include natural gas, liquefied natural gas, or synthetic gas of pipeline quality for mixtures of natural gas and synthetic gas). For purposes of Subpart F of this Plan, the term pollutant or contaminant means any pollutant or contaminant which may present an imminent and substantial danger to public health or welfare (40 CFR 300.6).

Release means any spilling, leaking, pumping, pouring, emitting, emptying, discharging, injecting, escaping, leaching, dumping, or disposing into the environment, including the abandonment or discarding of barrels, containers, and other closed receptacles containing any hazardous substance or pollutant or contaminant, but excludes (A) any release which results in exposure to persons solely within a workplace, with respect to a claim which such persons may assert against the employer of such persons, (B) emissions from the engine exhaust of a motor vehicle, rolling stock, aircraft, vessel, or pipeline pumping station engine, (C) release of source, byproduct, or special nuclear material from a nuclear incident, as those terms are defined in the Atomic Energy Act of 1954, if such release is subject to requirements with respect to financial protection established by the Nuclear Regulatory Commission under section 170 of such Act, or for the purposes of section 104 of this title or any other response action, any release of source byproduct, or special nuclear material from any processing site designated under section 102(a)(1) or 302(a) of the Uranium Mill Tailings Radiation Control Act of 1978, and (D) the normal application of fertilizer [CERCLA §101(22)].

Relevant and Appropriate Requirements are those Federal requirements that, while not "applicable," are designed to apply to problems sufficiently similar to those encountered at CERCLA sites that their application is appropriate. Requirements may be relevant and appropriate if they would be "applicable" but for jurisdictional restrictions associated with the requirement (40 CFR 300.6).

Remedial Investigation is a process undertaken by the lead agency (or responsible party if the responsible party will be developing a cleanup proposal) which emphasizes data collection and site characteristics. The remedial investigation is generally performed concurrently and in an interdependent fashion with the feasibility study. However, in certain situations, the lead agency may require potentially responsible parties to conclude initial phases of the remedial investigation prior to initiation of the feasibility study. A remedial investigation is undertaken to determine the nature and extent of the problem presented by the release. This includes sampling and monitoring, as necessary, and includes the gathering of sufficient information to determine the necessity for and proposed extent of remedial action.

Part of the remedial investigation involves assessing whether the threat can be mitigated or minimized by controlling the source of the contamination at or near the area where the hazardous substances or pollutants or contaminants were originally located (source control remedial actions) or whether additional actions will be necessary because the hazardous substances or pollutants or contaminants have migrated from the area of their original location (management of migration) (40 CFR 300.6).

Remedial Project Manager (RPM) means the Federal official designated by EPA (or the USCG for vessels) to coordinate, monitor, or direct remedial or other response activities under Subpart F of this Plan; or the Federal official DOD designates to coordinate and direct Federal remedial or other response actions resulting from releases of hazardous substances, pollutants, or contaminates from DOD facilities or vessels (40 CFR 300.6).

Remedy or *Remedial Action* means those actions consistent with permanent remedy taken instead of or in addition to removal actions in the event of a release or threatened release of a hazardous substance into the environment, to prevent or minimize the release of hazardous substances so that they do not migrate to cause substantial danger to present or future public health or welfare or the environment. The term includes, but is not limited to, such actions at the location of the release as storage; confinement; perimeter protection using dikes, trenches, or ditches; clay cover; neutralization; cleanup of released hazardous substances or contaminated materials; recycling or reuse; diversion; destruction; segregation of reactive wastes; dredging or excavations; repair or replacement of leaking containers; collection of leachate and runoff; on-site treatment or incineration; provision of alternative water supplies; and any monitoring reasonably required to assure that such actions protect the pubic health and welfare and the environment. The term includes the costs of permanent relocation of residents and businesses and community facilities where the President determines that, alone or in combination with other measures, such relocation is more cost-effective than and environmentally preferable to the transportation, storage, treatment, destruction, or secure disposition off-site of hazardous substances, or may otherwise be necessary to protect the public health or welfare; the term includes off-site transport and off-site storage, treatment, destruction, or secure disposition off-site of hazardous substances and associated contaminated materials [CERCLA §101(24)].

Remove or *Removal* means the cleanup or removal of released hazardous substances from the environment, such actions as may be necessary taken in the event of the threat of release of hazardous substances into the environment, such actions as may be necessary to monitor, assess, and evaluate the release or threat of release of hazardous substances, the disposal of removed material, or the taking of such other actions as may be necessary to prevent, minimize, or mitigate damage to the public health or welfare or to the environment, which may otherwise result from a release or threat of release. The term includes, in addition, without being limited to, security fencing or other measures to limit access, provision of alternative water supplies, temporary evacuation and housing of threatened individuals not otherwise provided for, action taken under section 104(b) of this Act, and any emergency assistance which may be provided under the Disaster Relief Act of 1974 [CERCLA §101(23)].

Respond or *Response* means remove, removal, remedy, and remedial action. All such terms (including the term "removal" and "remedial action") include enforcement activities related thereto [CERCLA §101(25)].

Source Control Action is the construction or installation and start-up of those actions necessary to prevent the continued release of hazardous substances or pollutants or contaminants (primarily from a source on top of or within the ground, or in buildings or other structures) into the environment (40 CFR 300.5).

Source Control Maintenance Measures are those measures intended to maintain the effectiveness of source control actions once such actions are operating and functioning properly such as the maintenance of landfill caps and leachate collection systems (40 CFR 300.5).

Territorial Sea and *Contiguous Zone* shall have the meaning provided in section 502 of the Federal Water Pollution Control Act [CERCLA §101(30)].

Transport or *Transportation* means the movement of a hazardous substance by any mode, including pipeline (as defined in the Pipeline Safety Act), and in the case of a hazardous substance which has been accepted for transportation by a common or contract carrier, the term transport or transportation shall include any stoppage in transit which is temporary, incidental to the transportation movement, and at the ordinary operating convenience of a common or contract carrier, and any such stoppage shall be considered as a continuity of movement and not as the storage of a hazardous substance [CERCLA §101(26)].

United States and *State* include the several States of the United States, the District of Columbia, the Commonwealth of Puerto Rico, Guam, American Samoa, the United States Virgin Islands, the Commonwealth of the Northern Marianas, and any other territory or possession over which the United States has jurisdiction [CERCLA §101(27)].

Vessel means every description of watercraft or other artificial contrivance used, or capable of being used, as a means of transportation on water [CERCLA §101(28)].

12

RELEASE REPORTING REQUIREMENTS

This chapter describes three different types of federal release reporting requirements: spill reporting, notification of previous hazardous waste management, and reporting of releases to the National Toxic Release Inventory.

SPILL REPORTING

Because there are five major federal laws addressing spill reporting, the national spill reporting program has overlap. Each of the five major laws covering spill reporting addresses different substances, modes, or releases. The five federal laws dealing with spill reporting are as follows:

- Comprehensive Environmental Response, Compensation, and Liability Act (CERCLA, or Superfund)
- Emergency Planning and Community Right-to-Know Act (EPCRA)
- Toxic Substances Control Act (TSCA)
- Hazardous Materials Transportation Act (HMTA)
- Clean Water Act (CWA)

The Clean Water Act was the original statute dealing with emergency notification of releases; however, it was limited to releases to water. The enactment of Superfund expanded the reporting requirements to releases into the environment, which includes land, water, and air. EPCRA, which was enacted as a direct result of the

1984 disaster in Bhopal, India, was designed to address releases presenting an immediate, acute hazard. The Toxic Substances Control Act covers only releases of PCBs. HMTA addresses releases that may cause a threat to public health during transportation-related incidents.

The regulatory requirements under these statutes overlap, but Superfund is considered the focal point for the emergency release notification regulations. Hence, the requirements under each of these statutes have been slowly modified and tied into Superfund, thus alleviating conflicts and overlapping requirements. It is anticipated that this process will continue. Table 12-1 presents a basic overview of the five major spill reporting laws.

Superfund

Superfund has established broad federal authority to deal with releases or threats of releases of hazardous substances from vessels and facilities. Superfund requires the reporting of a release of a hazardous substance into the environment at or above the designated reportable quantity. *Hazardous substances* are basically a compilation of substances regulated under other specified federal environmental statutes (40 CFR 300.5). The referenced substances include the following:

- Hazardous air pollutants under §112 of the Clean Air Act
- Hazardous wastes under RCRA
- Toxic pollutants (priority pollutants) under §307(a) of the Clean Water Act
- Hazardous substances under §311(b)(2)(A) of the Clean Water Act
- Substances designated as an imminent hazard under §7 of TSCA. (To date, there have been no such designations.)

It is important to note that petroleum is specifically excluded from the definition of hazardous substance under Superfund, but must be reported if it is released into waters of the United States (as defined) under the Clean Water Act, which is discussed in its own section of this chapter. However, releases of oil onto most lands or into the air are not necessarily subject to reporting under federal law.

Reportable Quantities. There are more than 700 listed hazardous substances. These substances and their designated reportable quantities (RQs) are listed in 40 CFR Part 302. Each newly listed hazardous substance automatically receives an RQ of 1 pound until it is superseded by regulation establishing a different RQ. EPA adjusts RQs based on a substance's intrinsic physical, chemical, and toxicological properties. The intrinsic properties include aquatic toxicity, mammalian toxicity (oral, dermal, and inhalation), ignitability, reactivity, chronic toxicity, and potential carcinogenicity. EPA ranks each intrinsic property (except for carcinogenicity potential) on a five-tier scale that has an RQ level for each tier. The five-tier scale uses RQ levels (1; 10; 100; 1,000; and 5,000 pounds), previously established by the Clean Water Act. Each substance is also evaluated and ranked for its carcinogenicity

Table 12-1. Overview of Federal Spill Reporting Laws

Law	Materials	Mode	Contact	Comments
Superfund	Hazardous substances	Into the environment	National Response Center	Encompasses all substances of other laws except for petroleum and those extremely hazardous substances not specifically listed as a Superfund hazardous substance.
HMTA	Hazardous substances	During transportation	National Response Center	Limited to a per-container basis, but Superfund may apply to multiple containers of less than the HMTA RQ.
CWA	Hazardous substances and petroleum	Into navigable waters	National Response Center	Essentially the same as Superfund except for petroleum, which Superfund excludes.
TSCA	PCBs	(1) Into surface water, sewers, grazing land, or vegetable gardens (2) As a result of a fire-related incident	National Response Center	This provision addresses PCBs at or above 50 ppm. If a spill also contains 1 lb. or more of PCBs, Superfund applies. For fire-related incidents, any amount requires notification.
EPCRA	Extremely hazardous substances and Superfund hazardous substances	Into the environment	State Emergency Response Commission, Local Emergency Planning Committee	Some extremely hazardous substances are also Superfund hazardous substances. Releases confined to within a facility's property boundary are excluded.

potential. A low, medium, or high carcinogenicity potential receives a corresponding RQ level of 1, 10, or 100 pounds. At the end of this ranking process, a substance is assigned the lowest RQ from the evaluation, provided that the RQ is protective of human health and the environment.

In addition to all listed hazardous substances, §102(a) authorizes EPA to designate a reportable quantity for any nondesignated hazardous substance that may present substantial danger when released into the environment.

Reporting Requirements. Any person in charge of a facility or vessel that has a release of a reportable quantity must report that release immediately to the National Response Center (NRC) [40 CFR 302.6(a)], which is a 24-hour hotline (800-424-8802) operated by the U.S. Coast Guard. When a call is received, the duty officer requests information, including the identity, location, and nature of the release; the identity of the transporter or owner of the facility or vessel; the nature of any injuries or property damage; and other relevant information. The National Response Center relays the release information directly to either the predesignated on-scene coordinator (OSC) at the appropriate EPA Regional Office, or to the appropriate U.S. Coast Guard District Office. This notification is a trigger for informing the government of a release so that appropriate federal personnel can evaluate the need for a federal removal action and, if warranted, pursue any action in a timely fashion. Federal personnel evaluate all reported releases, but they will not necessarily initiate a removal or remedial action in response to all reported releases because a release of a hazardous substance does not necessarily pose a hazard to public health or the environment. Many considerations other than the quantity released affect the government's decision concerning whether and how it should undertake a response to a particular release. The location of the release, its proximity to drinking-water supplies, the likelihood of exposure to nearby populations, response actions taken by responsible parties, and other factors must be assessed by the federal OSC on a case-by-case basis.

It is important to note that, unless specifically exempted from Superfund, a party responsible for a release of a hazardous substance is liable for the costs of cleaning up that release and for any natural resource damages caused by the release, even if the release is not subject to notification requirements of 40 CFR Part 302. The fact that a release of a hazardous substance is properly reported, or that it is not subject to the notification requirements of 40 CFR Part 302, does not preclude EPA or other appropriate government agencies from undertaking response actions, seeking reimbursement from responsible parties, or pursuing an enforcement action against responsible parties. It is the position of EPA that Superfund clearly establishes liability whenever there is a release of a hazardous substance regardless of the quantity. Thus, liability applies to releases of hazardous substances below reportable quantities.[1]

[1] EPA memorandum from Robert M. Perry, Associate Administrator and General Counsel to all regional counsel, "Liability under the Comprehensive Environmental Response, Compensation, and Liability Act of 1980 for Releases of Hazardous Substances in Amounts Less than Reportable Quantities," December 15, 1982.

Note: It is strongly recommended that the National Response Center be contacted concerning a release without necessarily determining if it meets an RQ because such determination can be very difficult and time-consuming, especially during an emergency. It is important to note that no facility has ever been cited for reporting a release that did not meet an RQ, whereas there have been numerous citations for failure to report a release that did meet an RQ.

Determination Period. An RQ is determined within a 24-hour period [40 CFR 302.6(a)]. Thus, if a substance with an RQ of 10 pounds were releasing 1 pound per hour, it would have to be reported as soon as it met the RQ because the spill would release an RQ within the 24-hour time period. If the release rate had been 0.25 pounds per hour, the release would not have to be reported because the spill would not equal the reportable quantity of 10 pounds within 24 hours. This does not mean that if there is a reportable spill, the notification can be delayed until the 24-hour time period elapses. The spill must be reported *immediately* upon meeting the reportable quantity.

Into-the-Environment Requirement. For a release to be reportable; it must be *into the environment* which "means the navigable waters, the waters of the contiguous zone, and the ocean waters . . . any other surface water, ground water, drinking water supply, land surface or subsurface strata, or ambient air . . ." (40 CFR 300.5). A spill in a contained building that met an RQ would not require reporting. However, if that same spill were then vented to the outside (or through a drain), and the amount of vented material met the designated reportable quantity, it would be reportable (50 *FR* 13462, April 4, 1985). A facility's boundaries are irrelevant for reporting requirements; the requirements depend on whether the release is into the environment.

Mixture Rule. EPA applies the *mixture rule,* previously developed under §311 of the Clean Water Act, to releases and solutions containing hazardous substances. This rule provides that releases of mixtures and solutions are subject to reporting only when a component hazardous substance is released in a quantity equal to or greater than its reportable quantity [40 CFR 302.6(b)]. For example, toxaphene has an RQ of 1 pound. If there is a formulation with 10 percent toxaphene and 90 percent inert ingredient, 10 pounds of the formulated pesticide would have to be released to constitute a 1-pound reportable toxaphene release. RQs of different substances are not additive under the mixture rule. Thus, releasing a mixture containing half an RQ of one substance and half an RQ of another substance does not require reporting. If the exact proportions are unknown, 1 pound of the formulation is the appropriate RQ.

Federally Permitted Releases. Reporting of a federally permitted release is not required under Superfund. Significant confusion surrounds this exemption, which stems from a reliance on vague statutory language. Currently, the only legal language addressing this provision is contained in the law itself [§101(10)]. Federally permitted releases include the following:

- Discharges in compliance with a permit under §402 of CWA.
- Continuous or anticipated intermittent discharges from a point source, identified in a permit or permit application under §402 of the CWA, which are caused by events relevant to operating or treatment systems.
- Discharges in compliance with a legally enforceable permit under §404 of the CWA.
- Releases in compliance with a legally enforceable hazardous waste permit under RCRA when such hazardous substance is specifically identified in the permit.
- Any emission into the air subject to a permit or control or approved State Implementation Plan.
- The introduction of any pollutant into a publicly owned treatment works (POTW), provided the pollutant is in compliance with the pretreatment standards.

To qualify for the reporting exclusion, the release must be in compliance with the facility's enforceable permit or conditions as specified in this list. What exactly constitutes "compliance" and with what, specifically, remain unresolved. EPA has twice proposed clarifications (48 *FR* 23552, May 25, 1983, and July 19, 1988, 53 *FR* 27268) in an attempt to define more precisely federally permitted release but no final EPA policy has been promulgated.

Continuous Releases. Superfund provides some relief from the reporting requirements for a release of a hazardous substance that is continuous and stable in quantity and rate if certain notifications have been given; that is, this provision allows for reduced reporting, but it is not an exemption. Per 40 CFR 302.8(b), a *continuous release* is a release that occurs without interruption or abatement or that is routine, anticipated, and intermittent and incidental to normal operations or treatment processes. *Stable in quantity and rate* is defined as a release that is predictable and regular in amount and rate of emission [40 CFR 302.8(b)]. The required notifications are as follows [40 CFR 302.8(c)]:

- Initial immediate telephone notification to the NRC.
- Initial written notification within 30 days of the initial telephone notification.
- Follow-up written notification within 30 days of the first anniversary date of the initial written notification. [The SARA Title III Form R may be used instead if specific information is contained in the report per 40 CFR 302.8(j).]
- Notification to NRC of a change in composition, source, or other information in the previous written notification.
- Notification to the NRC any time a statistically significant increase in the hazardous substance above the "established" reporting level occurs during a 24-hour period.

Public Notice. Any owner or operator who has a release of a hazardous substance must notify the potentially affected parties by publication in the local newspaper. This is a statutory requirement [§111(g)] of CERCLA. EPA has not provided any further clarification on this provision.

Penalties. Section 103(b) authorizes penalties, including criminal sanctions, for persons in charge of vessels or facilities who fail to report releases of hazardous substances that equal or exceed an RQ. Section 109 of SARA amended Superfund by increasing the maximum penalties and years of imprisonment. Any person having knowledge of a reportable release who fails to report the release immediately, or who knowingly submits any false or misleading information, shall upon conviction be subject to fines of not more than $250,000 for individuals (or $500,000 for an organization) and/or imprisonment for not more than three years.

EPCRA Spill Reporting

In 1986, Congress passed the Emergency Planning and Community Right-to-Know Act (EPCRA) as Title III of the Superfund Amendments and Reauthorization Act (SARA). It is commonly referred to as Title III because it appears as the third part of that other law, SARA. Even though Title III is part of Superfund, it is also a free-standing act.[2]

A release of an *extremely hazardous substance* (40 CFR Part 355, Appendix A) at or above its reportable quantity must be reported immediately to the State Emergency Response Commission (SERC) and the community emergency coordinator of the Local Emergency Planning Committee (LEPC).[3] If there is no LEPC, notification to a local emergency official will suffice (e.g., police, hazmat team, fire department). The information required to be reported for such releases, to the extent practicable, includes the following:

- Name and location of the facility
- Name of the chemical
- Whether the chemical is defined as an extremely hazardous chemical
- Estimated quantity released
- Time and duration of release
- Media (e.g., air, water, soil) in which release occurred
- Known or anticipated health risks resulting from the release
- Precautions that should be taken as a result of the release

[2] Facilities managing hazardous materials and/or wastes may have additional reporting and record-keeping requirements under EPCRA that are not addressed in this book, including emergency planning and hazardous material storage.

[3] It is important to note that 138 designated *extremely hazardous substances* are also listed *Superfund hazardous substances* (see Figure 12-1) that would require notification to the National Response Center as well if a reportable quantity was released.

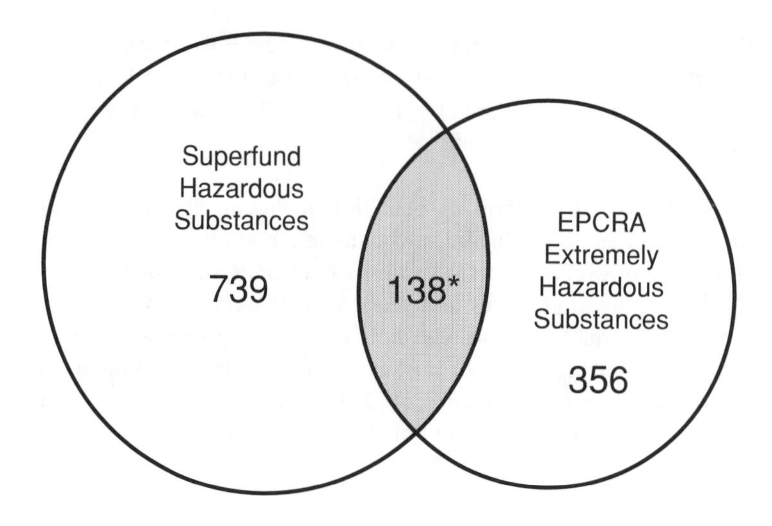

Figure 12-1. Universe of Extremely Hazardous Substances and Superfund Hazardous Substances

- Names and telephone numbers of points of contacts

However, a release of an extremely hazardous substance that results in exposure to persons solely within the boundaries of the facility is not reportable under EPCRA. This differs from Superfund, which requires notification (to federal authorities) of releases of *hazardous substances*, regardless of the facility's property boundary. A release of a Superfund hazardous substance (40 CFR 302.4) at or above the reportable quantity must be reported to the National Response Center [40 CFR 355.40(a)]. It is important to note that some of the Superfund hazardous substances also are SARA Title III extremely hazardous substances. As a result, a facility may have a multiple reporting requirement to local, state, and federal governments.

Follow-Up Notification. The facility must, as soon as practicable, submit a written statement providing details of the incident for any notification [40 CFR 355.40(b)(3)]. The report must contain information on actions taken to respond, the contents of the release, and any known or anticipated health risks as a result of the release.

TSCA/PCB Spill Reporting

Any spill of a material containing PCBs with a concentration of 50 parts per million (ppm) or greater into surface waters, drinking-water supplies, sewers, grazing land, or vegetable gardens must be reported immediately to the designated EPA Regional PCB Coordinator. The report must be made immediately after discovery, but in no case later than 24 hours [40 CFR 761.125(a)(1)(i)]. In addition, any spill of a material containing PCBs of 50 ppm that is released to a location not addressed in the previous sentence and that equals 10 pounds of PCBs must be reported to the designated EPA Regional PCB Coordinator.

It is important to note that regardless of the PCB spill requirements under TSCA, a release of 10 pounds or more of "pure" PCBs into the environment must be reported to the National Response Center. (*Pure* meaning that a release of PCB-containing material at a concentration of 500 ppm would require a release of 200 pounds, or about 20 gallons.) This is required because PCBs are a Superfund-listed hazardous substance. (If the concentration is higher, the less is required to meet the reporting level.)

In addition to these PCB spill notification requirements, any fire-related incident involving a PCB transformer (i.e., any transformer with a PCB concentration of 500 ppm or greater) must be reported to the NRC immediately [40 CFR 761.30(a)(1)(xi)]. A *fire-related incident* is defined as any incident involving a PCB transformer that entails the generation of sufficient heat and/or pressure (by any source) to result in the rupture (violent or nonviolent break) of a transformer and the release of PCBs.

HMTA Spill Reporting

Under the Hazardous Materials Transportation Act (HMTA), as amended, a release of a designated *hazardous substance* is subject to reporting. For a substance to be designated as a *hazardous substance* under HMTA (49 CFR 171.8), it must meet the following requirements:

- It is listed as a hazardous substance in the Appendix to 49 CFR 172.101.
- It is in a quantity, in *one* package, that equals or exceeds the reportable quantity (RQ) listed in the Appendix to 49 CFR 172.101. (Thus, its determination is based on a per-package basis.)
- If the substance is in a mixture or solution, its concentration (by weight) must equal or exceed the concentration corresponding to the RQ identified for the substance in the Appendix to 49 CFR 172.101.

If a transporter is involved in an incident where a release of an RQ of a hazardous substance occurs, the transporter must notify the National Response Center immediately. If a transporter has multiple containers of a "hazardous substance" but no single container with an RQ, it is possible that a collective release from multiple containers could meet or exceed the Superfund reportable quantity and thus would have to be reported.

CWA Spill Reporting

The Clean Water Act (CWA) was the first statute addressing the reporting of releases of substances that may present a threat to the environment. The act addressed the release of designated *hazardous substances* and *oil* into or upon the navigable waters of the United States [40 CFR 117.11(a)]. (These hazardous substances, listed in Table 117.3 of 40 CFR 117.3, also are Superfund listed hazardous substances [40 CFR 302.4)]. *Navigable waters* means water of the United States, including the territorial seas. This term includes all of the following:

- Waters that are currently used, were used in the past, or may be susceptible to use in interstate or foreign commerce, including all waters subject to the ebb and flow of the tide.
- Interstate waters, including interstate wetlands.
- All other waters, such as intrastate lakes, rivers, streams (including intermittent streams), mud flats, sand flats, and wetlands, the use, degradation, or destruction of which would affect or could affect interstate or foreign commerce, including any such waters
  - That are or could be used by interstate or foreign travelers for recreation or other purposes.
  - From which fish or shellfish are or could be taken and sold in interstate or foreign commerce.
- All impoundments of waters otherwise defined as navigable waters.
- Tributaries of waters identified in the above example, including adjacent wetlands.
- Wetlands adjacent to the above-identified waters. *Wetlands* are those areas that are inundated or saturated by surface water or groundwater at a frequency and duration sufficient to support, and that under normal conditions do support, a prevalence of vegetation typically adapted for life in saturated soil conditions. Wetlands generally include swamps, marshes, bogs, and similar areas such as sloughs, prairie potholes, wet meadows, prairie river overflows, mud flats, and natural ponds.

Petroleum. Whereas Superfund specifically excludes petroleum from the reporting requirements (40 CFR 300.5), CWA specifically requires the reporting of certain oil releases. *Oil* means oil of any kind or in any form, including but not limited to petroleum, fuel oil, sludge, oil refuse, and oil mixed with wastes other than dredged spoil [40 CFR 101.1(a)].

Under 40 CFR part 110, EPA has established requirements for certain discharges of oil. Oil discharges, into or upon U.S. waters, that must be reported to the NRC include those that

- Cause a sheen to appear on the surface of the water;
- Violate applicable water quality standards; or
- Cause a sludge or emulsion to be deposited beneath the surface of the water or upon the adjoining shorelines.

NOTIFICATION OF PREVIOUS HAZARDOUS WASTE MANAGEMENT

Whereas §103(a) of Superfund addresses the reporting of current releases, §103(c) addresses past releases, including the discovery of contamination at an existing facility.

Section 103(c) requires any person who owns or operates—or who at the time of disposal owned or operated—a facility at which RCRA hazardous wastes are or have been stored, treated, or disposed of, to notify EPA of the facility's existence. This requirement is for RCRA hazardous wastes only; it does not include other hazardous substances. This requirement is applicable regardless of when the activity occurred. There are, however, nine specific exclusions from reporting under §103(c) as follows:

1. A generator accumulating hazardous wastes for less than 90 days in compliance with 40 CFR 262.34.
2. Totally enclosed treatment facilities as defined in 40 CFR 260.10.
3. Wastewater treatment tanks as defined in 40 CFR 260.10.
4. Farmers disposing of waste pesticides from their own use.
5. Facilities that disposed of less than 55 gallons on-site.
6. Incidental spillage and leakage (*de minimis* losses).
7. Inactive storage facilities that no longer store wastes.
8. Persons filing or who have filed a RCRA §3010 notification as a hazardous waste management facility.
9. Any person who has filed a Part A permit application under RCRA.

Note: Exclusions 8 and 9 are not applicable if a contaminated area is found that was not identified on the forms.

The statutory language appears to indicate that after June 9, 1981, a facility would no longer be subject to this provision. However, EPA's Office of General Counsel has consistently provided the RCRA/Superfund hotline with the interpretation that §103(c) is an ongoing reporting requirement.

In summary, the presence of hazardous wastes at a facility means that unless the facility is specifically exempted, a notification is required. Thus, it is necessary to determine (1) if the material is a RCRA hazardous waste; (2) if the facility has previously made a notification; and (3) if the facility meets any of the exclusions. If the facility must notify EPA, notification must be done as soon as possible using the prescribed form contained in 46 *FR* 22155, April 15, 1981.

Information received pursuant to §103(c) notifications is placed in the Comprehensive Environmental Response, Compensation, and Liability Information System (CERCLIS). CERCLIS is a database with an inventory of potential Superfund sites that currently numbers about 16,000.

REPORTING TO THE TOXIC RELEASE INVENTORY

Section 313 of SARA Title III (also known as the Emergency Planning and Community Right-to-Know Act), requires certain facilities that manufacture, process, or otherwise use the more than 650 listed toxic chemicals to report certain information about such chemicals at the facilities, including the annual quantities of the listed chemicals entering the environment. Subject facilities also must report source reduction and recycling data for the listed chemicals pursuant to §6607 of the Pollution Prevention Act.

Under §313 of SARA Title III, facilities required to report must meet all of the following three criteria:

1. Have 10 or more full-time equivalent employees (i.e., 20,000 annual worker hours).
2. Are in any of the following Standard Industrial Classification (SIC) codes:

Major Group Codes:

- 10 (metal mining, except 1011 and 1081),
- 12 (coal mining, except 1241), or
- 20 through 39

Industry Codes:

- 4911 (electric services),
- 4931 (electric and other services combined),
- 4939 (limited to coal or combustion facilities generating commercial power),
- 4953 (limited to RCRA regulated facilities),
- 5169 (wholesale nondurable goods—chemicals and allied products not elsewhere specified),
- 5171 (petroleum bulk stations and terminals), or
- 7389 (limited to facilities primarily engaged in solvent recovery on a contract fee basis).

3. *Manufacture* 25,000 lbs., *process* 25,000 lbs., or *otherwise use* 10,000 lbs. of any chemical or chemical category listed in 40 CFR 372.65 (these are known as *threshold quantities*).

Note: It is important to distinguish and define the terms *manufacture*, *process*, and *otherwise use* to determine if reporting is required based on the threshold quantity determination. (Examples of each activity at a RCRA TSDF are listed in Table 12-2):

Manufacture—to produce, prepare, import, or compound a toxic chemical. Manufacture also applies to a toxic chemical that is produced coincidentally during the manufacture, processing, use, or disposal of another chemical or mixture of chemicals, including a toxic chemical that is separated from that chemical or mixture of chemicals as a by-product, and a toxic chemical that remains in that other chemical or mixture of chemicals as an impurity. (40 CFR 372.3)

Process—to prepare a toxic chemical, after its manufacture, for distribution in commerce: (1) in the same form or physical state as, or in a different form or physical state from, that in which it was received by the person so preparing such sub-

Table 12-2. Section 313 Activities and RCRA TSDF* Examples

Activity	Examples
Manufacture	• Creation of a new listed chemical through treatment of another chemical
Process	• Recovery of a listed chemical for distribution in commerce (e.g., solvents, metals)
Otherwise Use	• Recovery of a listed chemical for use on-site (e.g., solvents, metals)
	• Treatment of a listed chemical in waste
	• Disposal of a listed chemical removed from waste
	• Use of a listed chemical to treat another chemical

*TSDF = treatment, storage, and disposal facility

For additional information, see *Emergency Planning and Community Right-To-Know Act*, section 313: *Guidance for RCRA Subtitle C TSD Facilities* and Solvent Recovery Facilities, EPA/745B-97-015.

stance, or (2) as part of an article containing the toxic chemical. Process also applies to the processing of a toxic chemical contained in a mixture or trade name product. (40 CFR 372.3)

Otherwise Use or *Use*—to use a toxic chemical in any way that is not covered by the terms "manufacture" or "process" and includes (1) uses of a toxic chemical, including a toxic chemical contained in a mixture or trade name product and (2) use through treatment for destruction, stabilization (without subsequent distribution in commerce), and disposal if the facility engaged in treatment for destruction, stabilization, or disposal of the toxic chemical receives materials from other facilities for purposes of further waste management activities. Relabeling or redistributing a container of a toxic chemical where no repackaging of the toxic chemical occurs does not constitute use of the toxic chemical. (40 CFR 372.3)

Section 313 provides for reporting exemptions for specific processing or use activities. Thus, a facility does not have to count the excluded amounts toward the threshold quantity determination. The major exemptions include the following:

- *De Minimis Materials.* Listed chemicals in mixtures (but not wastes) are exempt from reporting if present at a concentration below 1 percent in noncarcinogens (as defined by OSHA) or 0.1 percent for carcinogens.
- *Articles.* Listed chemicals present in solids that undergo no shape change and do not release a listed chemical.
- *Structural Components.* Listed chemicals present in materials used to construct, repair, or maintain a plant building.
- *Maintenance Supplies.* Listed chemicals used for routine cleaning and grounds maintenance.
- *Laboratory Chemicals.* Listed chemicals used in laboratory activities (i.e., sampling and analysis, research and development, and QA/QC activities).

Each facility meeting the threshold quantities and not exempted by the criteria mentioned must complete a Form R for each listed chemical covering the previous calendar year by July 1 of the following year. (The Form R can be obtained from the RCRA/Superfund Hotline or EPA's internet home page, which are listed in Appendix E.) It is important to note that the amount of a chemical released does not affect the reporting requirements. Thus, even if there are no releases of a listed chemical, a facility must report if it meets the requirements regarding the SIC, number of employees, and the threshold quantities.

Reportable Information

Facilities subject to the reporting requirements and that have listed chemicals at or above the threshold quantity level must complete and submit a Form R for each chemical. (Records are required to be maintained for three years.) The completed form must be submitted to both the EPA and the designated state official. The Form R requires basic and specific information on the activity and release. The required information includes the following:

- Name and location of the facility
- Identity of the listed chemical
- Whether the chemical is manufactured, processed, or otherwise used
- Maximum quantity of the chemical on-site an any time during the year
- Quantities of the chemical released during the year to the environment on-site (by media), including routine releases and accidental spills from—
  - — stacks
  - — source releases to water
  - — underground injection
  - — waste sent to landfills
  - — routine releases
  - — spills
  - — fugitive emissions
- Quantities of the chemical subjected to on-site and waste management, including recycling, energy recovery, or treatment;
- Off-site locations where waste contains the chemical and the quantities of the chemical sent to those locations
- Source reduction activities
- Treatment and/or disposal methods used for wastes containing the chemical

Data from each Form R are incorporated into the national Toxic Release Inventory (TRI) data base. The TRI database contains all data reported on Form R since 1987, as well as chemical fact sheets containing health and environmental effects information for TRI toxic chemicals. Information on the TRI database is available from EPA's EPCRA hotline (see Appendix E).

SPILL REPORTING COMPLIANCE CHECKLIST

Superfund and EPCRA

Is the substance defined as a hazardous substance?	Yes___ No___
If yes, has the substance been released to the environment?	Yes___ No___
If yes, has the release met or exceeded its RQ in 40 CFR 302.4?	Yes___ No___
If yes, has the RQ level been met within a 24-hour period?	Yes___ No___
If yes, has the National Response Center been notified?	Yes___ No___
If yes, does the release qualify as a federally permitted release?	Yes___ No___
If yes, does the release constitute a continuous release?	Yes___ No___
If yes, has a newspaper notice be given to potentially affected persons?	Yes___ No___
Is the substance an extremely hazardous substance (40 CFR part 355)?	Yes___ No___
If yes, have the LEPC and SERC also been notified?	Yes___ No___

PCB Spill Reporting

Did the spill contain 50 ppm PCB or greater?	Yes___ No___
If yes, was the spill into surface water, drinking water supplies, sewers, grazing land, or vegetable gardens?	Yes___ No___
(Have the National Response Center and EPA Regional PCB Coordinator been notified?)	Yes___ No___
If no, did the material released contain 10 pounds of "pure" PCBs?	Yes___ No___
If yes, have the National Response Center and EPA Regional PCB Coordinator been notified?	Yes___ No___
If no, did the spill involve 10 pounds or more of a material that contains PCBs at 50 ppm or greater?	Yes___ No___
*If yes, has the EPA Regional PCB Coordinator been notified?	Yes___ No___
*Has a PCB Transformer been involved in a *fire-related incident*?	Yes___ No___
If yes, have the National Response Center and EPA Regional PCB Coordinator been notified?	Yes___ No___

*This requirement is discussed in Chapter 18, which addresses additional requirements under TSCA.

HMTA

Do any of the packages hold a designated hazardous substance?	Yes___ No___

If yes, has a release occurred from such containers?	Yes___ No___
If yes, has the National Response Center been notified?	Yes___ No___

CWA

Has a release occurred into navigable waters?	Yes___ No___
If yes, is the substance defined as oil?	Yes___ No___
Has the release caused a sheen, violated applicable water standards, or caused a sludge or emulsion to be deposited beneath the surface or upon adjoining shorelines?	Yes___ No___
If yes, has the National Response Center been notified?	Yes___ No___

13

RESPONSE ACTIONS UNDER SUPERFUND

Superfund provides the federal government with the authority to respond directly to, or to compel potentially responsible parties to respond to releases or threatened releases of hazardous substances or pollutants or contaminants.

RESPONSE AUTHORITY

Section 104 of Superfund explicitly authorizes federal government responses to a release or threatened release of *hazardous substances* or *pollutants* or *contaminants*. How the federal government responds to various releases is governed by a set of regulations known as the National Contingency Plan (40 CFR part 300), which is discussed in the next section.

Petroleum Exclusion

The definitions of hazardous substances (40 CFR 300.5) and pollutants or contaminants (40 CFR 300.5) specifically exclude "petroleum, including crude oil or any fraction thereof," unless specifically listed. There is no definition of *petroleum* in Superfund. However, EPA interprets the *petroleum exclusion* provision to include crude oil and fractions of crude oil, including the hazardous substances, such as benzene, that are indigenous in those petroleum substances. Because these hazardous substances are found naturally in crude oil and its fractions, they are included in the term *petroleum.* The term also includes hazardous substances that are normally mixed with or added to crude oil or crude oil fractions during the refining process, including hazardous substances whose levels are increased during refining. These substances are also part of *petroleum* because their addition is part of the normal oil separation and processing operations at refineries that produce the product commonly understood to be petroleum. However, hazardous substances that are added to petroleum (e.g., mixing of solvents with used oil) or that increase in concentration solely as a result of contamination of the petroleum during use are not part of the petroleum and thus are not excluded from Superfund.[1]

Prohibited Response Actions

Congress has prohibited the federal government from responding to certain releases unless there is a public or an environmental health emergency, or if no other person can respond to that release. These releases are as follows:

- A release of a naturally occurring substance in its unaltered form, or altered through natural conditions at a location where it is naturally found (e.g., radon gas).
- A release from building products contained in residential, community, or commercial structures where the exposure resulting from the products is within these structures (e.g., urea formaldehyde foam insulation, asbestos).
- A release into a public or private water supply due to the deterioration of the water supply system through ordinary use (e.g., asbestos or lead pipes).
- Workplace exposures covered by the Occupational Safety and Health Act (OSHA).
- Normal application of fertilizer.

NATIONAL CONTINGENCY PLAN

Superfund requires that all actions taken in response to releases of hazardous substances shall, to the greatest extent possible, be in accordance with the provisions of the National Contingency Plan (NCP), which is codified in 40 CFR part 300. The NCP outlines the steps the federal government must follow when responding to situations in which hazardous substances are released or are likely to be released into the environment. Response actions that are developed and implemented by private parties under Superfund also must be consistent with the NCP. The NCP implements the response authorities and responsibilities created by Superfund and the Clean Water Act. Actually a predecessor of Superfund, the NCP was originally written to implement provisions of §311 of the Clean Water Act, which addresses spills of oil and hazardous substances into waters of the United States. Pursuant to §105 of Superfund and Executive Order 12316, EPA is responsible for promulgating revi-

[1] U.S. EPA memorandum from the Office of General Counsel concerning the CERCLA Petroleum Exclusion, July 31, 1987.

sions to the NCP, which it has done periodically. This chapter discusses only the hazardous substance response provisions of the NCP.

The NCP has the following basic components:

- Methods for discovering sites at which hazardous substances have been disposed.
- Methods for evaluating releases to determine potential impact to public health and the environment.
- Methods and criteria for determining the appropriate response and extent of cleanup.
- Means of assuring that remedial action measures are cost-effective.

The process established by the NCP for addressing hazardous substances is triggered by the identification of potential hazardous waste sites under Superfund's §103 notification program, as discussed in Chapter 12. The process begins with a preliminary determination of whether there is an emergency requiring immediate action at a particular site. If there is, the next step is to act as quickly as possible to remove or stabilize the threat. Even after the necessary action has been taken to control the immediate threat, contamination may remain at the site. A more detailed analysis of the contamination may be needed to determine if further decontamination is required. If long-term action is deemed necessary, a decision is made regarding the relative national priority of responding to the threat at that site. If it is warranted, the site will enter the remedial action process, as explained in Chapter 14.

RESPONSE ACTIONS

Whether an action is managed by the government or by potentially responsible parties, two categories of response actions are identified by the NCP: removal and remedial actions.

Under §104(e), a Superfund-financed initial response (removal) cannot take more than 12 months or cost more than $2 million. Beyond these limits, the action is considered a continued response (remedial). A continued response cannot receive federal funding unless the site has been finalized on the National Priorities List (NPL). However, there is a waiver available to allow a continued response if it is appropriate and consistent with future remedial actions.

Removal Response Action

A removal action involves cleanup or other actions that are taken in response to emergency conditions (e.g., spills) on a short-term or temporary basis. Examples of removal response actions include installation of security fences or other measures to limit access, installation of alternative water supplies, evacuation and housing of threatened individuals, segregation of incompatible wastes, and construction of temporary containment systems. A removal action generally requires far less preliminary planning than a remedial action because the emphasis is on finding a rapid, short-term solution to a serious problem.

The following factors are considered in determining the appropriateness of a removal action at a particular site [40 CFR 300.415(b)(2)]:

- Actual or potential exposure of nearby populations, animals, or the food chain to hazardous substances or pollutants or contaminants.
- Actual or potential contamination of drinking-water supplies or sensitive ecosystems.
- Hazardous substances or pollutants or contaminants in drums, barrels, tanks, or other bulk storage containers that may pose a threat of release.
- High levels of hazardous substances or pollutants or contaminants in soils, largely at or near the surface, that may migrate.
- Weather conditions that may cause hazardous substances or pollutants or contaminants to migrate or be released.
- Threat of fire or explosion.
- The availability of other appropriate federal or state response mechanisms to respond to the release.
- Other situations or factors that may pose threats to public health or welfare or the environment.

The evaluation of the appropriateness of a removal action is done by a removal site evaluation (40 CFR 300.410). If a removal action is deemed appropriate under Superfund, there are three categories of removal actions. However, it should be noted that Superfund requires all removal actions to be conducted so as to contribute to the efficient performance of long-term remedial measures that EPA deems practicable. The response action categories are as follows:

- *Emergency,* which generally refers to a release that requires that removal activities begin on-site within hours of the lead agency's determination that a removal action is appropriate.
- *Time-critical,* where the lead agency determines that a removal action is appropriate and that there is a period of less than six months available before removal activities must begin on the site.
- *Non-time-critical,* where the lead agency determines that a removal action is appropriate and that there is a planning period of more than six months before on-site removal activities must begin. (It is important to note that an engineering evaluation/cost analysis is required to be prepared for all non-time-critical removal actions.)

An engineering evaluation/cost analysis (EE/CA) is a document that provides a framework for selecting alternative technologies at non-time-critical removal actions that do not warrant a full-blown remedial action response. The EE/CA document identifies removal action alternatives, analyzes removal action alternatives, and analyzes the feasibility and cost of each alternative.

Remedial Response Action

A remedial action tends to be long-term in nature and involves response actions that are consistent with a permanent remedy taken in lieu of or in addition to a removal action. The remedial action process is explained in detail in Chapter 14.

LIABILITY

Section 107 of Superfund sets forth the liabilities involved with releases or threatened releases of hazardous substances. The following entities are outlined as potentially liable under §107(a):

- The current owner or operator of a site that contains hazardous substances.
- Any person who owned or operated the site at the time when hazardous substances were disposed.
- Any person who arranged for the treatment, storage, or disposal of the hazardous substances at the site.
- Any generator who disposed of hazardous substances at the site.
- Any transporter who transported hazardous substances to the site.

These persons are liable for the following:

- All costs of removal or remedial action incurred by the government.
- Any other necessary costs of response incurred by any other person consistent with the National Contingency Plan.
- Damages for injury to, destruction of, or loss of natural resources, including the reasonable costs for assessing them.
- The costs of any health assessment or health effects study carried out under §104(i).

Section 107(e)(1) prohibits a potentially responsible party from transferring or contracting away its liability under §107 to another party by means of an indemnification, hold harmless, or similar contract agreement. However, the provisions of the section do not bar an agreement to insure, hold harmless, or indemnify a party to such an agreement for liability under §107.

Miscellaneous Liability Provisions

The following sections outline the various doctrines of liability contained or practiced under Superfund and other liability issues. The doctrines used depend on whether the material in question is a hazardous substance or a pollutant or contaminant, as depicted in Figure 13-1.

Hazardous Substance Liability. Under Superfund, statutory liability is imposed on hazardous substances. Therefore, EPA does not have to prove that the defendant's actions were the cause of the alleged environmental or health threat. Statutory liability is not imposed on pollutants or contaminants or any other substance that does not meet the definition of a hazardous substance (40 CFR 300.5). However, if cleanup costs are incurred by the federal government from responding to a release involving pollutants or contaminants, cost recovery is possible, although the government would have to file a civil action (lawsuit) and show *negligence, trespass,* or *nui-*

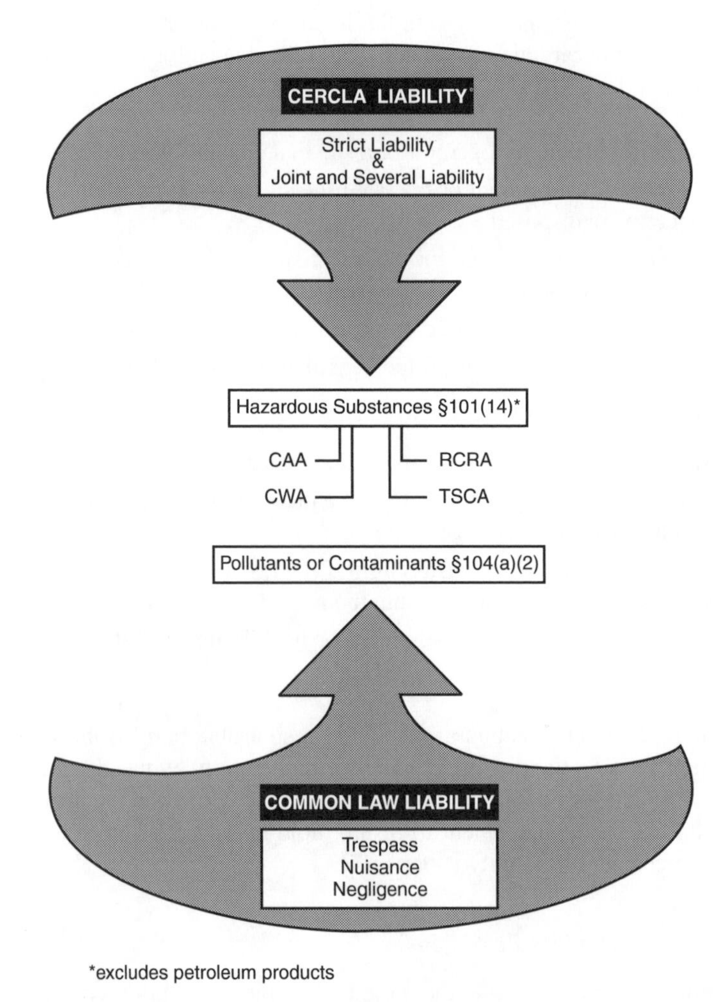

Figure 13-1. Liability under CERCLA

sance under common law doctrine on the part of the responsible party to obtain reimbursement for the response actions and potential punitive damages.

Strict Liability. Under Superfund, hazardous substances are governed under strict liability as established by the courts. Strict liability in tort law is liability without fault. Therefore, one who engages in an activity that has an inherent risk of injury is liable for all injuries proximately caused by the enterprise, even without showing negligence. Hence, one who handles hazardous substances is liable for all resulting injuries even if the utmost care is exercised.

Joint and Several Liability. Superfund does not statutorily impose joint and several liability. This has been left to the courts to decide. However, the Department of Justice (DOJ) and EPA have applied a standard of joint and several liability in appropriate cases, and this position has been upheld by the courts.[2] The term *joint and several* refers to the sharing of liabilities among a group of people collectively and also individually. Thus, if the government holds potentially responsible parties (PRPs) as jointly and severally liable for response costs, the government may sue some or all of the defendants together, or each one separately, and may collect equal or unequal amounts from each party. (*A potentially responsible party* is an individual or entity who may be liable under CERCLA.) For example, at a site with two PRPs, if one party disposed of 1,000 tons of hazardous waste and the other party disposed of 1 ton, under joint and several liability, each party could be liable for the same amount, or one party may be liable for the entire amount, even though the parties did not dispose of equal amounts. This is also known as the "deep pockets" approach. It basically enables the government to recover costs from whichever party can most afford the cleanup costs regardless of that party's degree of fault. The PRP who pays all or part of the response costs, however, has the right to sue other PRPs.

Liability Limits

Under Superfund, liability for damages for each release of a hazardous substance is limited to $50 million for owners and operators of vessels or facilities plus all costs of the response. However, there is no limit to liability when the following conditions apply:

- There was willful misconduct or willful negligence.
- The primary cause of the incident was a violation of applicable safety, construction, or operating standards or regulations.
- The responsible party failed to provide assistance when requested to by a public official in compliance with the NCP.

[2] See EPA memorandum from Frank Shepard, Associate Administrator for Legal Counsel and Enforcement, to Anne M. Gorsuch, Administrator, "Joint and Several Liability under the Comprehensive Environmental Response, Compensation, and Liability Act of 1980," July 27, 1981.

Exemptions to the Liability Provisions

There are four exemptions to the liability provisions:

- Secured creditor
- Service station dealers
- State and local governments
- Response action contractors

Secured Creditor. A person or institution whose ownership rights are held primarily to protect a security interest (e.g., loan, equipment, or other debt) is excluded from liability under §101(2)(A) of CERCLA.

Service Station Dealers. Service station dealers managing recycled/recyclable oil are exempt from some liability provisions provided certain requirements are met as set forth in §114(c) of CERCLA. This exclusion covers the role of a generator and transporter only. A service station dealer still may be liable as an owner or operator if oil is released on the property.

State and Local Governments. Except for gross or willful negligence, state and local governments are not liable for costs or damages resulting from a response to an emergency release of a hazardous substance.

Response-Action Contractors. In general, no person is liable for costs or damages as a result of actions taken or omitted in the course of rendering care, assistance, or advice in accordance with the National Contingency Plan or at the direction of an appointed on-scene coordinator. Under §119 of Superfund, any person who is a response-action contractor with respect to any release or threatened release is not liable under any federal law to any person for injuries, costs, damages, expenses, or other liabilities that result from such a release. This does not apply in the case of negligence, gross negligence, or willful misconduct. Also, this §119 does not apply to any PRP at the site or to a facility regulated under RCRA.

Section 104 of Superfund allows qualified PRPs (as determined by EPA) to carry out the response action, conduct the remedial investigation, or conduct the feasibility study in accordance with the Superfund settlement provisions (§122). It is important to note that if a PRP is involved in a remedial action, that party's liability is not lessened by the response action contractor liability provisions.

Defenses to CERCLA Liability

Section 107(b) of Superfund also provides four affirmative defenses to the liability provisions:

- an act of God,
- an act of war,

- an act or omission of a third party, and
- any combination of the foregoing.

The Third-Party Defense. To prove the third-party defense set forth in §107(b)(3), a person must establish by a preponderance of evidence that

- "The release or threat of release and . . . damages resulting there from were caused solely by . . . an act or omission of a third party other than an employee or agent of the defendant, or that one whose act or omission occurs in connection with a contractual relationship, existing directly or indirectly with the defendant . . . ";
- Due care was exercised with respect to the hazardous substance concerned, taking into consideration the characteristics of such hazardous substance, in light of all relevant facts and circumstances; and
- Precautions were undertaken against foreseeable acts or omissions of any such third party and the consequences that could frequently result from such acts or omissions.

Innocent Landowner Liability. The CERCLA statute of 1980 imposed *innocent landowner liability;* that is, a person who acquired property that contained hazardous substances may be liable for all response costs for a release regardless of whether the owner had knowledge of the substances. To address concerns that the strict liability standard under Superfund could cause inequitable results with respect to landowners who had not been involved in hazardous substance disposal activities, SARA included a new provision that created limited circumstances under which landowners who acquire property without the knowledge of the existence of hazardous substances may avail themselves of the third-party defense to liability contained in §107(b)(3).

Persons otherwise liable under Superfund are not liable if the release in question was caused by the act or omission of a third party, unless the third party's act or omission occurred in connection with a contractual relationship with the defendant. The protection afforded the innocent landowner under SARA is clarified with the definition of the term *contractual relationship* under §101(35): a person acquiring property through a contract, deed, or other such instrument will not be deemed to have a contractual relationship with the previous landowner if the subsequent owner can demonstrate that the person did not know and had no reason to know that any hazardous substance was present at the site at the time the property was acquired. To show that the landowner had no reason to know of the hazardous substances at the time of acquisition, a person must satisfy *due diligence* requirements to establish the third-party defense.

Due Diligence. Section 101(35) extends the third-party defense to persons who acquired the property after the disposal or placement of the hazardous substance only if, at the time of acquisition, the defendant "did not know and had no reason to know that any hazardous substance which is the subject of the release . . . was disposed of . . . at the facility." This section expressly provides that for a defendant to prove that he or she had "no reason to know" of the disposal of hazardous substances, the defendant must demonstrate by a preponderance of the evidence that, before acquisition, he or she conducted all appropriate inquiry into the previous ownership and uses of the property consistent with good commercial or customary practice. A landowner who demonstrates that "all appropriate inquiry" has been conducted will not be deemed to have constructive knowledge under §122(g)(1)(B).

Under §101(35)(B), the following factors must be considered when determining whether "all appropriate inquiry" has been made:

- Any specialized knowledge or experience on the part of the defendant.
- Commonly known or reasonably ascertainable information about the property.
- Relationship of the purchase price to the value of the property if uncontaminated.
- Obviousness of the presence or likely presence of contamination at the property.
- Ability to detect such contamination by appropriate inspection.

These factors clearly indicate that a determination of what constitutes "all appropriate inquiry" under all circumstances must be made on a case-by-case basis. Generally, when determining whether a landowner has conducted all appropriate inquiry, EPA will require a more comprehensive inquiry for those involved in commercial transactions than for those involved in residential transactions for personal use. For example, an investigation following the American Society for Testing and Materials (ASTM) procedures for prepurchase assessments (Phase I and, if necessary, Phase II Assessments) would be appropriate for some commercial transactions, whereas this type of inquiry would not typically be recommended for the purchaser of personal residential property. Thus, the determination will be made on the basis of what is reasonable under all of the circumstances. Guidance on landowner liability and on the type of investigation a buyer should perform to demonstrate "due care" is found in 54 *FR* 34235 (August 18, 1989).

Discretionary Policies on Liability. EPA has set forth criteria to use discretionary authority in determining whether to pursue certain parties who might otherwise feel the full brunt of liability under CERCLA. These parties include

- Residential homeowners
- Owners of property above contaminated aquifers
- Sewage sludge handlers

Residential Homeowners—Regardless of whether a person had knowledge about contamination at a residential property, owners of residential properties will not be subject to an enforcement action unless they caused the release (OSWER Directive 9834.6).

Owners of Property above Contaminated Aquifers—If a landowner did not contribute to a release, but owns property that has become contaminated solely as a result of off-site migration of hazardous substances, the owner will not be liable. (See EPA's Policy toward Owners of Property Containing Contaminated Aquifers, May 24, 1995.)

Sewage Sludge Handlers—Generators and transporters of municipal sewage sludge generally will not be held liable for releases of the sewage sludge subject to conditions. However, this does not cover entities that contributed hazardous substances to the sewage or its sludge (OSWER Directive 9843.13).

Natural Resource Damages

Pursuant to §107 of Superfund, natural resource trustees may be compensated for damages or injury to natural resources resulting from the release of a hazardous substance. *Natural resources* are defined as land, fish, wildlife, biota, air, water, groundwater, drinking-water supplies, and other resources (40 CFR 300.5). Upon discovery of injury to, destruction of, loss of, or threat to natural resources, trustees may take the following actions [40 CFR 300.700(c)]:

- Conduct a preliminary survey of the affected area.
- Coordinate assessments, planning, and investigation.
- Carry out damage assessments. (A *damage assessment* is the process of determining the extent of injury to or destruction or loss of a natural resource.)
- Devise and carry out a plan for restoration, rehabilitation, replacement, or acquisition of equivalent natural resources.

Federal trustees may also request the U.S. Attorney General to seek compensation from liable parties for the damages assessed and for the costs of an assessment and for restoration planning. Trustees may also participate in negotiations with PRPs concerning natural resource damages.

Natural Resources Trustees. Section 107(d) of Superfund requires EPA and the governor of each state to designate officials to act on behalf of the public as trustees for natural resources under Superfund and the Clean Water Act (for releases of oil). EPA is required to publish the identity of the designated federal trustees in the NCP (e.g., U.S. Fish and Wildlife Service, U.S. Forestry Service, Bureau of Reclamation, Bureau of Land Management). The governor of each state is required to notify EPA of the designated state trustees (e.g., State Fish and Wildlife, State Forestry Department, Environmental Protection) who are required to assess damages for injury to, destruction of, or loss of natural resources under their trusteeship. Also, federal trustees are authorized to assess damages for natural resources under a state's trusteeship upon request and reimbursement from a state and at the federal officials' discretion.

Any funds recovered by a federal or a state trustee must be retained by the trustee for use only to restore, replace, or acquire the equivalent of such natural resources. However, liability for damages to natural resources under §107 is not limited to the cost of restoring, replacing, or acquiring such natural resources.

Natural Resource Damage Assessments. The Department of the Interior is responsible for implementing natural resource damage assessments, which typically occur during the remedial investigation/feasibility study process. Procedures to conduct natural resource damage assessments (assessing injury to, destruction of, or loss) are codified in 43 CFR Part 11. Two types of assessments have been promulgated: standard procedures for simplified assessments requiring minimal field observation (Type A), and site-specific procedures for detailed assessments of specific sites (Type B).

Brownfields Economic Redevelopment Inititative

On January 25, 1995, EPA announced the Brownfields Action Agenda, an outline of EPA's activities and future plans dealing with "brownfields." EPA's Brownfields Economic Redevelopment Initiative is an economic redevelopment program intended to prevent, assess, clean up, and reuse brownfields. *Brownfields* are abandoned, idled, or underused industrial and commercial facilities where expansion or redevelopment is complicated by real or perceived environmental contamination.

Section 104(a) of Superfund grants EPA broad authority to take response actions to address releases and threatened releases of hazardous substances, pollutants, and contaminants. As specified in §104(a), these response activities are undertaken by EPA consistent with the NCP, and may be taken at any site at which a release or threatened release occurs, not just those that are listed on the NPL. Because brownfields projects involve sites at which either an actual or threatened release is present, response actions authorized by §104 may be taken. Under §111(a)(1), Superfund monies may be used to pay for response actions undertaken pursuant to §104, which would include response actions taken in connection with brownfields projects.

Liability for Brownfield Property. A significant barrier to assessing, cleaning up, and redeveloping brownfields is the public's apprehension about becoming involved with a site for fear of inheriting cleanup liabilities for contamination they did not create. EPA has attempted to address these by clarifying the following relevant liability issues:

- *The Guidance on Agreements with Prospective Purchasers of Contaminated Property.* This guidance states the situations under which EPA may enter into an agreement to not file a lawsuit against a prospective purchaser of a contaminated property for contamination that existed before the purchase. Such an agreement in effect eliminates much of the "retroactive liability" concern associated with purchasing contaminated or previously contaminated property where some evidence of federal environmental interest exists, and it encourages parties to purchase, assess, clean up, and redevelop brownfields they might otherwise avoid because of a reasonable fear of incurring Federal liability.

- *EPA's Policy for Owners of Property Containing Contaminated Aquifers.* This policy reassures landowners that EPA does not plan to sue them for groundwater contamination if they did not cause or contribute to the contamination.
- *The Land Use in the CERCLA Remedy Selection Process.* This policy promotes the use of remedies tailored to the anticipated future land use of a Superfund site and the use of presumptive remedies at certain site types.
- *The Policy on CERCLA Enforcement against Lenders and Government Entities that Acquire Property Involuntarily.* EPA and the Department of Justice (DOJ) jointly issued a memo explaining their policy on Superfund enforcement against lenders and government entities that acquire property involuntarily. The memo states that EPA and DOJ intend to apply as guidance the provisions of the "Lender Liability Rule" promulgated in 1992. EPA and DOJ will not pursue cleanup costs from those lenders that provide money to an owner or developer of a contaminated property, but do not actively participate in daily management of the property. Superfund releases from liability governmental units that involuntarily take ownership of property through the operation of federal, state, or local law. EPA clarified which actions would be considered "involuntary" and would therefore not subject the governmental unit to potential liability.

Comfort Letters. One of the most significant barriers to purchasing or developing brownfield sites is the uncertainty about potential contamination or Superfund liability. In an attempt to allay the fear of potential federal pursuit of parties for cleanup of brownfields, EPA may provide varying degrees of comfort by communicating EPA's intentions toward a particular piece of property. Comfort may range from a formal legal agreement containing a covenant not to sue (discussed in Chapter 16 under settlements), which releases a party from liability for cleanup of existing contamination, to site-specific statements known as *comfort letters*.

EPA has established four model "comfort letters" that address the most common situations at contaminated or potentially contaminates sites.[3] The four comfort letters are as follows:

- *No Previous Federal Superfund Interest.* This letter may be provided to parties where there is no historical evidence of federal Superfund program involvement with the property/site in question (i.e., the site is not listed in the Com-

[3] See EPA memorandum from Steven A. Herman, Assistant Administrator, Office of Enforcement and Compliance Assurance to Addressees, "Policy on the Issuance of Comfort/Status Letters," November 8, 1996.

prehensive Environmental Response, Compensation, and Liability Act Information System, or CERCLIS).

- *No Current Federal Superfund Interest.* This letter may be provided when the property/site either (1) has been archived and is no longer part of the CERCLIS inventory of sites, has been deleted from the NPL, or (2) is situated near but not within the defined properties of a CERCLIS site.
- *Federal Interest.* This letter may be provided where EPA either plans to respond in some manner or already is responding at a site. This letter is intended to inform the recipient of the status of EPA's involvement at the property.
- *State Action.* This letter may be provided when a state has the lead for day-to-day activities and oversight of a response action.

FEDERAL FACILITIES

Section 120(a) of Superfund affirms that, with few exceptions, federal facilities (e.g., military installations, Department of Energy weapons sites, Department of the Interior mining sites) are subject to the provisions, guidelines, rules, regulations, and criteria of Superfund in the same manner as any nongovernmental entity. Section 120(a) also outlines the process by which federal agencies are required to undertake remedial actions.

The selection of a remedy for a release is a joint determination by EPA and the affected federal agency. If the parties do not agree on the selection, EPA determines the remedy. EPA is required to evaluate federal facilities for possible inclusion on the NPL, and review the results of completed federal agency remedial investigations and feasibility studies.

Schedule for Response Activities

Section 120 also establishes a schedule for response actions at federal facilities based on the following. Federal agencies must

- Commence the RI/FS for a facility within 6 months of its listing on the NPL.
- Enter into an interagency agreement with EPA within 6 months of EPA's review of RI/FS concerning the review of alternatives. If the parties are unable to agree on the joint selection of a remedy, EPA will then select the remedy, schedule, and operation and maintenance arrangements.
- Begin remedial action within 15 months of completing RI/FS.
- Report annually to Congress regarding the progress achieved in implementing this section.

Miscellaneous Federal Facilities Provisions

There are a number of other statutory requirements affecting federal facilities that including the following:

- Listing on the NPL
- Hazardous waste compliance docket
- Preremedial activities
- Real-property transfers

These are detailed in the following sections.

Listing on the NPL. Because federal agencies are subject to the same requirements as nongovernmental entities, federal agencies may be listed on a modified National Priorities List. It is a modified list because federal facilities are not eligible for Superfund monies for response actions. If a site is listed on the "regular" NPL, it is eligible for fund money for a remedial response action. In response, EPA has established a separate section on the NPL that lists appropriate federal sites that are not eligible for Superfund monies.

Hazardous Waste Compliance Docket. Section 102(c) of Superfund requires EPA to establish a Federal Facilities Hazardous Waste Compliance Docket. The purpose of this docket is to identify federal facilities that may be contaminated with hazardous substances, to compile and maintain information pertaining to Superfund, and to ensure the availability of this information to the public. EPA periodically publishes the docket in the *Federal Register.* In addition, EPA has established information repositories in every EPA Regional Office, which contains docket information on federal facilities within the EPA region.

Preremedial Activities. The federal agency that owns the site or has primary responsibility for the site is responsible for conducting the preliminary assessment or the site inspection (the preremedial activities). EPA, however, is responsible for ensuring that the activity was conducted in accordance with the National Contingency Plan.

Real-Property Transfers. Each time the federal government enters into any contract for the sale or transfer of real property at which a hazardous substance was released (in amounts greater than its reportable quantity), disposed of, or stored at for a year or more, the sales contract must include a disclosure notice pertaining to the substance and activity (40 CFR 373.1). Additionally, any federal agency transferring federally owned real property must include in the deed a covenant assuring the transferee that all remedial action was taken before the transfer, and that the government will conduct any remedial action necessary after the transfer.

For further information on this subject, see EPA memorandum from Elliot P. Laws, Assistant Administrator, to Regional Administrators, "Federal Property Transfer: Guidance on EPA Evaluation of Federal Agency Demonstrations that Remedial Actions Are 'Operating Properly and Successfully' under CERCLA §120(h)(3)."

14

REMEDIAL RESPONSE

A Superfund-authorized action at a site involves either a short-term removal action or a long-term remedial response. Whereas a *removal action* is a relatively quick action taken over the short term to address a release, a *remedial action* is a long-term action that is intended to stop and clean up a release of hazardous substances so that they no longer present a threat to public health or the environment.

Because of the unanticipated excessive costs and time required to conduct response actions under the Superfund program, EPA initiated the Superfund Accelerated Cleanup Model (SACM). The main goals of SACM are to achieve immediate risk reduction at Superfund sites, perform more efficient and cost-effective cleanups, and avoid duplication of effort. SACM is an internal process that uses lessons learned from other response actions to better predict the needs and objectives appropriate for a particular site (OSWER Directive 92031-05I).

REMEDIAL RESPONSE ACTION PROCESS

There is a basic nine-step process (which is not necessarily sequential) for a remedial response at a Superfund site, as outlined in Figure 14-1. The steps are as follows:

1. Site discovery
2. Preliminary assessment
3. Site inspection
4. Hazard ranking analysis

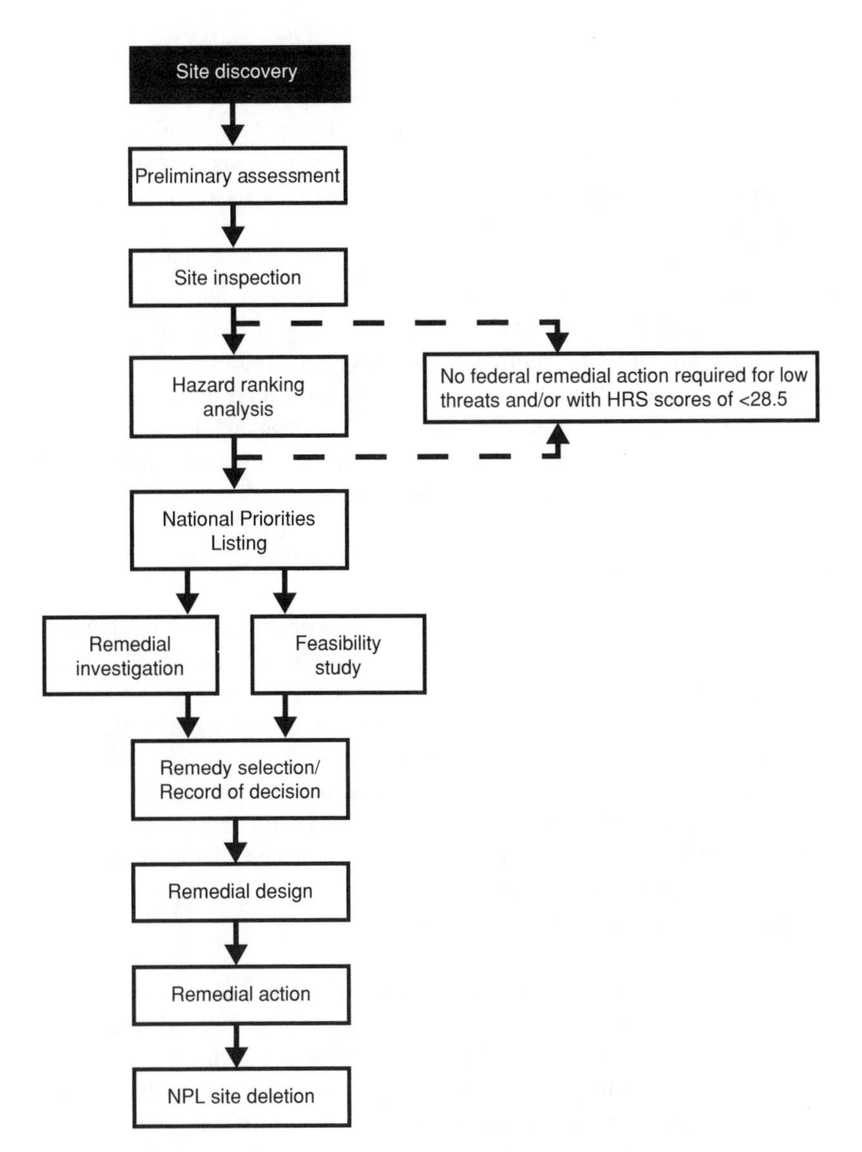

Figure 14-1. The CERCLA Remedial Response Process

5. National Priorities List determination
6. Remedial investigation and feasibility study
7. Remedy selection/record of decision
8. Remedial design
9. Remedial action
10. Deletion from the NPL

Site Discovery

Releases of hazardous substances at or above the designated reportable quantity and sites where hazardous substances have been managed must be reported as discussed in Chapter 12. Sites are also discovered by citizen complaints, formal citizen petitions authorized by 40 CFR 300.420(b)(5), employees, state records, and compliance inspections.

All sites where releases or potential releases have been reported are entered into the Comprehensive Environmental Response, Compensation, and Liability Information System (CERCLIS), which is the national reporting, tracking, and management tool for Superfund. CERCLIS currently contains about 16,000 sites. EPA recently archived approximately 24,000 sites, out of a previous total of 40,000 sites, from CERCLIS. Many of these sites were found to be clean, while others are being addressed by state cleanup programs. The federal government is unlikely to have any further Superfund interest in these archived sites.

Preliminary Assessment

A *preliminary assessment* is the first phase in the process of determining whether a site is releasing, or has the potential to release, hazardous substances or pollutants or contaminants into the environment, and whether it requires response action that is authorized by Superfund. The preliminary assessment includes an initial analysis of available information to determine if a release of hazardous substances may be serious enough to require additional investigation or action. All sites listed on CERCLIS must undergo a preliminary assessment [40 CFR 300.420(b)]. Approximately half of all CERCLIS sites are eliminated from further consideration based on the results of the preliminary assessment.

The preliminary assessment has four specific goals:

1. *Eliminate sites where remedial action is not required.* The first goal of the preliminary assessment is to screen out those sites in the CERCLIS inventory that are statutorily ineligible for a Superfund remedial response, that pose no threat to public health or the environment, or at which no further action is warranted or authorized under the remedial program.
2. *Identify sites that require emergency response.* Superfund removal authority allows EPA to take immediate action at a site regardless of whether the site is on the National Priorities List (NPL). The preliminary assessment can assist in determining if the site, or a portion of it, may qualify for a removal action, thereby warranting referral to the removal program. This allows cleanup activities in advance of a determination about whether the site qualifies for the NPL.
3. *Compile information necessary to develop preliminary and projected HRS scores.* If the site may pose a threat that warrants remedial action, the preliminary assessment collects data to develop preliminary and proposed

hazard ranking system (HRS) scores (discussed later in this chapter under "Hazard Ranking Analysis").

4. *Set priorities for site inspections.* The fourth goal of the preliminary assessment is to set the priority of the site for a site inspection. Because more sites are typically referred for further action than resources can accommodate, EPA has to establish priorities for further investigation, based on the degree of potential impact.

The preliminary assessment is a multistep process consisting of the following:

- Review of available information (e.g., state records, manifests, permit applications, historical land use records, deeds, etc.)
- Site reconnaissance
- Development of preliminary and projected HRS scores
- Application of qualitative criteria
- Prioritization for site inspection
- Report preparation
- Documentation
- CERCLIS tracking

During a preliminary assessment, the investigator compiles and evaluates available information about a site and its surrounding environment, including information on potential waste sources, migration pathways, and receptors. At the conclusion of the preliminary assessment, a brief report (EPA Form 2070-12) with formal recommendations is prepared. Although the preliminary assessment is used in an attempt to establish whether the site has the potential to adversely affect the environment, it is not intended to determine the exact magnitude of the release. This is accomplished when the site is scored under the hazard ranking system analysis after completion of a site inspection and, more comprehensively, during the subsequent remedial investigation. However, should a site warrant no further action, the site is designated as No Further Response Action Planned (NFRAP), and the site is removed from CERCLIS and archived.

Preliminary Assessment Petitions. Any person may formally petition EPA (or other federal agency if the site of concern is a federal facility) to conduct a preliminary assessment [40 CFR 300.420(b)(5)]. The petition must pertain to a release or threatened release that affects or may affect the petitioner.

There is no specified format for the preliminary assessment petition. However, a petition should contain basic information about the release and the petitioner. This information should include the following:

- The name, address, telephone number, and signature of the petitioner.

- The exact location of the actual or potential release. This may include marking on a street map or on a U.S. Geological Survey topographical map.
- A description of how the petitioner is, or may be, affected by the actual or potential release (e.g., contaminated groundwater, air releases, or unusual health affects).

The petition must be sent to the EPA Regional Administrator responsible for the site. After receiving a preliminary assessment petition, EPA must determine if there is reason to believe that an actual or potential site exists, and if EPA has the legal authority to respond to the site. Within 12 months after a petition is received, EPA is required to review it and prepare a report. This report will state whether the petition was approved or not, and the reasons for the decision. A copy of this report must be sent to the petitioner within the one-year time frame. If a petition is approved, EPA will conduct a preliminary assessment and provide a copy of the preliminary assessment to the petitioner.

Site Inspection

If the preliminary assessment indicates a suspected release of hazardous substances that may threaten human health or the environment, EPA may initiate a *site inspection* [40 CFR 300.420(c)], which is conducted to ascertain the extent of the problem and to obtain information needed to determine whether a removal action is warranted at the site, or whether it should be included on the NPL for a remedial action. In addition to sampling, inspections usually include a reconnaissance of the site's layout, surrounding topographical features, and the location of nearby populations to document any risks the site may pose. Site inspections are divided into two tiers: focused site inspections and expanded site inspections.

The goals of the focused site inspection are to

- Collect additional data to calculate a better preliminary hazard ranking score.
- Establish priorities among sites most likely to qualify for the National Priorities List.
- Identify the most critical data requirements for a hazard ranking analysis.

Those sites that are most likely to qualify for the National Priorities List are candidates for expanded site inspections, which address the data requirements for a hazard ranking analysis, and may cover air, groundwater, soil, and surface-water. Expanded site inspections are also used to calculate a preliminary hazard ranking score and to support the scoping phase of the remedial investigation and development of the remedial work plan.

Following completion of the site inspection, a report is prepared that includes the following information:

- A description of the site conditions and activities necessitating the site inspection.

- Description of known contaminants.
- Descriptions of contaminant migration pathways.
- Identification and description of human and environmental receptors.
- Discussion of further action.

Hazard Ranking Analysis

Because there are not enough Superfund federal monies to clean up all of the nation's hazardous waste sites, a system of establishing priorities was developed. The primary criteria for establishing priorities for remedial action are based on the Hazard Ranking System (HRS), which was designed to help evaluate the relative risk to public health and the environment and to allow for the quick ranking of sites. (The HRS procedures are contained in Appendix A of 40 CFR part 300.) EPA developed the HRS scoring system to help evaluate the relative risks to public health and the environment posed by different sites. Each site submitted by the states and EPA is scored according to the HRS. The HRS score reflects the potential harm to human populations and the environment from the migration of hazardous substances involving groundwater, surface water, air, and soil. A composite score, based on separate, weighted scores calculated for each of the possible contaminant migration routes, is determined. The score for each migration route is obtained by assigning a numerical value based on predetermined guidelines, contained in the HRS, to a set of factors that characterize the potential of the release to cause harm. An HRS numerical score for potential inclusion on the NPL is then determined. The HRS score is based on 100 points, and any site that receives a score of 28.5 is eligible for inclusion on the NPL. For EPA to compile a listing of 400 sites on the first-ever NPL, as mandated by CERCLA, it set the HRS score at 28.5 for NPL inclusion. Although the score was somewhat arbitrarily determined, it has remained the same.

Responsible agencies (e.g., states, EPA, or other federal agencies) have the primary responsibility for identifying potential sites, computing HRS scores, and submitting candidate sites to their respective EPA Regional Office. The EPA regional offices then conduct a quality control review of the state's candidate sites and also may assist in investigating, monitoring, and scoring sites. EPA headquarters conducts further quality assurance audits to ensure accuracy and consistency among the various EPA and state offices involved in the scoring. EPA then proposes the new sites that meet the listing requirements in the *Federal Register* and solicits public comments on the proposal. After completion of the public comment period and a review of the comments (which takes approximately eight months from the time of proposal), the final rulemaking is published in the *Federal Register.*

National Priorities List Determination

The primary purpose of the National Priorities List (NPL) is to identify those facilities and sites that appear to present the most significant threat to human health or the environment and that need long-term remedial response actions. A site listed on the NPL is eligible for federal monies for a remedial response action. The NPL does not determine priorities for removal actions. EPA may pursue removal actions at any site, whether listed on the NPL or not, provided that the action is consistent with the National Contingency Plan (NCP). Likewise, EPA may take enforcement actions, as well as conduct remedial investigations or feasibility studies, regardless of whether the site is on the NPL. However, a site's listing on the NPL must be finalized for it to receive fund monies for remedial action (construction). Because the NPL is part of the NCP, it is subjected to standard rulemaking procedures, including proposal in the *Federal Register,* public comment, and a final determination and subsequent publication in the *Federal Register.* In addition, listing of a site on the NPL does not require any action by a private party, nor does it determine the liability of any party for the cost of cleanup at the site.

As mentioned in the previous secton, for a site to become eligible for inclusion on the NPL, it must receive a Hazard Ranking System (HRS) score of 28.5 or greater. However, listing on the NPL does not mean that the site necessarily represents an immediate threat to the public health, although to get on the list by the HRS score, each site must represent some significant, long-term threat to public health.

A site may be placed on the NPL regardless of its HRS score if it meets the following conditions [40 CFR 300.425(c)]:

- If the Agency for Toxic Substances and Disease Registry (ATSDR) issues a health advisory for a site.
- If EPA anticipates that it will be more cost-effective to use its remedial authority than to use its removal authority to respond to a release.
- If EPA determines that a release poses a significant threat to the public health.

NPL Listing Policies. Because of the unique conditions or statutory implications at certain sites, EPA has established policies to address these sites, which include the following:

- Federal facilities
- RCRA facilities
- Mining sites

Federal Facilities. Although §111(e)(3) of Superfund prohibits the use of fund monies for remedial actions at federal facilities, EPA lists federal facilities on the NPL. EPA has established this policy because §120(a) affirmed that federal facilities are subject to Superfund and because listing federal facilities on the NPL is consistent with the NPL's purpose of providing information to the public with respect to sites that present potential hazards (March 13, 1989, 54 *FR* 10512).

RCRA Facilities. EPA published its policy for including sites on the NPL that are subject to Subtitle C of RCRA on June 10, 1986 (51 *FR* 21054). Under this policy, facilities not subject to RCRA Subtitle C corrective action requirements (discussed in Chapter 9) will remain eligible for NPL inclusion. Examples of RCRA Subtitle C, NPL-eligible facilities include the following:

- Facilities that ceased treating, storing, or disposing of hazardous wastes before November 19, 1980.
- Sites at which only materials exempted from the statutory or regulatory definition of solid waste or hazardous waste are managed.
- Hazardous waste generators or transporters not required to have interim status or a RCRA permit.

Sites with releases that can be addressed under the RCRA Subtitle C corrective action provisions will not be placed on the NPL. However, RCRA facilities may be listed if they meet all of the other criteria for listing, and if they fall within one of the following categories:

- Facilities owned by persons who are bankrupt.
- Facilities that have lost authorization to operate, and for which there are additional indications that the owner or the operator will not be willing to undertake corrective action.
- Sites, analyzed on a case-by-case basis, whose owners or operators have shown an unwillingness to undertake corrective action.

Mining Sites. EPA has affirmed that mining sites are eligible for inclusion on the NPL. This policy was published in the *Federal Register* on September 18, 1985 (50 *FR* 37950), and June 10, 1986 (51 *FR* 21054).

Remedial Investigation and Feasibility Study

Information obtained from the preliminary assessment, site inspection, and other sources is used to determine the general types of response actions applicable to the site for use in planning the remedial investigations. The NCP requires that a detailed remedial investigation (RI) and a feasibility study (FS) be conducted for sites listed on the NPL and targeted for remedial response under §104. The RI/FS process is the methodology that the Superfund program has established for characterizing the nature and extent of risks posed by uncontrolled hazardous waste sites and for evaluating potential remedial options. In short, the RI/FS process is the remedy evaluation and selection process.

Determination of the Lead Agency. A lead agency that will have the primary responsibility for coordinating a response action must be selected. The selection of the lead agency is typically done after a site is placed on the National Priorities List. The lead agency, which is represented by the remedial project manager (RPM), has the primary responsibility for coordinating a response action and will oversee all technical, enforcement, and financial aspects of a remedial response. EPA, a state environmental agency, or another federal agency (e.g., Department of Defense, Department of Energy, Department of the Interior) may serve as the lead agency; however, EPA has final authority regarding remedy selection where it is not the lead agency. Remedy selection is done through interagency negotiations involving EPA, states, and other federal agencies.

State as Lead Agency. Where EPA and a state are involved in remedial activities, the lead agency is identified in a Superfund Memorandum of Agreement (SMOA), a Cooperative Agreement (CA), or a State Superfund Contract (SSC). The SMOA is a general agreement that specifies the nature and extent of interaction between EPA and the state for one or more sites. The CA is a site-specific agreement that establishes EPA and state responsibilities for a specific response action. The SSC is an agreement that documents any required cost shares and assurances necessary from a state but does not involve the disbursement of federal monies (40 CFR 300.500).

Other Federal Agency as Lead Agency. A federal agency other than EPA also may assume the roles and responsibilities of the lead agency. The division of authority and responsibility for the federal agency as the lead (and support agency) is specified in a Federal Facility Agreement (FFA). Because federal agencies conducting response activities are expected to comply with the NCP, CERCLA, and other EPA guidance as mandated by §120 of CERCLA, the FFA is used to delineate the lead agency's responsibilities.

Responsible-Party-Initiated Action. Pursuant to §104 of Superfund, a potentially responsible party (PRP) may also conduct response actions, provided that the PRP is qualified and otherwise capable. For a PRP-initiated response action, EPA or the state is the lead agency for overseeing the PRP's activities. PRPs may participate in the remedy selection process by recommending their own preferred alternative to the lead agency at the conclusion of the feasibility study and by submitting comments during the formal comment period held before the final remedy selection for the site.

The RI/FS Process. As shown in Figure 14-2, the RI and the FS are interdependent. Although the RI/FS process is presented in a fashion that makes the steps appear sequential and distinct, in practice the process is highly interactive. In fact, the RI and the FS are usually conducted concurrently. Whereas the RI emphasizes data collection and site characterization, the FS focuses on data analysis and evaluation of alternatives. Data collected in the RI influence the development of remedial alternatives in the FS, which in turn affects the data needs and scope of treatability studies and field investigations. In addition, because of the complex nature of many sites, new site characterization information may be compiled as the RI progresses, which may require reassessment of the types of response actions identified. In turn, this may require expanding the RI to obtain the data necessary to evaluate the new alternatives.

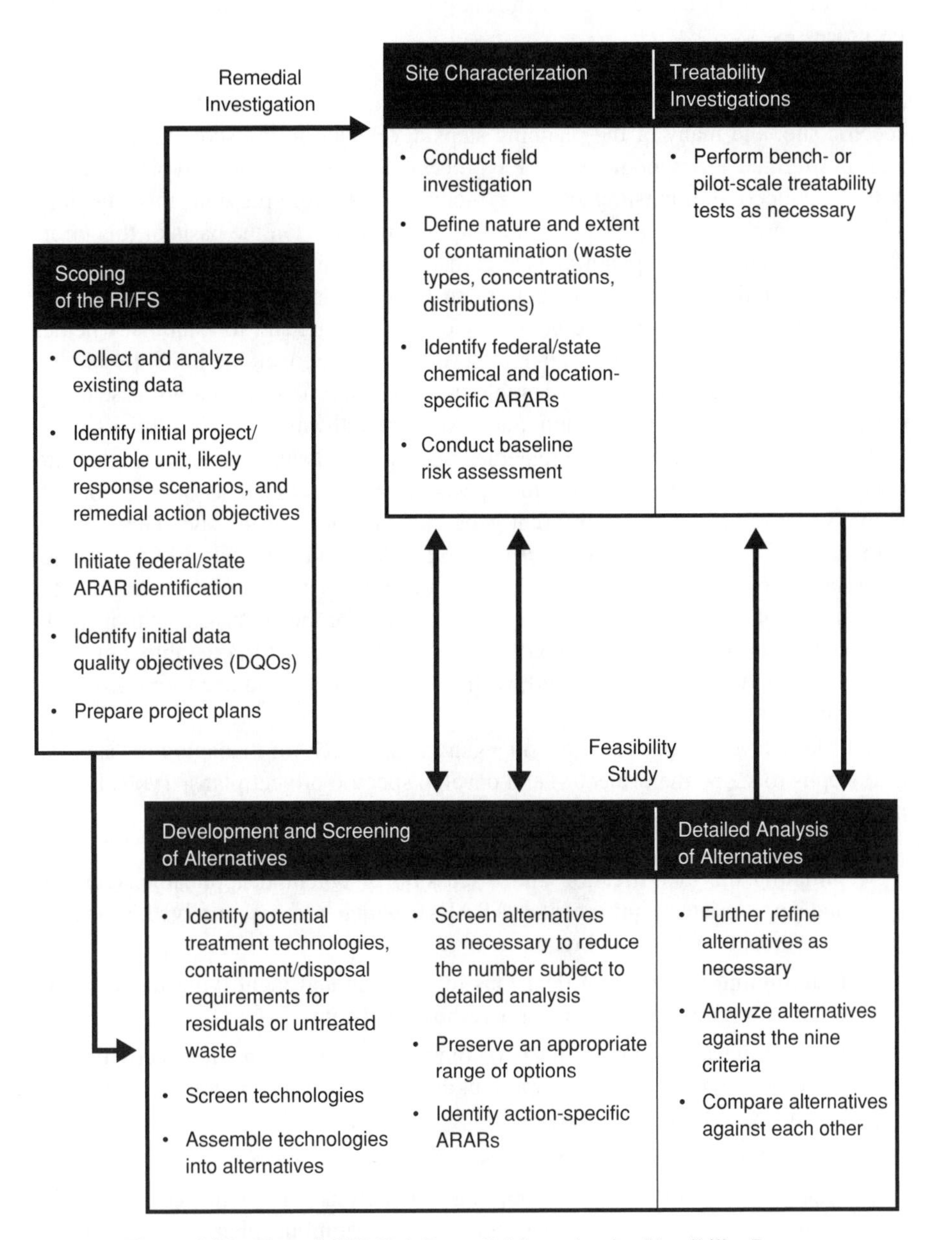

Figure 14-2. Phased CERCLA Remedial Investigation/Feasibility Process

Management and coordination of RI/FS activities will affect the resources, timing, and completeness of the RI and the FS reports. Site-specific conditions will govern the extent of data collection and analysis for each level of both the RI and the FS processes..

Scoping of the RI/FS. Scoping is the initial planning stage for the RI/FS process at a specific site, and many of the planning steps developed during scoping are continued and refined throughout the RI/FS process. Scoping activities typically begin with the collection of existing site data, including data from previous investigations such as the preliminary assessment and site investigation. On the basis of this information, site management planning is undertaken for preliminary identification of the boundaries of the study area, to determine likely remedial action objectives and whether interim actions may be necessary or appropriate, and to establish whether the site may best be remedied as one or several separate, operable units. (*Remedial action objectives* are media- or operable-unit-specific goals, which are essentially cleanup goals based on the contaminants, exposure pathways, and health risks.)

Current practices in designing remedies for Superfund sites often divide sites into operable units that address discrete aspects of the site (such as source control, groundwater remediation) or different geographic portions of the site. This is especially true for complex sites. An *operable unit* is defined in 40 CFR 300.5 as "a discernible part of the entire response action that decreases a release, threat of release, or pathways of exposure." RI/FSs may be conducted for the entire site, and operable units may be broken out during or after the feasibility study. Or, operable units may be treated individually from the start, with focused RI/FSs conducted for each operable unit.

When the involved parties agree on a general approach for managing the site, the next step is to scope the project(s) and develop specific project plans. Typical scoping activities include the following:

- Initiating the identification and discussion of potential applicable, relevant, and appropriate requirements (ARARs), which are essentially the cleanup standards, based on legal requirements.
- Determining the types of decisions to be made and identifying the data and other information needed to support those decisions.
- Assembling a technical advisory committee to assist in these activities, to serve as a review board for important deliverables, and to monitor progress, as appropriate, during the study.
- Establishing data quality objectives.
- Preparing the work plan, the sampling and analysis plan (which consists of the quality assurance project plan and the field sampling plan), the health and safety plan, and the community relations plan.

The scoping process is critical to the development of a sampling plan, which describes the sampling studies that will be conducted, including sample types, analyses, locations, and frequency. Planning needs, such as sampling operational plans, materials, record keeping, personnel needs, and sampling procedures, also are developed or identified for the investigation.

Risk Assessment. The primary purpose of Superfund is to protect human health and the environment from current and potential threats posed by uncontrolled hazardous substances releases. To assist in this mandate, EPA has developed a Superfund risk assessment program as part of its remedial response program. EPA issued guidance regarding the increased consideration of anticipated future land uses in remedy selection decisions at NPL sites.[1] The guidance encourages discussions among local land-use-planning authorities, other officials, and the community as early as possible in the site assessment process.

The Superfund risk assessment program is composed of three separate parts:

- Human health and environmental evaluations
- Health assessments
- Endangerment assessments

Human Health and Environmental Evaluations. The goal of the Superfund human health and environmental evaluation is to provide a framework for obtaining the risk information needed for decision making at Superfund sites. Specifically, the process provides the following:

- An analysis of baseline risks.
- A basis for determining levels of chemicals that can remain on-site without threatening public health and the environment.
- A basis for comparing potential health impacts of various remedial alternatives.
- A consistent process for evaluating and documenting public health and environmental threats at sites.

The risk information generated by the human health and environmental evaluation process is designed to be used in the RI/FS. The baseline risk assessment contributes to site characterization and the subsequent development, evaluation, and selection of appropriate response alternatives. (*Baseline risks* are existing or potential risks to human health and the environment that might exist if no remedy or institutional control were applied at a site.) Such an assessment consists of data collection and analysis, exposure assessment, toxicity assessment, and risk characterization. The results of the baseline risk assessment are used to

- Help determine whether a response action is necessary.
- Modify preliminary remediation goals.

[1] *Land Use in the CERCLA Remedy Selection Process*, May 1995.

- Help support selection of the no-action alternative.

  Note: The "no-action alternative" is developed as the baseline for the comparison of other alternatives and is required to be carried throughout the entire remedial screening and selection process.

- Document the magnitude of risk at a site, and the primary causes of that risk.

Health Assessment. A public health assessment must be conducted by the Agency for Toxic Substances and Disease Registry (ATSDR) for each site proposed or included on the NPL before the completion of the RI/FS [40 CFR 300.400(f)]. The ATSDR health assessment, which is relatively qualitative in nature, is separate from the human health evaluation, which is more quantitative than the ATSDR assessment.

EPA human health evaluations include quantitative, substance-specific estimates of the risk that a site poses to human health. These estimates depend on statistical and biological models using data from human epidemiological investigations and animal toxicity studies. The information generated from a human health evaluation is used in risk management decisions to establish cleanup levels and select a remedial alternative. ATSDR health assessments, although they may also employ quantitative data, are more qualitative in nature than the EPA evaluations, as the health assessments not only focus on the possible health threats posed by chemical contaminants attributable to a site but consider *all* health threats, both chemical and physical, to which residents near a site may be subjected. These health assessments focus on the medical and public health concerns associated with exposures at a site and discuss especially sensitive populations, toxic mechanisms, and possible disease outcomes. EPA considers the information in a health assessment along with the results of the baseline risk assessment to give a complete picture of health threats.

Endangerment Assessments. Before taking enforcement action against PRPs, EPA must make a determination that a site presents an imminent and substantial endangerment to public health or the environment. Such a legal determination is called an *endangerment assessment.* In the past, an endangerment assessment was often prepared as a study separate from the baseline risk assessment (human health and environmental evaluation). However, with the enactment of SARA and changes in EPA practices, the need to perform a detailed endangerment assessment as a separate effort from the baseline risk assessment has been eliminated. Elements included in the baseline risk assessment conducted at a Superfund site during the RI/FS process satisfy the requirements of the endangerment assessment.

Remedial Investigation. The remedial investigation (RI) is composed of two phases: site characterization and treatability investigations. The data developed from these two phases support the evaluation of remedial alternatives. The following should be noted:

- The RI must be consistent with the National Contingency Plan.

- Data needs differ between enforcement-led, fund-led (federally financed), and private-party-led remedial investigations (the data collection process must be tailored to meet specific investigative needs and objectives, including data quality and sufficiency).
- All supporting files and supporting documentation must be collected and retained for public viewing.

Site Characterization. The site characterization process involves field sampling and laboratory analyses. Field sampling typically is phased so that results of the initial sampling efforts can be used to refine plans developed during scoping to better focus subsequent sampling efforts. Data quality objectives are revised as appropriate, based on an improved understanding of the site, to facilitate the efficient and accurate characterization of the site.

A preliminary site characterization summary is prepared to provide information on the site early in the process and before preparation of the full RI report. This summary is useful in determining the feasibility of potential technologies and in assisting with the initial identification of ARARs.

Treatability Investigations. If existing site and/or treatment data are insufficient for adequate evaluation of alternatives, treatability tests may be necessary to analyze a particular technology for specific site wastes. Treatability investigations generally use bench-scale studies, but occasionally pilot-scale studies are necessary.

Bench- or pilot-scale studies may be needed in the RI for investigators to obtain enough data to select a remedial alternative. The bench and pilot studies of the RI are specifically concerned with waste treatability, startup of innovative technologies, technology application issues, and evaluation of specific alternatives. Bench and pilot studies may also be conducted during remedial alternative design or construction to more fully evaluate the specific requirements of the selected alternative. However, these studies are outside the RI/FS process. In general, bench-scale studies are appropriate for the RI stage, whereas pilot-scale studies, if required, may be conducted during the final remedial design phase.

Feasibility Study. The feasibility study also is composed of two phases: development and screening of alternatives, and then detailed analysis of those alternatives.

Development and Screening of Alternatives. The development of alternatives generally begins soon after the RI/FS scoping process, when likely response scenarios are first identified. The development of alternatives requires the following:

- Identification of remedial action objectives.
- Identification of potential treatment, resource recovery, and containment technologies that will satisfy these objectives.
- Screening of the technologies based on their effectiveness, implementability, and cost.

- Assembling technologies and their associated containment or disposal requirements into alternatives for managing the contaminated media at the site or for the operable unit.

Alternatives can be developed to address a contaminated medium (e.g., soil), a specific area of the site (e.g., a surface impoundment), an operable unit (e.g., a series of surface impoundments), or the entire site. Alternatives for specific media, site areas, and operable units can be studied separately in the FS process or combined into comprehensive alternatives for the entire site.

As practicable, a range of treatment alternatives is developed, varying primarily in the extent to which they rely on the long-term management of residuals and untreated wastes. The upper bound of the range would be an alternative that would eliminate, to the extent feasible, the need for any long-term management (including monitoring) at the site. The lower bound would consist of an alternative involving treatment as a principal element, although some long-term management of portions of the site that did not constitute "principal threats" would be required. Between the upper and lower bounds of the treatment range, alternatives varying in the type and degrees of treatment and associated containment disposal requirements should be included as appropriate, and a no-action alternative must be developed.

After the development of the alternatives, the next step is to screen them. In this step, the universe of potentially applicable technology types and process options is reduced by evaluating the options with respect to their technical implementability. During screening, process options and entire technology types are eliminated from further consideration on the basis of technical implementability. This is accomplished by using readily available information (from the RI site characterization of contaminated types and concentrations and on-site characteristics) to screen out technologies and process options that cannot be effectively implemented at the site. As with all decision making during an RI/FS, the screening of technologies must be documented.

Detailed Analysis of Alternatives. The detailed analysis of alternatives consists of analyzing and presenting relevant information to help decision makers select a site remedy. However, the detailed analysis of alternatives is not the decision making process itself. During this analysis, each alternative is assessed against the evaluation criteria described in the following list. The results of the assessment are organized so alternatives can be compared and key tradeoffs among them identified. This approach is designed to provide decision makers with enough information to compare alternatives adequately, select an appropriate remedy for a site, and demonstrate satisfaction of the Superfund remedy selection requirements in the record of decision (ROD).

Section 121(b)(1)(A) of Superfund requires an evaluation of the long-term effectiveness and related considerations for each of the alternative remedial actions. Nine evaluation criteria have been developed to address the statutory requirements and to address the additional technical and policy considerations EPA has determined to be important for selecting among remedial alternatives, as outlined in Figure 14-3 [40

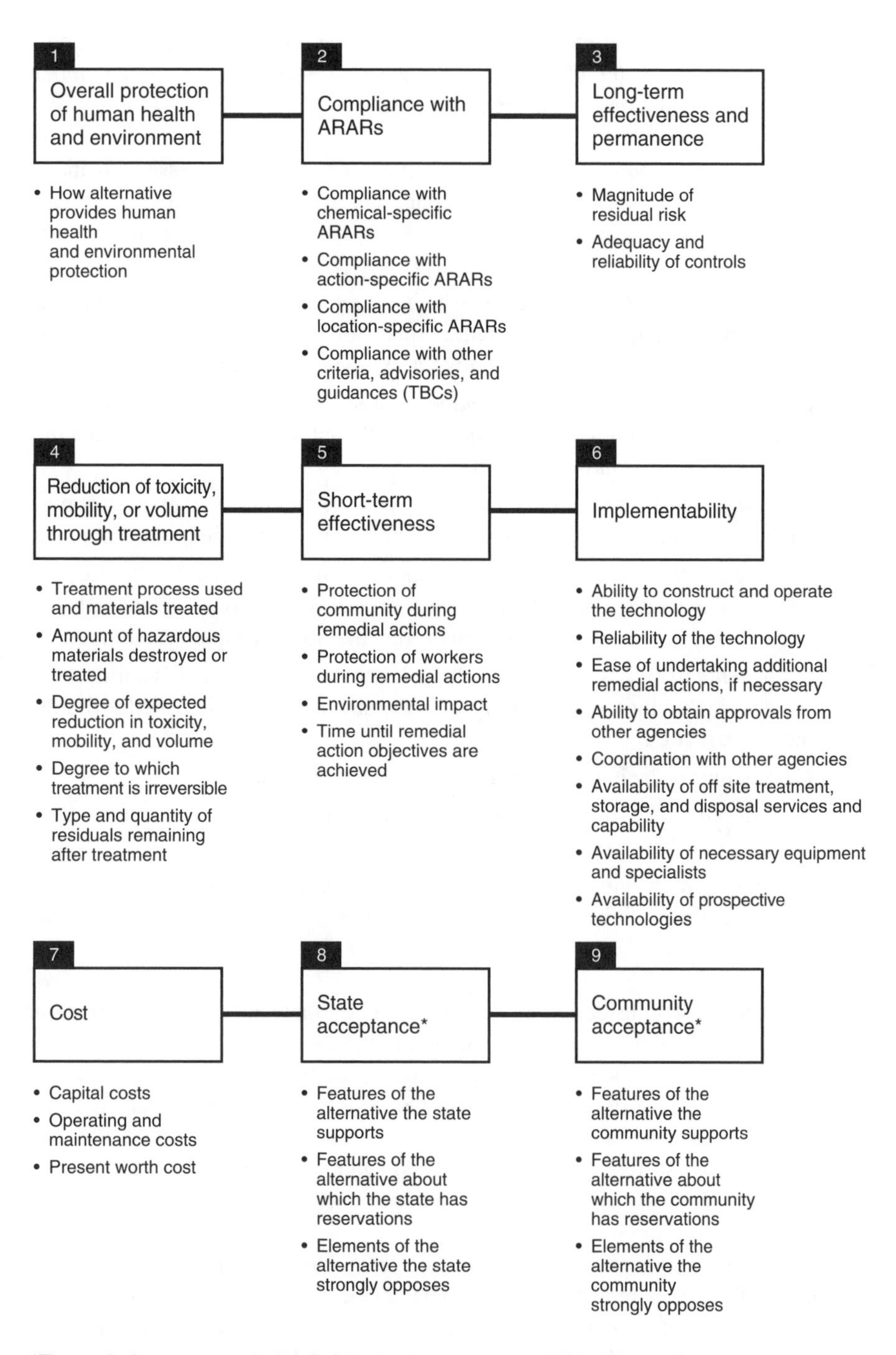

Figure 14-3. The Nine CERCLA Criteria for Evaluating a Remedy

CFR 300.430(e)(9)]. These evaluation criteria serve as a basis for conducting detailed analyses during the FS and for subsequently selecting an appropriate remedial action. The evaluation criteria are as follows:

- *Overall protection of human health and the environment* assesses whether a remedy provides adequate protection and describes how risks posed through each pathway are eliminated, reduced, or controlled through treatment, engineering controls, or institutional controls.
- *Compliance with ARARs* assesses whether a remedy will meet all the applicable or relevant appropriate requirements of federal and state environmental standards and/or provide grounds for involving a waiver.
- *Long-term effectiveness and permanence* assesses the ability of a remedy to maintain reliable protection of human health and the environment over time, once cleanup goals have been met.
- *Reduction of toxicity, mobility, or volume through treatment* assesses the anticipated performance of the treatment technologies that a remedy may employ.
- *Short-term effectiveness* assesses how long it will take during the construction and implementation period to provide protection and prevent any adverse impacts on human health and the environment until cleanup goals are met.
- *Implementability* assesses the technical and administrative feasibility of a remedy, including the availability of materials and services needed to implement a particular option.
- *Cost* estimates capital and operation and maintenance costs.
- *State acceptance* evaluates the technical or administrative issues and concerns the state may have regarding each alternative.
- *Community acceptance* assesses the issues and concerns the public may raise regarding each of the alternatives and evaluates the degree of community acceptance.

Concerning cost considerations, the NCP states that "an alternative that far exceeds (for example, by an order of magnitude) the costs of other alternatives evaluated and does not provide substantially greater public health or environmental benefits should usually be excluded from further consideration." If the site is fund-financed, it must also be *fund-balanced.* Fund balancing means that the cost required for action at the site must be weighed against other priority sites to determine if there is adequate money in the fund to respond to those sites as well.

Cleanup Requirements. Superfund does not specify cleanup standards per se; however, the NCP requires remedial actions to be undertaken in compliance with *applicable* or *relevant* and *appropriate* environmental and public health requirements (ARARs) [40 CFR 300.400(g)]. As a result, ARARs have become the de facto cleanup standards. The requirements that must be complied with are those that are applicable or relevant to the hazardous substances or pollutants or contaminants at a site. Any such requirements may be waived under specified conditions, as outlined under the "Statutory Waivers" heading that follows, provided that protection of human health and the environment is still assured.

Applicable requirements are those cleanup standards, standards of control, and other environmental protection requirements, criteria, or limitations promulgated under federal or state law that specifically address a hazardous substance, pollutant or contaminant, remedial action, location, or other circumstances at a Superfund site. If a requirement is not applicable, it still may be relevant and appropriate [40 CFR 300.400(g)].

Relevant and *appropriate requirements* are those cleanup standards, standards of control, and other environmental protection requirements, criteria, or limitations promulgated under federal or state law that, while not applicable to a hazardous substance, pollutant, or contaminant, remedial action, location, or other circumstances at a Superfund site, address problems or situations similar to those encountered at the site, and whose use is well suited to the particular site. The relevance and appropriateness of a requirement can be judged by comparing a number of factors, including the characteristics of the remedial action, the substances in question, and the physical circumstances of the site [40 CFR 300.400(g)(2)].

In addition to ARARs, EPA considers the use of nonpromulgated, to-be-considered (TBC) guidelines and controls. TBCs, which are not legally binding and do not have the same status as ARARs, are used when ARARs are not fully protective. TBCs typically include health or exposure advisories to specific chemicals or classes of chemicals.

Whereas the applicability determination is a legal one, the determination of relevancy and appropriateness relies on professional judgment that considers environmental and technical factors at the site. There is some flexibility in the relevance and appropriateness determination: a requirement may be relevant, in that it covers situations similar to that at the site, but it may not be appropriately applied for various reasons, and therefore may not be well suited to the site. In some situations, only portions of a requirement or regulation may be judged relevant and appropriate, but if a requirement is applicable, all substantive parts must be followed (OSWER Directive 9234.01).

For example, if closure requirements under Subtitle C of RCRA (e.g., the unit received hazardous waste after 1980) are applicable, the unit must be closed in compliance with one of the closure options available under Subtitle C. These closure options are closure by removal (clean closure) or closure with wastes or waste residues in place (landfill-type closure). However, if Subtitle C is not applicable, then a hybrid system, which could include other types of closure designs, also could be used. The hybrid closure option arises from a determination that only certain closure requirements in the two Subtitle C closure alternatives are relevant and appropriate.

ARARs are divided into three general types, which are chemical-specific, action-specific, and location-specific requirements:

- *Chemical-Specific ARARs* are usually health- or risk-based numerical values or methodologies that, when applied to site-specific conditions, result in the establishment of numerical values. Thus, these values establish an "acceptable" amount or concentration of a chemical that may be found in, or discharged to, the ambient environment.

- *Action-Specific ARARs* are usually technology- or activity-specific requirements or limitations on actions taken with respect to hazardous waste.
- *Location-Specific ARARs* are usually restrictions placed on the concentration of hazardous substances or the conduct of activities solely because they occur in special locations.

On-Site versus Off-Site Actions. Under Superfund, any on-site action is exempt from federal, state, or local permitting requirements (e.g., RCRA, TSCA) [40 CFR 300.400(e)]. On-site actions generally need to comply only with the technical aspects of the requirements, not with the administrative aspects; that is, neither permit applications nor other administrative reviews are considered ARARs for actions conducted on-site, and therefore they should not be pursued. However, the RI/FS, record of decision, and design documents should demonstrate full compliance with all technical requirements that are considered applicable or relevant and appropriate.

Off-site actions do, however, require full compliance with all aspects of the applicable regulations. Thus, off-site actions only need to comply with applicable requirements, not with relevant and appropriate requirements. The off-site transfer of hazardous wastes is allowed only if the transfer is made to a facility operating in compliance with RCRA (or in compliance with TSCA or other federal laws where applicable) and applicable state requirements. As soon as a hazardous waste is removed off-site from a Superfund site, appropriate administrative requirements become applicable (e.g., permits, manifests) [40 CFR 300.440(b)].

In addition, any off-site transfer of a RCRA hazardous waste is authorized only if the facility has no releases of hazardous waste and the facility has no significant violations (if the facility has had a significant release, it must be controlled by an enforceable agreement for corrective action).

Using ARARs. ARARs are identified on a site-specific basis and depend on the specific chemicals at a site, the proposed remedy selection, and the site characteristics. The ARARs for any site should be identified and considered at various points in the remedial planning stages, specifically, as follows:

- During the scoping process of RI/FS, chemical-specific and location-specific ARARs should be identified on a preliminary basis.
- During the site characterization phase of the remedial investigation, when the risk assessments are conducted to assess risks at a site, the chemical-specific and location-specific ARARs should be identified more comprehensively than they first were, and then used to help determine the cleanup goals.
- During the development of remedial alternatives in the feasibility study, action-specific ARARs should be identified for each of the proposed alternatives and considered along with other ARARs.

- During the detailed analysis of alternatives, ARARs for each alternative should be examined as a group to determine what is needed to comply with other laws and to be protective of human health and the environment.
- When an alternative is selected, it must be able to attain the ARARs unless one of the six statutory waivers discussed in the following subsection is invoked.
- During the remedial design phase, the technical specifications of construction must ensure attainment of the ARARs.

Statutory Waivers. EPA is authorized to select a remedial action that protects human health and the environment but does not meet the ARARs for on-site actions if any of the following applies [40 CFR 300.430(f)(1)(ii)(C)]:

1. The remedial action is an interim measure, and the final remedy will attain the ARARs upon its completion.
2. Compliance with the ARARs will result in greater risk to human health and the environment than other options.
3. Compliance with the ARARs is technically impracticable.
4. An alternative remedial action will attain the equivalent of the ARARs.
5. In the case of state requirements, the state has not consistently applied state requirements in similar circumstances.
6. For remedial actions under §104, compliance with the ARARs will not provide a balance between protecting public health, welfare, and the environment at the facility and the availability of fund money for response at other facilities *(fund-balancing).*

If a remedial action results in any hazardous substances remaining on-site, EPA is required to review the remedy at least every five years to assure that human health and the environment are being protected. If the review indicates that additional action is needed, EPA is required to take the action and report to Congress on the sites and the actions taken.

ARARs for Contaminated Water. When determining the applicable or relevant and appropriate actions involving contaminated surface water or groundwater, the most important factors to consider are the uses and potential uses of the water. The actual or potential use of water, and the manner in which it is used, will determine what requirements may be applicable or relevant and appropriate. For groundwater that is or may be used for drinking, the maximum contaminant level goals (MCLGs) and maximum contaminant levels (MCLs) established under the Safe Drinking Water Act are generally the applicable or relevant and appropriate standard. If MCLs do not exist for contaminants identified at a site, then health advisories established by EPA's Office of Drinking Water, reference doses (RFDs), and carcinogenic slope factors (CSFs) are applicable or relevant and appropriate.

State ARARs. The requirement that ARARs must be attained also applies to any state requirement promulgated under a state environmental or facility-citing law that is more stringent than any federal requirement, and that has been identified to EPA in a timely manner.

The application of a state requirement that results in a statewide ban on land disposal is prohibited unless each of the following considerations is met:

- The state requirement is generally applicable and was adopted by formal means.
- The state requirement was adopted on the basis of hydrologic, geologic, or other relevant considerations and was not adopted to preclude on-site remedial actions or other land disposal for reasons unrelated to protection of human health and the environment.
- The state arranges for and pays the incremental costs of utilizing another treatment or disposal facility.

If the proposed remedial action does not attain an ARAR (federal or state), EPA must provide the state with an opportunity to concur with the remedy before EPA accepts a settlement with responsible parties. If the state concurs, it may become a signatory to the consent decree. (A *consent decree* is an administrative order, having the force of law, that outlines the requirements for the responsible parties concerning a site's remedial action.) If the state does not concur and insists the ARAR be met, it can intervene, as a matter of right, in the enforcement action before joining the consent decree to ensure the remedial action conforms to the ARAR. The remedial action must be made to conform to the ARAR if the state establishes, on the basis of the administrative record, that EPA's decision to waive the ARAR was not supported by substantial evidence. However, if the court determines that the remedial action does not need to conform to the ARAR, and the state insists the ARAR be met, the consent decree must be modified to incorporate the ARAR if the state agrees to pay the additional costs associated with meeting it.

Land Disposal Restrictions. The temporary or permanent placement of restricted hazardous wastes on land can trigger the RCRA land disposal restrictions. According to OSWER Directive 9347.3-05FS, placement, which is the applicability trigger for land disposal restriction, does not occur when wastes are moved or treated within an area of contamination, but may occur when hazardous wastes are placed on the land outside of areas of contamination. (An *area of contamination* is essentially a discrete zone of continuous contamination, such as a Superfund site.)

Post-RI/FS Action. After the RI/FS is completed, the results of the detailed analyses, combined with the risk management judgments made by the decision maker, become the rationale for selecting a preferred alternative and preparing the proposed plan, as explained in the following discussion on remedy selection. Therefore, the results of the detailed analysis, or more specifically the comparative analysis, should highlight the relative advantages and disadvantages of each alternative so that the key tradeoffs can be identified. These tradeoffs, coupled with risk management decisions, will serve as the rationale for action, providing a transition between the RI/FS report and the development of the proposed plan and, ultimately, the record of decision.

Remedy Selection/Record of Decision

The remedy selection process, outlined in Figure 14-4, involves the selection of the preferred alternative, the issuance of a proposed plan, a public comment period for the preferred alternative, the final remedy selection, and, documentation of the final remedy selection in the record of decision (ROD).

Preferred Alternative Selection. After the RI/FS is complete, the lead agency (e.g., EPA, the state) identifies a *preferred alternative*: the protective, ARAR-compliant approach judged to provide the best balance of tradeoffs with respect to the nine evaluation criteria described previously in the "Detailed Analysis of Alternatives" subsection of this chapter and depicted in Figure 14-3.

Proposed Plan. The preferred alternative for a site is presented to the public in a *proposed plan,* the purpose of which is to facilitate public participation in the remedy selection process by [40 CFR 300.430(f)(2)]

- Identifying the preferred alternative for a remedial action at a site or an operable unit and explaining the reasons for the preference.
- Describing other remedial options that were considered in detail in the RI/FS report.
- Soliciting public review and comment on all the alternatives described.
- Providing information on how the public can be involved in the remedy selection process.

The proposed plan contains the "identified" preferred alternative, chosen on the basis of available information, but it does not present a remedy "selected" for implementation. An important function of the proposed plan is to solicit public comment on *all* of the alternatives considered in the RI/FS detailed analysis because a remedy other than the preferred alternative may be selected.

Public Comment. The lead agency must provide a reasonable opportunity for submission of written and oral comments and an opportunity for a public meeting, at or near the facility at issue, regarding the proposed plan and any proposed findings. The lead agency must make the relevant documents (e.g., proposed plan, RI/FS) available to the public when the public comment period begins. This period must last a minimum of 30 days to allow time for the public to comment on the informa-

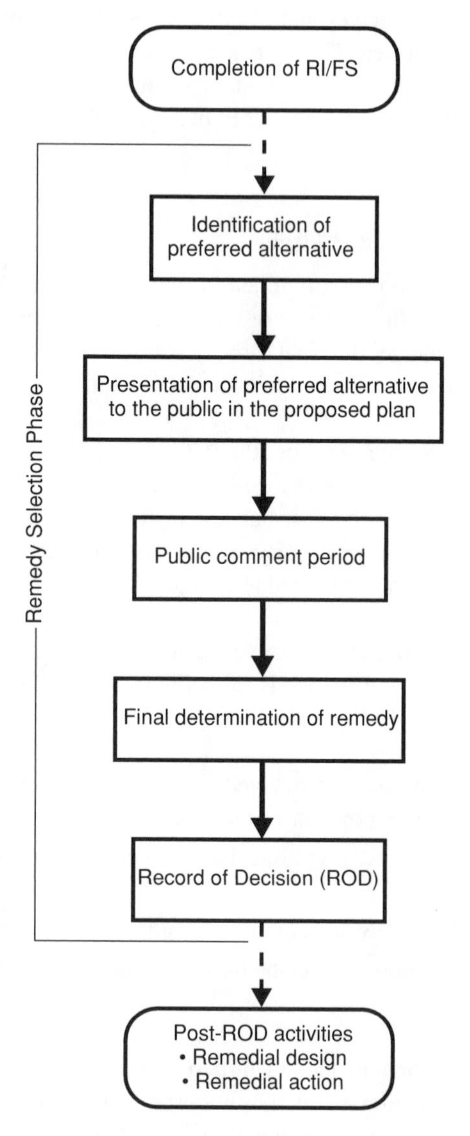

Figure 14-4. CERCLA Remedy Selection Process

tion contained in the RI/FS report. However, upon timely request, an additional 30 days of public comment may be requested [40 CFR 300.430(f)(3)(C)].

Final Remedy Selection. In this step, the lead agency makes the final remedy selection based on any new information or points of view expressed by the state or the public [40 CFR 300.430(f)(4)]. The proposed remedy selection may be adopted as is, modified, or replaced by a more appropriate alternative. Selecting a final remedy at federal facilities requires a joint selection by the federal agency and EPA. If an agreement is not reached, EPA selects the final remedy.

Record of Decision. The record of decision (ROD) documents the remedy selection and serves as the final remedial action plan for a site or an operable unit [40 CFR 300.430(f)(5)]. The ROD has the following purposes:

- It serves a legal function, in that it certifies that the remedy selection process was carried out in accordance with the NCP.
- It serves as a technical document that outlines the engineering components and remediation goals of the selected remedy.
- It provides the public with a consolidated source of information about the history, characteristics, and risks posed by the conditions at the site, as well as a summary of the cleanup alternatives considered, their evaluation, and the rationale behind the selected remedy.

The ROD consists of three basic components:

- declaration,
- decision summary, and
- responsiveness summary.

The *declaration* is an abstract of the key information contained in the ROD and is the section of the ROD signed by the EPA Regional Administrator. It provides a brief description of the selected remedy for the site and a formal statement explaining that the selected remedy complies with CERCLA and is consistent, to the extent practicable, with the NCP.

The *decision summary* provides an overview of the site characteristics, the alternatives evaluated, and the analysis of those options. The decision summary also identifies the selected remedy and explains how the remedy fulfills statutory requirements.

The *responsiveness summary* addresses public comments received on the proposed plan, the RI/FS report, and other information in the administrative report.

Remedial Design

The *remedial design phase* is the engineering phase. The purpose of the remedial design is to develop a detailed set of plans and specifications for conducting the selected remedial action based on the ROD, the cleanup levels to be attained, and site characteristics. At this point, for EPA-led sites, EPA may turn over the management of the cleanup action to the U.S. Army Corps of Engineers because of its experience in major construction projects. The Corps then hires its own contractors to perform the work; however, the EPA Remedial Project Manager (RPM) provides environmental oversight at the site. If a state or responsible party has the lead, it continues to manage the design and construction for the site under the oversight of the EPA RPM.

Remedial Action

Following completion and approval of the remedial design, the remedial action phase is implemented. Otherwise known as the "construction" phase, the *remedial action phase* is the process in which remedial activities, including treatment, removal, and all other necessary tasks, are undertaken. The time required for this phase depends on the complexity of the site. When all phases of the remedial activities at a site have been completed and no further action is warranted, the site will be deleted from or recategorized on the NPL.

The remedial action must conform to the remedy set forth in the ROD and all other decision documents. If the remedial action phase differs from the ROD, an explanation of significant differences must be published or the ROD must be revised [40 CFR 300.435(b)(2)].

At any site where the remedial action does not completely remove all contamination, operation and maintenance (O&M) is conducted. (The site may be eligible for evental deletion from the NPL.) For federal fund-lead sites, the state must assume responsibility for any operation associated with the site remedy. EPA shares the costs of O&M for a period not to exceed one year, based on the date of project completion [40 CFR 300.435(f)].

Deletion from the NPL

When no further response action is appropriate, a site may be deleted from the NPL or recategorized.

The criteria that must be met for a site to be deleted from the NPL include the following [40 CFR 300.425(e)]:

- EPA, in consultation with the state, must have determined that the responsible or other parties have implemented all appropriate response actions.
- All appropriate fund-financed responses must have been implemented, and EPA, in consultation with the state, must have determined that no further response is appropriate.
- Based on a remedial investigation, EPA, in consultation with the state, must have determined that no further response is appropriate.

In addition, because the modification of the NPL means the modification of the NCP, the process is subject to standard rulemaking procedure. Thus, EPA must notify the public via the *Federal Register* and provide the opportunity for public comment.

EPA has established a new category for the NPL, the "construction completion" category. Sites may be categorized as "construction complete" only after remedies have been implemented and are operating properly. Such sites may include those awaiting deletion, sites awaiting five-year review or deletion, and sites undergoing long-term remedial actions.

EPA has established a policy on the partial deletion of sites listed on the NPL (November 1, 1995, 60 *FR* 55466). This policy allows for portions of an NPL site where no further response action is appropriate to be deleted from the NPL.

It is important to note that, for those sites where waste remains, deletion from the NPL (even after cleanup is complete) cannot occur until at least one five-year review has been conducted. When a site is deleted from the NPL, it is technically closed out with respect to the federally funded remedial response. However, any site deleted from the NPL is eligible for further federal fund-financed remedial response actions if further conditions warrant such actions.

15

MISCELLANEOUS PROVISIONS UNDER SUPERFUND

A number of provisions established under Superfund are indirectly part of the response action activities. This chapter discusses these provisions. The "miscellaneous" provisions that are addressed in this chapter are as follows:

- Enforcement
- Public participation
- Worker safety and health program

ENFORCEMENT

A major goal of Superfund is to encourage responsible parties to finance and conduct appropriate response actions and to recover the costs for response actions that were financed with fund money. Sections106 and 107 of Superfund form the basis of this authority. The enforcement provisions of §106 authorize the use of administrative orders or the pursuit of civil actions to compel responsible parties to undertake cleanup action. Section 107 makes responsible parties liable for the costs incurred by federal or state governments or private parties that conduct response actions consistent with the NCP, as discussed in Chapter 14.

The Superfund enforcement process normally used by EPA to promote involvement of the potentially responsible party (PRP) typically includes the following:

- EPA attempts to identity PRPs as early as possible. Once they have been identified, EPA notifies the parties of their potential liability.

- In the course of identifying the response work that must be done, EPA encourages PRPs to do work on the site.
- If EPA believes the PRPs to be willing and able to do the work, EPA attempts to negotiate an enforcement agreement with them. The enforcement agreement may be an agreement entered in court (e.g., a judicial consent decree) or a mutual agreement outside of court (e.g., an administrative order).
- If a settlement is not reached, EPA can use its authority to issue a unilateral administrative order against the PRPs. If the PRPs do not respond to an administrative order, EPA can file a lawsuit to compel performance.
- If the PRPs do not perform the response action and EPA undertakes the work, EPA may file suit against the PRPs and recover up to triple the government's response costs.

Identification of PRPs

After site discovery, the first step is to search for and identify the potentially responsible parties. Information from a local county clerk or registrar can be used to identify a person or party paying property taxes, previous title holders, or business license applications. In addition to an informational search, a field investigation may aid in the identification of responsible parties. An on-site visual investigation might include inspections of trash receptacles, drum labels, abandoned vehicles and license plates, and names of contract companies (e.g., for fences, trash removal, landscaping). If a PRP is located, especially a solvent one, that party is usually made aware that he or she could be liable for all response costs incurred and that identifying additional parties would help distribute the liability, an approach that often facilitates the search.

After a PRP has been identified, EPA issues a *general notice letter* stating that the government has identified the party as a PRP liable for any cleanup costs incurred. This notice procedure allows the PRPs, if there is more than one, to organize and select representatives to meet with EPA in subsequent negotiations. PRPs who do not participate may be sued either by the government or by other PRPs for cost reimbursement after cleanup costs have been incurred (40 CFR 300.700).

Special Notices and Enforcement Moratorium. EPA may issue a *special notice* calling for a temporary enforcement moratorium if negotiations would facilitate an agreement with PRPs to either undertake or finance a response action [Superfund §122(e)(2)(A)]. EPA is required to notify all parties as well as the state of such negotiation procedures, and to provide them with appropriate information when available. This information includes the names and addresses of other PRPs, the volume and nature of substances contributed by each PRP, and a ranking by volume of the substances. The PRPs who receive special notice have 60 days to submit a proposal to undertake or finance the response action. During this 60-day period, EPA may not initiate the RI/FS; however, additional studies or investigations authorized under Superfund §104(b) may be initiated. Nothing precludes EPA from undertaking response or enforcement actions if a site poses a significant threat to human health or the environment. If the PRPs do not submit a good-faith proposal within 60 days of special notice, EPA may proceed with the RI/FS. If PRPs submit a good-faith effort within 60 days of special notice, the enforcement moratorium continues until 90 days past the date of notice while EPA evaluates the proposal.

Negotiations. Negotiation with potentially responsible parties is an integral part of the enforcement process. Negotiations are conducted either to secure responsible party action to fund or conduct cleanup or to recover federal funds already expended.

Negotiations for removal actions usually are of short duration, whereas negotiations for remedial actions generally last longer and may take place before, during, or after a response action, depending on site circumstances as well as the nature and immediacy of the threat.

Section 122 of Superfund sets forth procedures for negotiating settlements with PRPs for conducting response actions. This section essentially formalizes the settlement process that had been established under EPA's existing settlement policy, with some additions.

Settlements

EPA is authorized to enter into agreements with PRPs to conduct response actions and to decide if the specified settlement procedures should be used. There are two types of settlement documents: administrative orders and consent decrees. An administrative order is not subject to judicial review. A *consent decree,* which is a legal document specifying an entity's obligations when that entity enters into a settlement option with the government, must be entered as an order by a federal district court or a state court. A consent decree must also be published in the *Federal Register* for public comment before its approval by a judge. (Remedial action agreements must be entered into as consent decrees, whereas administrative orders are allowed for both removal and non-RI/FS remedial actions.) EPA is not required to make a finding of imminent and substantial endangerment, and entering into a consent decree is not an acknowledgment of an imminent and substantial endangerment or liability. Participation by a PRP in a settlement is not admissible in any judicial or administrative proceeding except as is consistent with federal rules of evidence.

There are a number of settlement options available to EPA and PRPs. These depend on a number of factors, including the extent of action by other PRPs and the amount of waste contributed. Available settlement options include the following:

- Nonbinding preliminary allocation of responsibility
- Covenants not to sue
- Mixed funding settlements
- *De minimis* waste settlements
- *De micromis* settlements

Nonbinding Preliminary Allocation of Responsibility. A nonbinding preliminary allocation of responsibility (NBAR) is a method used by EPA to allocate percentages of total response costs at a facility among PRPs, primarily based on the volume of waste contributed by a PRP. NBARs are used to expedite a settlement if PRPs fail to determine allocations of responsibilities among themselves. An NBAR is not admissible as evidence in any proceeding including citizen suits, is not subject to court review, and does not constitute an apportionment or other statement of divisibility of harm or causation. Generally, an NBAR is prepared during the RI/FS. The costs incurred by EPA in preparing the NBAR must be reimbursed by the PRPs. The guidelines for preparing NBARs were promulgated on May 28, 1987 (52 *FR* 19919).

Covenants Not to Sue. Section 122(f) of Superfund authorizes EPA to issue covenants not to sue for liability, including future liability, if certain conditions are met (e.g., the covenant is in the public interest and would expedite a response action approved by EPA). The criteria for such a covenant are the same as those provided in EPA's *settlement policy* (50 *FR* 5034, February 5, 1985). Such covenants are required to contain a reservation of rights, known as a reopener, for unknown conditions, except in extraordinary circumstances. (A *reopener* is a provision that reserves EPA's right to require settling parties to take further response action in addition to cleanup measures already provided for in a settlement agreement, notwithstanding the covenant not to sue.) EPA also will include in such covenants a second reopener to cover situations in which additional information reveals that a remedy no longer protects public health or the environment. This second reopener is triggered by a threshold of protection of public health or the environment rather than the imminent and substantial endangerment threshold.

EPA must provide a covenant not to sue under Superfund for future releases under certain circumstances; for example, if an on-site remedial action that meets the NCP is not selected and the hazardous substances are transported off-site to a facility that meets the requirements of RCRA, or if a remedial action involves treatment so the hazardous substances will no longer present any current or currently foreseeable future risks to public health, welfare, or the environment.

A covenant not to sue concerning future liability to the federal government does not take effect until EPA certifies that the remedial action has been completed in accordance with Superfund at the site subject to such covenant. The date of completion of the remedial action is the date at which remedial construction has been completed. The exact point when EPA can certify completion of a particular remedial action depends on the particular requirements of that remedial action; however, in general, the operation and maintenance activities are not included as the remedial action if they successfully attain the requirements set forth in the record of decision and the remedial design.

EPA considers a number of factors in determining whether to issue a covenant not to sue. These factors include the following:

- The effectiveness and reliability of the remedy

- The nature of the risks remaining at the facility
- The extent to which performance standards are included
- The extent to which the response action provides a complete remedy
- The extent to which the technology has been demonstrated to be effective
- Whether the fund would be available for any additional remedial action
- Whether the remedial action will be carried out, in whole or in part, by the responsible parties.

Mixed Funding Settlements. Because of the number and economic status of PRPs at a typical Superfund site, EPA was granted authority under §122(b) to settle with some PRPs at a site while pursuing cost recovery with nonsettled PRPs. On March 14, 1988 (53 *FR* 8279), EPA promulgated its policy on mixed funding settlements. This policy includes a "cash out" option where a PRP pays for a portion of the response costs up front and EPA performs the response action. Or, the PRP may agree to conduct discrete portions of the response action.

De Minimis Settlements. When practicable and in the public interest, §122(g) of Superfund authorizes EPA to reach final settlements with PRPs if the settlement involves a minor portion of the response costs and the wastes sent to the site by the PRP are minimal compared to the other hazardous substances at the facility in terms of volume and toxicity *(de minimis).* Final settlements may also be negotiated with landowner PRPs if the landowner did not conduct or permit the disposal of hazardous waste at the site, did not contribute to the release of hazardous substances by an act or omission, and did not buy the property with the knowledge that hazardous substances had been disposed of at the site. Potentially responsible parties claiming a defense to liability, as opposed to being less culpable than other PRPs, must meet the requirements of §101(f) to establish that defense.

To qualify as a *de minimis* generator or transporter (contributor), the PRP must have contributed an amount of hazardous substances that is minimal compared to the total amount of hazardous substances at the facility. Also, the PRP must have contributed hazardous substances that do not have a significantly greater hazardous effect than others at the facility. For example, if all of the PRPs at a site disposed of the same solvent wastes, those PRPs who had contributed a minimal amount in relation to the total could qualify for *de minimis* status. If, however, a PRP disposed of a minimal amount of waste that was more toxic or that exhibited other more serious hazardous effects than the other wastes at the site, that PRP, despite the minimal volume of the contribution, would probably not qualify as a *de minimis* contributor. Even if a waste contributor meets the volume and toxicity requirements for *de minimis* contributor status, a possible settlement with a *de minimis* PRP must be determined by EPA to be "practicable and in the public interest."

The goal of negotiations with *de minimis* parties is to achieve quick and standardized agreements with the minimal expenditure of enforcement resources and transaction costs. To attain this goal, the *de minimis* settlement does not require a commitment to perform work but rather a payment to be made to the fund. However, in appropriate cases, EPA can consider entering into *de minimis* settlements under which the settling parties agree to perform a discrete portion of the response action needed for the site, such as preparing a RI/FS or the cleanup of an operable unit.

De minimis settlers may be granted a covenant not to sue to protect them from civil claimants seeking injunctive relief under §§106 and 107 of Superfund if EPA determines that such a covenant is consistent with the public interest. Natural resource damage claims may not be released, however, and are expressly reserved unless the natural resource trustee has agreed in writing to such a covenant not to sue.

De Micromis Settlements. A *de micromis* waste contributor, a subset of a *de minimis* contributor, is a generator or transporter who is responsible for only a miniscule amount of waste at a Superfund site. Similar to *de minimis* contributors, *de micromis* contributors can "cash out" with a settlement early in the response action process (OSWER Directive 9834.17).

Cost Recovery

If the parties involved do not reach settlement(s), EPA will conduct a response action, using monies provided by Superfund. EPA will subsequently pursue cost recovery from the liable parties. Private parties may also seek reimbursement for response costs from other liable parties. EPA can attempt to recover costs through a variety of legal and administrative tools including negotiations, demand letters, arbitration, alternative dispute resolution, administrative settlements, judicial settlements, and litigation. Also, Superfund §§107(l) and (m) authorize the imposition of a lien against the PRP or subject site.

Citizen Suits

Superfund allows any person to take civil action against EPA for failure to undertake a nondiscretionary duty under Superfund or against any person (including personnel at federal facilities) for violation of any standard, regulation, condition, requirement, or order under Superfund.

A citizen suit cannot begin until 60 days after a plaintiff has given notice of violation to EPA, to the state where the alleged violation occurred, and to the alleged violator. An action cannot be commenced for a violation if EPA is "diligently" prosecuting the action under RCRA or Superfund.

PUBLIC PARTICIPATION

Because of the site-specific nature of Superfund response actions, and the program's underlying intent, a comprehensive public participation program has been developed. The NCP mandates community involvement at specific points in the response process. Pertaining to public participation, response actions include all removal, remedial, and enforcement actions.

EPA's general route of action is to pursue legal means to compel responsible parties to pay for or conduct the cleanup. Thus, in an enforcement-led site, EPA may be limited in the amount and nature of the information that it can make available to the public because there might be sensitive or confidential information whose public disclosure could damage the government's legal case. However, if an enforcement action is embodied in a consent decree, including other proposed settlement agreements, the consent decree must be available for public comment [40 CFR 300.430(c)].

As provided for in 40 CFR 300.700(a), private parties may undertake a response action when a hazardous substance is released. Although private-party response actions must comply with most of the public participation requirements under CERCLA, they are not required to establish and maintain an administrative record file or an information repository, which is done by the lead agency [40 CFR 300.700(c)(6)].

Administrative Record File

An administrative record file containing all information and documentation used in the selection of a response action must be established [40 CFR 300.800(a)]. The administrative record file must be available for public review, and a copy must be maintained near the site. In addition, a duplicate record must be maintained in a central location, such as a state office or the lead agency's office. For remedial response actions, the administrative record file must be made available to the public when the remedial investigation begins. For removal response actions, the administrative record file must be established no later than 60 days after initiation of an on-site removal activity for emergency and time-critical actions, or, for non-time-critical actions, when the engineering evaluation/cost analysis is made publicly available.

The administrative record file must contain all relevant and appropriate documentation, public comments, site-specific data, decision documents, enforcement orders, guidance documents, and technical documents and references used by the lead agency in considering decisions involving the response action (40 CFR 300.810). In addition, an inventory of all the file's contents must be included in the file.

Community Relations Plan

Except for emergency and time-critical removal actions, a site-specific community relations plan must be developed [40 CFR 300.415(m)(3)(1) and 300.430(c)(2)(ii)]. The content and scope of the plan are developed by interviewing residents, local officials, and community groups to ascertain the community's primary concerns and the extent of the community's desire to participate in the response action decision-making process. This information is used to prepare a site-specific community relations plan. The plan outlines in detail the activities that will be conducted to ensure that members of the community can present their opinions and concerns about the site as well as be kept informed throughout the response process. Typically, one of the first steps in the community relations program is to establish an information file at a local public building known as an information repository, which is separate from the administrative record file. The information repository is to contain site-related material such as news releases, technical reports, and fact sheets related to site activities. Each site also has a designated Community Involvement Coordinator, usually an EPA staff person located at a EPA Regional Office, who is available to answer questions regarding the site.

Response-Action Specific Requirements

Because response action—removal and remedial—are different, the following sections summarize the public participation requirements for each.

Removal Actions. A removal action usually involves an immediate response designed for a quick cleanup. In such cases, the community relations effort is limited to informing the public about the response action and its potential health and environmental effects [40 CFR 300.415(m)]. However, because removal actions can differ based on the time necessary to complete the action (i.e., less than six months and more than 120 days), there are the following different public participation opportunities:

- *Removal Actions Less Than Six Months.* For these short-term response actions, an administrative record file must be maintained, the public must be notified, and the public must be provided an opportunity to comment on the removal response action [40 CFR 300.415(m)(2)].
- *Removal Actions More Than 120 Days.* For these longer-term removal actions, the public must be notified; the public, local officials, and affected parties must be interviewed to determine the extent of their concern; an administrative record file must be established; a community relations plan must be prepared; and the public must be provided an opportunity to comment on the response action [40 CFR 300.415(m)(3)]. If an engineering evaluation/cost analysis (EE/CA) will be prepared for the response action as discussed in Chapter 14, the public will also have an opportunity to specifically comment on the EE/CA.

Remedial Actions. A remedial action can take years to develop and implement and can include a number of steps. Because of the significance of its potential impact, a more detailed public participation program and community relations plan is required for a remedial action. For example, because the feasibility study requires a detailed analysis of the level of community support for each alternative, more substantial public involvement is necessary. In addition, the plan has to include provisions for the additional opportunities and degree of involvement necessitated by the remedial response process. For example, the public must be involved in the proposed listing of a site on the NPL, the RI/FS scoping, the feasibility analysis, selection of the preferred alternative, the remedial design and action phase, the eventual proposed deletion of the site from the NPL, and consent decrees. The specific requirements of public participation include the following:

- Before the adoption of any remedial action plan, the public must have access to a published notice and a brief analysis of the proposed action, as well as an opportunity for a public meeting.
- The final remedial action plan must be made available to the public before the commencement of any remedial action. If there are any significant changes to the proposed plan, a discussion of the reasons for these changes must be included.
- Responses to all significant public comments, criticisms, and new data submitted in written comments or during oral presentations must accompany the data.
- After the adoption of a final remedial action plan, if there is remedial action, enforcement action, or settlements that differ in any significant aspects from the final plan, an explanation of the reasons for such changes must be published in a major local newspaper.

Although EPA tries to include the community's preferences in selecting a remedy for a site, it may be forced to select a response action that is not the community's choice because Superfund requires that the chosen remedy be cost-effective, fund-balanced, and selected as part of a permanent solution.

Technical Assistance Grants

Qualified citizen's groups are eligible to receive Technical Assistance Grants (TAGs) of up to $50,000. These grants may be used to hire independent technical advisors to help citizens understand, review, and comment on technical factors and activities related to response actions. To be eligible, a group must be a citizen's association or an environmental or health advocacy group that demonstrates a genuine interest in the site and must also be incorporated as a nonprofit organization. A TAG grant may be renewed for an additional $50,000 in most circumstances.

WORKER SAFETY AND HEALTH PROGRAM

The NCP requires response actions to comply with the requirements of 29 CFR 1910.120 as promulgated and administered by the U.S. Department of Labor (DOL; 40 CFR 300.150). Executive Order 12196 mandates that federal agencies are required to comply with these standards, and 40 CFR 311.1 states that DOL requirements also apply to state and local government employees in states that do not have OSHA-approved programs.

Program Components

The worker safety and health program requires each employer to develop and implement a written safety and health program that identifies, evaluates, and controls safety and health hazards and that provides emergency response procedures for each hazardous waste site or treatment, storage, and/or disposal facility. This written program must include specific and detailed information on the following:

- An organizational workplan
- Site evaluation and control
- A site-specific safety and health plan
- An information and training program
- A personal protective equipment program
- Monitoring
- A medical surveillance program
- Decontamination procedures
- Engineering controls and work practices
- Handling and labeling drums and containers
- Record keeping

The written safety and health program must be periodically updated and made available to all affected employees, contractors, and subcontractors. The employer must also inform contractors and subcontractors, or their representatives, of any identifiable safety and health hazards or potential fire or explosion hazards before they enter the work site.

Each of the components of the safety and health program is discussed in the following sections.

Organizational Workplan. The workplan must support the overall objectives of the control program and provide procedures for its implementation, and it must incorporate the employer's standard operating procedures for safety and health [29 CFR 1910.120(b)(1)]. A chain of command must be established to specify employer and employee responsibilities in carrying out the safety and health program.

The workplan should include the following:

- A list of supervisor and employee responsibilities and means of communication.
- The name of the person who supervises all of the hazardous waste operations.
- The name of the site supervisor with both the responsibility and the authority to develop and implement the site safety and health program and to verify compliance.

In addition to this organizational structure, the plan should define the tasks and objectives of site operation as well as the logistics and resources required to fulfill those tasks. The following topics should be addressed:

- The anticipated cleanup and/or operating procedures.
- A definition of work tasks and objectives and methods of accomplishment.
- The established personnel requirements for implementing the plan.
- The procedures needed to implement training, informational programs, and medical surveillance requirements.

Site Evaluation and Control. A trained person must conduct a preliminary evaluation of an uncontrolled hazardous waste site before entering the site [29 CFR 1910.120(c)(2)]. The evaluation must include all suspected conditions that are immediately dangerous to life or health, or that may cause serious harm to employees (e.g., confined space entry, potentially explosive or flammable situations, visible vapor clouds, physical threats). As available, the evaluation must include information on the location and size of the site, site topography, site accessibility by air and roads, dispersal pathways for hazardous substances, a description of worker duties, and the time needed to perform given tasks.

The use of a buddy system also is required for the rescue of any employee who may become unconscious, trapped, or seriously disabled on-site. In such a system, two employees must keep an eye on each other, and only one should be in a dangerous area. If one gets into trouble, the second can call for help.

Site-Specific Safety and Health Plan. The site-specific safety and health plan must not only include all the basic requirements of the overall safety and health program but also focus attention on characteristics unique to the particular site [29 CFR 1910.120(b)(4)]. For example, the site-specific plan may outline procedures for confined-space entry, air and personal monitoring and environmental sampling, and a spill containment program.

The site-specific safety and health plan must identify the hazards of each phase of the site's operations and must be kept on the work site. Pre-entry briefings must be conducted before employees enter the site and at other times as necessary to ensure that employees are aware of the site safety and health plan and its implementation. The employer must also ensure that safety and health inspections of the site are performed periodically and that all known deficiencies are corrected before any work is done at the site.

Information and Training Program. As part of the safety and health program, employers are required to develop and implement a program to inform workers (including contractors and subcontractors) performing hazardous waste operations of the level and degree of exposure they are likely to encounter [29 CFR 1910.120(e) and (i)].

Employers are also required to develop and implement procedures for introducing effective new technologies or equipment that can improve worker protection in hazardous waste operations.

The employer must develop a training program for all employees exposed to safety and health hazards during hazardous waste operations. Both supervisors and workers must be trained to

- Recognize hazards and prevent them.
- Select, care for, and use respirators properly, as well as other types of personal protective equipment.
- Understand engineering controls and their use.
- Use proper decontamination procedures.
- Understand the emergency response plan, medical surveillance requirements, confined-space entry procedures, the spill containment program, and any appropriate work practices.

Employers may not perform any hazardous waste operations unless they have been trained to the level required by their job function and responsibility and have been certified by their instructor to have completed the necessary training. All applicable workers must receive refresher training that is sufficient for them to maintain or demonstrate their competency annually. These requirements may include recognizing and knowing the hazardous wastes they will encounter and their risks, knowing how to select and use appropriate personal protective equipment, and knowing the appropriate control, containment, or confinement procedures and how to implement them. Training does not have to be repeated if the employee goes to work at a new site unless conditions warrant otherwise. Employees who receive the specified training must be given a written certificate upon successful completion of that training.

Personal Protective Equipment Program. The employer is required to develop a written *personal protective equipment program* for all employees involved in hazardous waste operations [29 CFR 1910.120(a)(5)]. This program must include an explanation of equipment selection and use, maintenance and storage, decontamination and disposal, training and proper fit testing, donning and doffing procedures, inspection, in-use monitoring, program evaluation, and equipment limitations.

The employer must also provide and require the use of personal protective equipment where engineering control methods are infeasible to keep worker exposures at or below the permissible exposure limit. Personal protective equipment must be selected that is appropriate to the requirements and limitations of the site, the task-specific conditions and task duration, and the hazards and potential hazards identified at the site.

Monitoring. The employer must conduct monitoring before site entry at uncontrolled hazardous waste sites to identify conditions immediately dangerous to life and health, such as oxygen-deficient atmospheres and areas where toxic substances exposures are above permissible limits [29 CFR 1910.120(c)(6)]. Accurate information on the identification and quantification of airborne contaminants is essential for the following:

- Selecting personal protective equipment.
- Delineating areas where protection and controls are needed.

- Assessing the potential health effects of exposure.
- Determining the need for specific medical monitoring.

After hazardous waste cleanup operations begin, the employer must periodically monitor those employees who are likely to have relatively high exposures to determine if they have been exposed to hazardous substances in excess of permissible exposure limits. The employer must also monitor for any potential condition that is immediately dangerous to life and health or for higher exposures that may occur as a result of new work operations.

Medical Surveillance Program. The employer must establish a medical surveillance program for the following [29 CFR 1910.120(f)]:

- All employees exposed or potentially exposed to hazardous substances or health hazards above the permissible exposure limits for more than 30 days per year.
- Workers exposed above the published exposure levels (if there is no permissible exposure limit for the hazardous substances) for 30 days or more per year.[1]
- Workers who wear respirators for 30 or more days per year on site.
- Workers who are exposed to unexpected or emergency releases of hazardous wastes above exposure limits (without wearing appropriate protective equipment), or who show signs, symptoms, or illness that may have resulted from exposure to hazardous substances.

All examinations must be performed under the supervision of a licensed physician, without cost to the employee, without loss of pay, and at a reasonable time and place. Examinations must include a medical and work history with special emphasis on (1) symptoms related to the handling of hazardous substances and health hazards and (2) fitness for duty, including the ability to wear any required personal protective equipment under conditions that may be expected at the work site. These examinations must be given as follows:

- Before job assignments and annually thereafter (or every two years if a physician determines that is sufficient).
- At the termination of employment.
- Before reassignment to an area where medical examinations are not required (that is, if the employee has not had an examination within the last six months).
- If the examining physician believes that a periodic follow-up is medically necessary.

[1] For published permissible exposure limits, consult the American Conference of Governmental and Industrial Hygienists, *Threshold Limit Values to Airborne Toxicants in the Workplace*.

- As soon as possible for employees that are injured or become ill from exposure to hazardous substances during an emergency, or who develop signs or symptoms of overexposure from hazardous substances.

The employer must give the examining physician a copy of the OSHA hazardous waste operations standard and its appendixes, a description of the employee's duties related to his or her exposure, the exposure level or anticipated exposure level, a description of any personal protective and respiratory equipment used or to be used, and any information from previous medical examinations. The employer must obtain a written opinion from the physician that contains the results of the medical examination and any detected medical conditions that would place the employee at an increased risk from exposure, any recommended limitations on the employee or on his or her use of personal protective equipment, and a statement that the employee has been informed by the physician of the results of the medical examination. The physician is not to reveal in the written opinion given to the employer any specific findings or diagnoses unrelated to employment.

Decontamination Procedures. Decontamination procedures are a component of the site-specific safety and health plan and, consequently, must be developed, communicated to employees, and implemented before workers enter a hazardous waste site [29 CFR 1910.120(k)]. As necessary, the site safety and health officer must require and monitor decontamination of the employee or decontamination and disposal of the employee's clothing and equipment, as well as the solvents used for decontamination, before the employee leaves the work area. If an employee's nonimpermeable clothing becomes grossly contaminated with hazardous substances, the employee must immediately remove that clothing and take a shower. Impermeable protective clothing must be decontaminated before it is removed by an employee.

Protective clothing and equipment must be decontaminated, cleaned, laundered, maintained, or replaced to retain its effectiveness. The employer must inform any person who launders or cleans such clothing or equipment of the potentially harmful effects of exposure to hazardous substances.

Engineering Controls and Work Practices. To the extent feasible, the employer must institute engineering controls and work practices to help reduce and maintain employee exposure at or below permissible exposure limits [29 CFR 1910.120(g)]. To the extent they are not feasible, engineering and work practice controls may be supplemented with personal protective equipment. Examples of suitable and feasible engineering controls include the use of pressurized cabs or control booths on equipment and/or the use of remotely operated materials-handling equipment. Examples of safe work practices include removing all nonessential employees from potential exposure when drums are opened, wetting down dusty operations, and placing employees upwind of potential hazards.

Handling and Labeling Drums and Containers. Before drums or containers are handled, the employer must ensure they meet the required OSHA, EPA, and DOT regulations and that they are properly inspected and labeled [29 CFR 1910.120(j)]. Damaged drums or containers must be emptied of their contents by using a device classified for the material being transferred, and they must be properly discarded. In areas where spills, leaks, or ruptures occur, the employer must furnish employees with salvage drums or containers, a suitable quantity of absorbent material, and approved fire-extinguishing equipment (in case of small fires). The employer must also inform employees of the appropriate hazard warnings for labeled drums and the removal of soil or coverings, as well as the dangers of handling unlabeled drums or containers without prior identification of their contents. To the extent feasible, the moving of drums or containers must be kept to a minimum, and a program must be implemented to contain and isolate hazardous substances being transferred into drums or containers. In addition, an approved EPA ground-penetrating device must be used to determine the location and depth of any improperly discarded drums or containers.

Record Keeping. OSHA requires employers to provide employees with certain information to help employees manage their own safety and health. The OSHA standard "Access to Employee Exposure and Medical Records" (29 CFR 1910.20) gives OSHA as well as employees exposed to hazardous wastes (or their designated representatives) direct access to these records. The standard applies to, but does not require, medical and exposure records maintained by the employer.

The employer must keep exposure records for 30 years. Medical records must be kept for at least the duration of employment plus 30 years. Records of employees who have worked for less than 1 year need not be retained after the employee leaves, but the employer must provide these records to the employee upon termination of such employment. First-aid records of one-time treatments need not be retained for any specified period. The employer must inform each employee of the existence, location, and availability of these records.

Under the OSHA hazardous waste operations standard, medical records must include, at a minimu, the following information [29 CFR 1910.120(j)(8)]:

- Employee's name and social security number.
- Physician's written opinions.
- Employee's medical complaints related to exposure to hazardous substances.
- Information provided to the treating physician.

PART III

THE TOXIC SUBSTANCES CONTROL ACT

The Toxic Substances Control Act (TSCA) was enacted on October 11, 1976, just 10 days before the enactment of RCRA. TSCA's authority allows EPA to prohibit or control the manufacture, importation, processing, use, and disposal of any *chemical substance* that presents an *unreasonable risk of injury to human health or the environment.* Certain substances are statutorily excluded from the definition of chemical substance including pesticides (active ingredients only), foods, food additives, drugs, cosmetics, tobacco, firearms, and nuclear material.

TSCA regulations apply to any person who manufactures, imports, processes, uses, or disposes of *chemical substances.* However, because this book focuses on "hazardous waste" regulation, this part addresses only the use, management, disposal, and cleanup of polychlorinated biphenyls (PCBs). Also, §13 of TSCA applies to the importation into the United States of any hazardous waste for disposal or reclamation.

TSCA does not contain any provisions for state authorization; therefore, EPA has sole TSCA enforcement authority (although a state may assist EPA upon mutual agreement). In addition, §18 of TSCA prohibits EPA from preempting state regulations pertaining to chemical substances except under certain conditions. Some states regulate PCBs as a hazardous waste or regulate materials containing less than 50 parts per million (ppm) PCBs. Thus, there may be more stringent or overlapping requirements at the state level.

TSCA's authority is very broad, but primarily focuses on controlling the chemical manufacturing and processing industry. A unique aspect of TSCA is that it has the authority to regulate the entire life cycle of a chemical substance, from its manufacture to its disposal. However, as discussed in the following sections pertaining to specific laws, there are statutory provisions that have limited the use of TSCA to regulate hazardous waste (e.g., economic impact analyses, individual chemical substances versus commingled waste streams). These statutory provisions did not affect the regulation PCBs because PCBs were explicitly singled out in the law, which mandated the control of PCBs, allowing the circumvention of these hurdles.

TSCA'S RELATIONSHIP TO OTHER MAJOR LAWS

The following sections describe the basic relationships among TSCA and other federal statutes pertaining to the control of hazardous waste. The specific statutes addressed are

- Resource Conservation and Recovery Act
- Superfund
- National Environmental Policy Act

RESOURCE CONSERVATION AND RECOVERY ACT

The authority concerning disposal under TSCA is significantly different from that of RCRA. TSCA focuses on individual chemical substances, whereas RCRA may address waste streams that can contain many chemical substances. EPA may regulate a chemical substance under TSCA only if it is found that the chemical substance "presents an unreasonable risk of injury to human health or the environment." To determine an unreasonable risk under TSCA, EPA must conduct an economic cost/benefit analysis, which includes an analysis on the availability of substitutes. Because of the difficulty in quantifying environmental benefits compared to the relative ease in quantifying costs, few chemicals have survived TSCA's cost/benefit analysis for regulation, thus relatively few chemicals are controlled under this law. RCRA does not explicitly require the use of economic considerations in its rulemaking.[1]

PCBs are not listed as a hazardous waste under RCRA, and unless they are mixed with other chemicals, will not exhibit a characteristic of hazardous waste under the RCRA definition of hazardous waste. PCBs are listed as an Appendix VIII (40 CFR part 261) hazardous constituent and an Appendix IX (40 CFR part 264) groundwater constituent. There have been periodic unsuccessful attempts to transfer the regulation of PCBs from TSCA to RCRA to eliminate duplication and improve consistency. However, the 1989 promulgation of the manifesting, notification, and tracking rule covering PCBs and an EPA-led effort to ensure consistency among the two programs should ensure that the PCB program will remain under TSCA because the requirements closely resemble RCRA-type requirements.

[1] Although RCRA itself does not require economic considerations, other laws, including the Regulatory Flexibility Act, the Paperwork Reduction Act, the Unfunded Mandates Reform Act, and Executive Orders 12866, 12875, and 12898, all require some form of economic analysis to be prepared.

RCRA Mixed Wastes. A *mixed waste* contains a RCRA hazardous waste and PCBs of 500 ppm or greater. For example, if a PCB item is decontaminated with a solvent (e.g., xylene or toluene) that is a RCRA hazardous waste, a mixed waste will be generated. PCB-containing dielectric fluids removed from electrical transformers, capacitors, and associated PCB electrical equipment that also exhibit the RCRA toxicity characteristic for organics (i.e., D004–D017) is excluded from RCRA and subject to TSCA (40 CFR 261.8 and 55 *FR* 11841, March 29, 1990). This means that certain solvents may be used to remove the dielectric and only TSCA need be followed, subject to state restrictions.

Importing Hazardous Wastes. When importing hazardous wastes into the United States for reclamation or disposal, the importer, who must be a U.S. citizen, must comply with §13 of TSCA. Section 13 requires all importers of chemical substances to certify that such shipments either are in compliance with all applicable rules and regulations of TSCA or are not subject to TSCA. Most hazardous wastes will meet the definition of a chemical substance and will therefore be subject to TSCA import control. Chemicals that are not subject to TSCA under §13 (i.e., materials that are not included in the definition of *chemical substance*) include active pesticide ingredients, drugs, cosmetics, and nuclear materials. In determining whether a material is in compliance with TSCA, of primary concern are the following requirements: (1) every chemical component in the waste stream must be an *existing* chemical [i.e., included in EPA's 8(b) TSCA Chemical Substances Inventory] and (2) none of the chemical components of the waste stream can be prohibited from importation or manufacture. Regulatory information pertaining to the import status of chemical substances under TSCA can be obtained from EPA's TSCA Industry Assistance Hotline at 202-554-1404.

Superfund

PCBs are designated as a hazardous substance under Superfund (40 CFR part 302) because of the Clean Water Act. Therefore, any person involved in a release or threatened release of PCBs is liable for any and all costs incurred for a cleanup of released or threatened release of PCBs.

Pursuant to §103 of Superfund, a spill of 1 pound or more of PCBs is a reportable quantity subject to immediate notification of the National Response Center (800-424-8802). As discussed in Chapter 12, the spill must be 1 pound of "pure" PCBs. For example, if a transformer containing dielectric fluid that has 100,000 ppm (10 percent) PCBs spills, 10 pounds of fluid must be released to constitute 1 pound of released PCBs. However, if the exact concentration of PCBs is unknown (assuming it is greater than 500 ppm), a spill of 1 pound or greater of fluid must be reported.

National Environmental Policy Act

EPA has concluded that many of its actions do not require an environmental impact statement under the National Environmental Policy Act because EPA's actions constitute functional equivalency, as discussed previously in Part I. The functional-equivalence exclusion that relates to TSCA actions has been tested in court only for cleanup actions involving PCBs; however, it appears that, as far as EPA is concerned, when conducting or overseeing an action concerning TSCA, EPA may use the functional-equivalence exclusion if the specified criteria are met as established by case law. [See *Environmental Defense Fund v. EPA,* 489 F.2d 1247, 1257 (D.C. Cir. 1973); *Twitty v. State of N.C.,* 527 F.Supp. 778 (1981); *Maryland v. Train,* 415 F. Supp. 116, 121–22 (1976); and *Warren County v. State of N.C.,* 528 F.Supp. 276 (1981).]

16

KEY PCB DEFINITIONS

This chapter contains key regulatory definitions of PCB-related TSCA terms used throughout Part III. These definitions are listed in 40 CFR 761.3 and 761.123 and are noted accordingly. Definitions appear here verbatim.

Annual document log means the detailed information maintained at the facility on the PCB waste handling at the facility. (40 CFR 761.3)

Annual report means the written document submitted each year by each disposer and commercial storer of PCB waste to the appropriate EPA Regional Administrator. The annual report is a brief summary of the information included in the annual document log. (40 CFR 761.3)

Basel Convention means the Basel Convention on the Control of Transboundary Movements of Hazardous Wastes and Their Disposal as entered into force on May 5, 1992. (40 CFR 761.3)

Capacitor means a device for accumulating and holding a charge of electricity and consisting of conducting surfaces separately by a dielectric. Types of capacitors are as follows:

(1) *Small capacitor* means a capacitor which contains less than 1.36 kg (3 lbs.) of dielectric fluid. The following assumptions may be used if the actual weight of the dielectric fluid is unknown. A capacitor whose total volume is less than 1,639 cubic centimeters (100 cubic inches) may be considered to contain less than 1.36 kg (3 lbs.) of dielectric fluid and a capacitor whose total volume is more than 3,278 cubic centimeters (200 cubic inches) must be considered to contain more than 1.36 kg (3 lbs.) of

dielectric fluid. A capacitor whose volume is between 1,639 and 3,278 cubic centimeters may be considered to contain less than 1.36 kg (3 lbs.) of dielectric fluid if the total weight of the capacitor is less than 4.08 kg (9 lbs.).

(2) *Large high-voltage capacitor* means a capacitor which contains 1.36 kg (3 lbs.) or more of dielectric fluid and operates at 2,000 volts (a.c. or d.c.) or above.

(3) *Large low-voltage capacitor* means a capacitor which contains 1.36 kg (3 lbs.) or more of dielectric fluid and which operates below 2,000 volts (a.c. or d.c.). (40 CFR 761.3)

Certification means a written statement regarding a specific fact or representation that contains the following language:

> Under civil and criminal penalties of law for the making or submission of false or fraudulent statements or representations (18 U.S.C. 1001 and 15 U.S.C. 2615), I certify that the information contained in or accompanying this document is true, accurate, and complete. As to the identified section(s) of this document for which I cannot personally verify truth and accuracy, I certify as the company official having supervisory responsibility for the person who, acting under my direct instructions, made the verification that this information is true, accurate, and complete.

Chemical substance, (1) except as provided in paragraph (2) of this definition, means any organic or inorganic substance of a particular molecular identity, including: Any combination of such substances occurring in whole or part as a result of a chemical reaction or occurring in nature, and any element or uncombined radical. (2) Such term does not include: Any mixture; any pesticide (as defined in the Federal Insecticide, Fungicide, and Rodenticide Act) when manufactured, processed, or distributed in commerce for use as a pesticide; tobacco or any tobacco product; any source material, special nuclear material, or byproduct material (as such terms are defined in the Atomic Energy Act of 1954 and regulations issued under such Act); any article the sale of which is subject to the tax imposed by §4181 of the Internal Revenue Code of 1954 (determined without regard to any exemptions from such tax provided by §4182 or §4221 or any provisions of such Code); and any food, food additive, drug cosmetic, or device (as such terms are defined in §201 of the federal Food, Drug, and Cosmetic Act) when manufactured, processed, or distributed in commerce for use as a food, food additive, drug, cosmetic, or device. (40 CFR 761.3)

Chemical waste landfill means a landfill at which protection against risk of injury to health or the environment from migration of PCBs to land, water, or the atmosphere is provided from PCBs and PCB Items deposited therein by locating, engineering, and operating the landfill as specified in §761.75. (40 CFR 761.3)

Commercial storer of PCB waste means the owner or operator of each facility that is subject to the PCB storage unit standards of Sec. 761.65(b)(1) or (c)(7) or meets the alternate storage criteria of Sec. 761.65(b)(2), and who engages in storage activities involving either PCB waste generated by others or that was removed while servicing the equipment owned by others and brokered for disposal. The receipt of a fee or any other form of compensation for storage services is not necessary to qualify as a commercial storer of PCB waste. A generator who only stores its own waste is subject to the storage requirements of Sec. 761.65, but is not required to obtain approval as a commercial storer. If a facility's storage of PCB waste generated by others at no time exceeds a total of 500 gallons of liquid and/or non-liquid material containing PCBs at regulated levels, the owner or operator is a commercial storer but is not required to seek EPA approval as a commercial storer of PCB waste. Storage of one company's PCB waste by a related company is not considered commercial storage. A "related company" includes, but is not limited to: a parent company and its subsidiaries; sibling companies owned by the same parent company; companies owned by a common holding company; members of electric cooperatives; entities within the same Executive agency as defined at 5 U.S.C. 105; and a company having a joint ownership interest in a facility from which PCB waste is generated (such as a jointly owned electric power generating station) where the PCB waste is stored by one of the co-owners of the facility. A "related company" does not include another voluntary member of the same trade association. Change in ownership or title of a generator's facility, where the generator is storing PCB waste, does not make the new owner of the facility a commercial storer of PCB waste. (40 CFR 761.3)

Designated facility means the off-site disposer or commercial storer of PCB waste designated on the manifest as the facility that will receive a manifested shipment of PCB waste. (40 CFR 761.3)

Disposal means to intentionally or accidentally discard, throw away, or otherwise complete or terminate the useful life of PCBs and PCB Items. Disposal includes spills, leaks, and other uncontrolled discharges of PCBs as well as actions related to containing, transporting, destroying, degrading, decontaminating, or confining PCBs and PCB Items. (40 CFR 761.3)

Disposer of PCB waste as the term is used in subparts J and K of this part (761), means any person who owns or operates a facility approved by EPA for the disposal of PCB waste which is regulated for disposal under the requirements of subpart D of this part (761). (40 CFR 761.3)

Double wash/rinse means a minimum requirement to cleanse solid surfaces (both impervious and nonimpervious) two times with an appropriate solvent or other material in which PCBs are at least 5 percent soluble (by weight). A volume of PCB-free fluid sufficient to cover the contaminated surface completely must be used in each wash/rinse. The wash/rinse requirement does not mean the mere spreading of solvent or other fluid over the surface, nor does the requirement mean a once-over wipe with a soaked cloth. Precautions must be taken to contain any runoff resulting from the cleansing and to dispose properly of wastes generated during the cleansing. (40 CFR 761.123)

Emergency situation for continuing use of a PCB Transformer exists when: (1) Neither a non-PCB Transformer is currently in storage for ruse or readily available (i.e., available within 24 hours) for installation. (2) Immediate replacement is necessary to continue service to power users. (40 CFR 761.3)

EPA identification number means the 12-digit number assigned to a facility by EPA upon notification of PCB waste activity under §761.205. (40 CFR 761.3)

Fluorescent light ballast means a device that electrically controls fluorescent light fixtures, and that includes a capacitor containing 0.1 kg or less of dielectric. (40 CFR 761.3)

Generator of PCB waste means any person whose act or process produces PCBs that are regulated for disposal under subpart D of this part (761), or whose act first causes PCBs or PCB Items to become subject to the disposal requirements of subpart D of this part, or who has physical control over the PCBs when a decision is made that the use of the PCBs has been terminated and therefore is subject to the disposal requirements of subpart D of this part. Unless another provision of this part specifically requires a site-specific meaning, "generator of PCB waste" includes all of the sites of PCB waste generation owned or operated by the person who generates PCB waste. (40 CFR 761.3)

High-concentration PCBs means PCBs that contain 500 ppm or greater PCBs, or those materials which EPA requires to be assumed to contain 500 ppm or greater PCBs in the absence of testing. (40 CFR 761.123)

High-contact industrial surface means a surface in an industrial setting which is repeatedly touched, often for long periods of time. Manned machinery and control panels are examples of high-contact industrial surfaces. High-contact industrial surfaces are generally of impervious solid material. Examples of low-contact industrial surfaces include ceilings, walls, floors, roofs, roadways and sidewalks in the industrial area, utility poles, unmanned machinery, concrete pads beneath electrical equipment, curbing, exterior structural building components, indoor vaults, and pipes. (40 CFR 761.123)

High-contact residential/commercial surface means a surface in a residential/ commercial area which is reportedly touched, often for relatively long periods of time. Doors, wall areas below 6 feet in height, uncovered flooring, windowsills, fencing, banisters, stairs, automobiles, and children's play areas such as outdoor patios and sidewalks are examples of high-contact residential/commercial surfaces. Examples of low-contact residential/commercial surfaces include interior ceilings, interior wall areas above 6 feet in height, roofs, asphalt roadways, concrete roadways, wooden utility poles, unmanned machinery, concrete pads beneath electrical equipment, curbing, exterior structural building components (e.g., aluminum/vinyl siding, cinder block, asphalt tiles), and pipes. (40 CFR 761.123)

Impervious solid surfaces means solid surfaces which are nonporous and thus unlikely to absorb spilled PCBs within the short period of time required for cleanup of spills under this policy [the PCB Spill Cleanup Policy]. Impervious solid surfaces include, but are not limited to, metals, glass, aluminum siding, and enameled or laminated surfaces. (40 CFR 761.123)

Importer means any person defined as an "importer" at 40 CFR 720.3(1) who imports PCBs or PCB Items and is under the jurisdiction of the United States. (40 CFR 761.3)

In or near commercial buildings means within the interior of, on the roof of, attached to the exterior wall of, in the parking area serving, or within 30 meters of a non-industrial non-substation building. Commercial buildings are typically accessible to both members of the general public and employees, and include: (1) Public assembly properties, (2) educational properties, (3) institutional properties, (4) residential properties, (5) stores, (6) office buildings, (7) transportation centers (e.g., airport terminals, buildings, subway stations, bus stations, or train stations). (40 CFR 761.3)

Incinerator means an engineered device using controlled flame combustion to thermally degrade PCBs and PCB Items. Examples of devices used for incineration include rotary kilns, liquid injection incinerators, cement kilns, and high-temperature boilers. (40 CFR 761.3)

Industrial buildings means a building directly used in manufacturing or technically productive enterprises. Industrial buildings are not generally or typically accessible to other than workers. Industrial buildings include buildings used directly in the production of power, the manufacture of products, the mining of raw materials, and the storage of textiles, petroleum products, wood and paper products, chemicals, plastics, and metals. (40 CFR 761.3)

Laboratory means a facility that analyzes samples for PCBs and is unaffiliated with any entity whose activities involve PCBs. (40 CFR 761.3)

Leak or *leaking* means any instance in which a PCB Article, PCB Container, or PCB Equipment has any PCBs on any portion of its external surface. (40 CFR 761.3)

Low-concentration PCBs means PCBs that are tested and found to contain less than 500 ppm PCBs, or those PCB-containing materials which EPA requires to be assumed to be at concentrations below 500 ppm (i.e., untested mineral oil dielectric fluid). (40 CFR 761.123)

Manifest means the shipping document EPA form 8700-22 and any continuation sheet attached to EPA form 8700-22, originated and signed by the generator of PCB waste in accordance with the instructions included with the form and subpart K of this part (761). (40 CFR 761.3)

Manned control center means an electrical power distribution control room where the operating conditions of a PCB Transformer are continuously monitored during the normal hours of operation (of the facility), and, where the duty engineers, electricians, or other trained personnel have the capability to deenergize a PCB Transformer completely within 1 minute of the receipt of a signal indicating abnormal operating conditions such as overtemperature condition or overpressure condition in a PCB Transformer. (40 CFR 761.3)

Manufacture means to produce, manufacture, or import into the customs territory of the United States. (40 CFR 761.3)

Mark means the descriptive name, instructions, cautions, or other information applied to PCBs and PCB Items, or other objects subject to these regulations. (40 CFR 761.3)

Marked means the marking of PCB Items and PCB storage areas and transport vehicles by means of applying a legible mark by painting, fixation of an adhesive label, or by any other method that meets the requirements of these regulations. (40 CFR 761.3)

Mineral Oil PCB Transformer means any transformer originally designed to contain mineral oil as the dielectric fluid and which has been tested and found to contain 500 ppm or greater PCBs. (40 CFR 761.3)

Nonimpervious solid surfaces means solid surfaces which are porous and are more likely to absorb spilled PCBs prior to completion of the cleanup requirements prescribed in this policy [the PCB Spill Cleanup Policy]. Nonimpervious solid surfaces include, but are not limited to, wood, concrete, asphalt, and plasterboard. (40 CFR 761.123)

Non-PCB Transformer means any transformer that contains less than 50 ppm PCB; except that any transformer that has been converted from a PCB Transformer or a PCB-Contaminated Transformer cannot be classified as a non-PCB Transformer until reclassification has occurred, in accordance with §761.30(a)(2)(v). (40 CFR 761.3)

Nonrestricted access means any area other than restricted access, outdoor electrical substations, and other restricted access locations, as defined in this section. In addition to residential/commercial areas, these areas include unrestricted access rural areas (areas of low density development and population where access is uncontrolled by either man-made barriers or naturally occurring barriers, such as rough terrain, mountains or cliffs). (40 CFR 761.123)

On-site means within the boundaries of a contiguous property unit. (40 CFR 761.3)

Other restricted access (nonsubstation) locations means areas other than electrical substations that are at least 0.1 kilometer (km) from a residential/commercial area and limited by man-made barriers (e.g., fences and walls) substantially limited by naturally occurring barriers such as mountains, cliffs, or rough terrain. These areas generally include industrial facilities and extremely remote rural locations. (Areas where access is restricted but that are less than 0.1 km from a residential/

commercial area are considered to be residential/commercial areas.) (40 CFR 761.123)

Outdoor electrical substations means outdoor, fenced-off, and restricted access areas used in the transmission and/or distribution of electrical power. Outdoor electrical substations restrict public access by being fenced or walled off as defined under §761.301(1)(1)(ii). For purposes of this TSCA policy [the PCB Spill Cleanup Policy], outdoor electrical substations are defined as being located at least 0.1 km from a residential/commercial area. Outdoor fenced-off and restricted access areas used in the transmission and/or distribution of electrical power which are located less than 0.1 km from a residential/commercial area are considered to be residential/commercial areas. (40 CFR 761.123)

PCB and *PCBs* means any chemical substance that is limited to the biphenyl molecule that has been chlorinated to varying degrees or any combination of substances which contain such substance. (Refer to §761.1(b) for applicable concentrations of PCBs.) PCB and PCBs as contained in PCB Items are defined in §761.3. For any purposes under this part [761], inadvertently generated non-Aroclor PCBs are defined as the total PCBs calculated following division of the quantity of monochlorinated biphenyls by 50 and dichlorinated biphenyls by 5. (40 CFR 761.3)

PCB Article means any manufactured article, other than a PCB Container, that contains PCBs and whose surface(s) has been in direct contact with PCBs. "PCB Article" includes capacitors, transformers, electric motors, pumps, pipes, and any other manufactured item (1) which is formed to a specific shape or design during manufacture, (2) which has end use function(s) dependent in whole or in part upon its shape or design during end use, and (3) which has either no change of chemical composition during its end use or those changes of composition which have no commercial purpose separate from that of the PCB Article. (40 CFR 761.3)

PCB Article Container means any package, can, bottle, bag, barrel, drum, tank, or other device used to contain PCB Articles or PCB Equipment, and those whose surface(s) has not been in direct contact with PCBs. (40 CFR 761.3)

PCB Container means any package, can, bottle, bag, barrel, drum, tank, or other device that contains PCBs or PCB Articles and whose surface(s) has been in direct contact with PCBs. (40 CFR 761.3)

PCB-Contaminated Electrical Equipment means any electrical equipment, including, but not limited to, transformers (including those used in railway locomotives and self-propelled cars), capacitors, circuit breakers, reclosers, voltage regulators, switches (including sectionalizers and motor starters), electromagnets, and cable, that contains PCBs at concentrations $\geq$50 ppm and $<$500 ppm in the contaminating fluid. In the absence of liquids, electrical equipment is PCB contaminated if it has PCBs at $>$10 μg/100 m^2 and $<$100 μg/100 m^2 as measured by a standard wipe test (as defined in §761.123) of a non-porous surface. (40 CFR 761.3)

PCB Equipment means any manufactured item, other than a PCB Container or a PCB Article Container, which contains a PCB Article or other PCB Equipment, and includes microwave ovens, electronic equipment, and fluorescent light ballasts and fixtures. (40 CFR 761.3)

PCB Item means any PCB Article, PCB Article Container, PCB Container, PCB Equipment, or anything that deliberately or unintentionally contains or has as a part of it any PCB or PCBs. (40 CFR 761.3)

PCB Transformer means any transformer that contains ≥500 ppm PCB. (40 CFR 761.3)

PCB wastes means those PCBs and PCB Items that are subject to the disposal requirements of subpart D of this part (761). (40 CFR 761.3)

Posing an exposure risk to food or feed means being in any location where human food or animal feed products could be exposed to PCBs released from a PCB Item. A PCB Item poses an exposure risk to food or feed if PCBs released in any way from the PCB Item have a potential pathway to human food or animal feed. EPA considers human food or animal feed to include items regulated by the U.S. Department of Agriculture or the Food and Drug Administration as human food or animal feed; this includes direct additives. Food or feed is excluded from this definition if it is used or stored in private homes. (40 CFR 761.3)

Qualified incinerator means one of the following:

(1) An incinerator approved under the provisions of §761.70. Any level of PCB concentration can be destroyed in an incinerator approved under §761.70.

(2) A high-efficiency boiler which complies with the criteria of §761.71(a)(1), and for which the operator has given written notice to the appropriate EPA Regional Administrator in accordance with the notification requirements for the burning of mineral oil dielectric fluid §761.71(a)(2).

(3) An incinerator approved under section 3005(c) of the Resource Conservation and Recovery Act) 42 U.S.C. 6925(c)) (RCRA).

(4) Industrial furnaces and boilers which are identified in 40 CFR 260.10 and 40 CFR 279.61(a)(1) and (2) when operating at their normal operating temperatures (this prohibits feeding fluids, above the level of detection, during either startup up or shutdown operations). (40 CFR 761.3)

Quanitifable Level/Level of Detection means 2 micrograms per gram from any resolvable gas chromatographic peak, i.e. 2 ppm. (40 CFR 761.3)

Requirements and standards means: (1) "Requirements" as used in this policy [the PCB Spill Cleanup Policy] refers to both the procedural responses and numerical decontamination levels set forth in this policy as constituting adequate cleanup of PCBs. (2) "Standards" refers to the numerical decontamination levels set forth in this policy. (40 CFR 761.123)

Residential/commercial areas means those areas where people live or reside or where people work in other than manufacturing or farming industries. Residential areas include housing and the property on which housing is located as well as playgrounds, roadways, sidewalks, parks, and other similar areas within a residential community. Commercial areas are typically accessible to both members of the general pubic and employees and include public assembly properties, institutional properties, stores, office buildings, and transportation centers. (40 CFR 761.123)

Retrofill means to remove PCB or PCB-contaminated dielectric fluid and to replace it with either PCB, PCB-contaminated, or non-PCB dielectric. (40 CFR 761.3)

Responsible party means the owner of the PCB equipment, facility, or other source of PCB or his/her designated agent (e.g., a facility manager or foreman). (40 CFR 761.123)

Rupture of a PCB Transformer means a violent or non-violent break in the integrity of a PCB Transformer caused by an overtemperature and/or overpressure condition that results in the release of PCBs. (40 CFR 761.3)

Sale for purposes other than resale means sale of PCBs for purposes of disposal and for purposes of use, except where use involves sale for distribution in commerce. PCB Equipment which is first leased for purposes of use any time before July 1, 1979, will be considered sold for purposes other than resale. (40 CFR 761.3)

Significant exposure means any exposure of human beings or the environment to PCBs as measured or detected by any scientifically acceptable analytical method. (40 CFR 761.3)

Small quantities for research and development means any quantity of PCBs:

(1) that is originally packaged in one or more hermetically sealed containers of a volume of no more than five (5.0) milliliters; and (2) that is used only for purposes of scientific experimentation or analysis, or chemical research on, or analysis of, PCBs, but not for research or analysis for the development of a PCB product. (40 CFR 761.3)

Soil means all vegetation, soils and other ground media, including but not limited to, sand, grass, gravel, and oyster shells. It does not include concrete and asphalt. (40 CFR 761.123)

Spill means both intentional and unintentional spills, leaks, and other uncontrolled discharges where the release results in any quantity of PCBs running off or about to run off the external surface of the equipment or other PCB source, as well as the contamination resulting from those releases. This policy [the PCB Spill Cleanup Policy] applies to spills of 50 ppm or greater PCBs. The concentration of PCBs spilled is determined by the PCB concentration in the material spilled as opposed to the concentration of PCBs in the material onto which the PCBs were spilled. Where a spill of untested mineral oil occurs, the oil is presumed to contain greater than 50 ppm but less than 500 ppm PCBs and is subject to the relevant requirements of this policy. (40 CFR 761.123)

Spill area means the area of soil on which visible traces of the spill can be observed plus a buffer zone of 1 foot beyond the visible traces. Any surface or object (e.g., concrete sidewalk or automobile) within the visible traces area or on which visible traces of the spilled material are observed is included in the spill area. This area represents the minimum area assumed to be contaminated by PCBs in the absence of precleanup sampling data and is thus the minimum area that must be cleaned. (40 CFR 761.123)

Spill boundaries means the actual area of contamination as determined by post-cleanup verification sampling or by precleanup sampling to determine actual spill boundaries. EPA can require additional cleanup when necessary to decontaminate all areas within the spill boundaries to the levels required in this policy [the PCB Spill Cleanup Policy] (e.g., additional cleanup will be required if postcleanup sampling indicates that the area decontaminated by the responsible party, such as the spill area as defined in this section, did not encompass the actual boundaries of PCB concentration). (40 CFR 761.123)

Standard wipe test means, for spills of high-concentration PCBs on solid surfaces, a cleanup to numerical surface standards and sampling by a standard wipe test to verify that the numerical standards have been met. The definition constitutes the minimum requirements for an appropriate wipe testing protocol. A standard-size template (10 centimeters (cm) x 10 cm) will be used to delineate the area of cleanup; the wiping medium will be a gauze pad or glass wool of known size which has been saturated with hexane. It is important that the wipe be performed very quickly after the hexane is exposed to air. EPA strongly recommends that the gauze (or glass wool) be prepared with hexane in the laboratory and that the wiping medium be stored in sealed glass vials until it is used for the wipe test. Further, EPA requires the collection and testing of field blanks and replicates. (40 CFR 761.123)

Storage for disposal means temporary storage of PCBs that have been designated for disposal. (40 CFR 761.3)

Totally enclosed manner means any manner that will ensure no exposure of human beings or the environment to any concentration of PCBs. (40 CFR 761.3)

Transfer facility means any transportation-related facility including loading docks, parking areas, and other similar areas where shipments of PCB waste are held during the normal course of transportation. Transport vehicles are not transfer facilities under this definition, unless they are used for the storage of PCB waste, rather than for actual transport activities. Storage areas for PCB waste at transfer facilities are subject to the storage facility standards of §761.65, but such storage areas are exempt from the approval requirements of §761.65(d) and the record keeping requirements of §761.180, unless the same PCB waste is stored there for a period of more than 10 consecutive days between destinations. (40 CFR 761.3)

Transporter of PCB waste means, for the purposes of subpart K of this part [761], any person engaged in the transportation of regulated PCB waste by air, rail, highway, or water for purposes other than consolidation by a generator. (40 CFR 761.3)

Transport vehicle means a motor vehicle or rail car used for the transportation of cargo by any mode. Each cargo-carrying body (e.g., trailer, railroad freight car) is a separate transport vehicle. (40 CFR 761.3)

Treatability study means a study in which PCB waste is subjected to a treatment process to determine:

(1) Whether the waste is amenable to the treatment process;
(2) What pretreatment (if any) is required;
(3) The optimal process conditions needed to achieve the desired treatment;
(4) The efficiency of a treatment process for the specific type of waste (i.e., soil, sludge, liquid, etc.); or
(5) The characteristics and volumes of residuals from a particular treatment process. A "treatability study" is not a mechanism to commercially treat or dispose of PCB waste. Treatment is a form of disposal under this part. (40 CFR 761.3)

Waste oil means used products primarily derived from petroleum, which include, but are not limited to, fuel oils, motor oils, gear oils, cutting oils, transmission fluids, hydraulic fluids, and dielectric fluids. (40 CFR 761.3)

17

USE, STORAGE, AND DISPOSAL OF PCBS

Section 6(e) of the Toxic Substances Control Act (TSCA) specifically requires EPA to regulate the manufacture, importation, use, and disposal of polychlorinated biphenyls (PCBs). Other provisions of TSCA direct EPA to regulate chemicals that present an "unreasonable risk of injury to health and the environment," but §6(e) is the only provision of TSCA that directly controls the manufacture, processing, distribution in commerce, use, and disposal of a specific chemical substance—PCBs.

INTRODUCTION

TSCA specifically bans the manufacture of new PCBs; prohibits the processing, distribution in commerce, and use of PCBs in any way other than a *totally enclosed manner;* and regulates the disposal of PCBs. Although the intentional manufacture of PCBs is banned, Congress gave EPA the authority to grant some limited exceptions to the ban. Limited exemptions may be granted only upon petition on a case-by-case basis through rulemaking.

What PCBS Are

Polychlorinated biphenyls are produced by substituting chlorine atoms for hydrogen atoms on a biphenyl (double benzene ring) molecule. The number and the location of the chlorine attachments determine the physical properties and characteristics of a PCB molecule. There are 10 positions on a biphenyl ring that, through various combinations, can yield 209 possible PCB compounds. Commercial PCB products are most often mixtures of many types of PCB molecules. (Commercial PCB mixtures are also known by their generic name, *Aroclor.*) For regulatory purposes, a biphenyl ring chlorinated to *any* degree is considered a PCB (40 CFR 761.3). This includes monochlorinated biphenyls. Generally, PCBs used in electrical equipment tend to be viscous and heavy (10 to 13 pounds per gallon), similar to a light honey, but they may also be solid and waxy. They are extremely stable, nonflammable, fat-soluble, and resistant to degradation. These properties are ideal in an industrial application, but are also thought to be a cause of their negative environmental effects. Because of their fat solubility, PCB molecules can move up the food chain, bioaccumulating in higher trophic-level organisms. Also, because of their solubility, they decompose very slowly.

PCBs are toxic to fish at very low exposure levels and can adversely affect their survival rate and reproductive success. Documented toxic effects in animals include adverse reproductive effects, liver lesions, cancer, developmental toxicity, and chloracne (skin lesions). The human toxicological properties of PCBs depend on the amount and location of the chlorination. PCBs and chlorobenzenes, which are often mixed together, can give rise to polychlorinated dibenzofurans and polychlorinated dibenzo-*p*-dioxins under the right conditions (i.e., certain fire-related incidents). Both of these compounds are believed to be more toxic to humans than PCBs themselves.

Between 1929 and 1977, PCBs were manufactured primarily for use in electrical equipment, and a significant amount of PCB equipment sold in the United States is still in service. PCBs that were intentionally manufactured for use as a dielectric fluid were often mixed with certain organic solvents, such as chlorinated benzenes; thus, the dielectric fluid present in electrical equipment containing PCBs generally is not 100 percent PCB except for dielectric in capacitors.

Polychlorinated terphenyls (PCTs) also were manufactured as a commercial product and are similar in properties to PCBs. Although PCTs are not specifically covered by the TSCA PCB regulations, most were contaminated with up to 10,000 ppm PCB, and so would be regulated because of the PCB contamination.

The two major categories of PCBs are those intentionally manufactured for use in electrical and other types of equipment, and those PCBs produced inadvertently as by-products and impurities. (Although this chapter addresses only PCBs in electrical equipment, EPA's overall strategy for controlling the environmental release of PCBs includes the regulation of both PCBs in electrical equipment *and* intentionally generated by-product PCBs.) The intentional manufacture of PCBs for purposes other than research is prohibited; however, PCBs are still contained in electrical equipment and other types of equipment that were manufactured before the ban.

OVERVIEW OF PCB REGULATION

On May 31, 1979, EPA promulgated the first major rule (44 *FR* 31512) covering the manufacture, processing, distribution in commerce, use, and disposal of PCBs. The rule did the following:

- Classified the use of PCBs in transformers, capacitors, and electromagnets as totally enclosed.
- Established requirements for the marking and disposal of PCBs in concentrations over 50 parts per million (ppm).
- Established a regulatory cutoff of 50 ppm for the manufacture, processing, distribution in commerce, and use of PCBs.
- Authorized the use of PCBs for 11 specific activities.

The Environmental Defense Fund (EDF) sought judicial review of Provisions 1, 3, and 4 in the U.S. Circuit Court of Appeals for the District of Columbia. The court ruled that the EPA lacked substantial evidence to support both the classification of transformers, capacitors, and electromagnets as totally enclosed and the regulatory cutoff of 50 ppm for the manufacture, processing, distribution in commerce, and use of PCBs.

If the court's decision had gone into effect, the use of all transformers, capacitors, and electromagnets containing PCBs would have been immediately banned. Because most electrical transformers and capacitors at the time contained PCBs, this ban would have had a disastrous economic impact on industry and consumers alike. Thus, EPA and EDF filed a joint motion with the court requesting a stay of the court's mandate until additional rulemaking could be completed. The court granted the request and compelled EPA to begin additional rulemaking.

Major rules subsequent to the May 1979 rule include the following:

- On August 25, 1982 (42 *FR* 37357), EPA promulgated the Electrical Equipment Use Rule in response to the EDF suit. It further defined the concept of a totally enclosed manner of PCB use and regulated exposure risks to food and feed.
- On July 17, 1985 (50 *FR* 29199), EPA promulgated the PCB Transformer Fire Rule. This rule regulates transformers *in or near commercial buildings* and is designed to prevent threats to human health or the environment from PCB transformers involved in fires.
- On April 2, 1987 (52 *FR* 10688), EPA promulgated the PCB Spill Cleanup Policy, which established cleanup methods and uniform cleanup levels for spill of PCBs at concentrations of 50 ppm or greater. (This rule is the focus of Chapter 18.)
- On December 21, 1989 (54 *FR* 52716), EPA promulgated the Notification and Manifesting Rule. This rule requires notification of PCB activity, manifesting of PCB waste shipments, permitting of commercial storage for disposal, and additional record-keeping requirements to parallel RCRA.
- On March 18, 1996 (61 *FR* 11096), EPA overturned the "closed border" policy and authorized the import and export of PCBs 50 ppm or greater for the purposes of disposal.

- On June 29, 1998 (63 *FR* 35384), EPA promulgated a variety of modifications to address the use, disposal, and cleanup of PCBs. The rule also codified many policies developed during the EPA/PCB management program. EPA also reinstated the "closed border" policy for imports and exports of PCBs ≥50 ppm.

The remainder of this chapter presents sections on the following specific topics:

- PCBs in electrical equipment
- PCB marking requirements
- PCB storage requirements
- PCB disposal requirements
- Exporting and importing PCBs for disposal
- Record keeping and reporting
- TSCA enforcement provisions

A compliance checklist is provided at the end of this chapter to help the reader understand the PCB requirements under TSCA.

PCBs IN ELECTRICAL EQUIPMENT

This section describes the PCB regulatory program for electrical equipment; specifically, it provides an overview of the PCB electrical equipment program and a discussion of transformers, capacitors, and other equipment.

PCB Electrical Equipment Regulatory Program Overview

The regulations governing PCBs in electrical equipment are contained in 40 CFR Part 761. These regulations are designed to ensure the proper disposal of PCBs and PCB items while minimizing risk to health and the environment during use, handling, storage, and disposal. The regulations, with some exceptions, apply to any substance, mixture, or item with a concentration of 50 ppm PCBs or greater, or contaminated by a source of PCBs 50 ppm or greater. Specific requirements (discussed later in this chapter) include the following:

- *Marking:* PCB items must be clearly identified.
- *Inspections:* PCB items in service, storage for reuse, or storage for disposal must be inspected regularly to ensure releases are either prevented or identified immediately.
- *Storage:* PCBs and PCB items must be stored in accordance with requirements designed to ensure safe storage prior to disposal.
- *Disposal:* Except as provided, PCBs and PCB items must be disposed of by high-temperature incineration.

- *Tracking*: Generators of waste PCBs and PCB items must notify EPA of their activities and must ship all PCB waste with a uniform hazardous waste manifest, and must receive certificates of disposal.
- *Record keeping:* Certain records (e.g., manifests, certificates of disposal, inspection logs) must be kept by facilities using, storing, and disposing of PCBs and PCB items.

The regulations also address prohibited and authorized uses and spill cleanup (discussed in Chapter 18). In general, EPA will hold the owners of equipment containing PCBs responsible for compliance with the TSCA regulations. However, in cases involving PCB use by a person who does not own the equipment or PCB equipment located on property owned by a third party, EPA will consider the facts of each case to determine whether the user or the landowner should be held responsible for compliance either in addition to or instead of the owner of the PCB items (TSCA Compliance Program Policy, No. 6-PCB-1, March 4, 1981).

PCB Classifications

Intentionally generated polychlorinated biphenyls (e.g., commercial products) are classified as follows, based on their PCB concentration:

Non-PCB: less than 50 ppm PCB

PCB-contaminated: 50 to 499 ppm PCB

PCB: 500 ppm or greater PCB

In addition to the PCB content, classification and regulations are also based on the type of equipment. The equipment classification includes

- Transformers
- Capacitors
- Hydraulic systems
- Miscellaneous equipment

Reclassification. PCB electrical equipment may be reclassified to PCB-contaminated or non-PCB status [40 CFR 761.30(a)(2)(v)]. Reclassifying involves the draining of all of the dielectric fluid, then filling the equipment with non-PCB dielectric fluid. Provided that the dielectric fluid is retested following a minimum of three months' *in-service* use after the date of filling, and it has a PCB concentration of less than 500 ppm or less than 50 ppm, the item may be reclassified and handled accordingly. During the in-service period, the transformer must be used under loaded electrical conditions that raise the temperature of the dielectric fluid to a minimum of 50°C.

Note: EPA does not, however, recognize the validity of field screening tests for reclassification purposes. Appropriate laboratory analysis must be used.

Dilution to Circumvent Disposal. Dilution to circumvent disposal requirements is prohibited. For example, if any amount of PCB fluid (500 ppm or greater) is introduced to PCB-contaminated or non-PCB fluid or equipment, the entire mixture or equipment must be classified and handled as PCB (i.e., 500 ppm or greater) regardless of the actual level [40 CFR 761.1(b)]. Likewise, no person may process liquid PCBs into non-liquid PCBs to circumvent the high-temperature incineration requirements [40 CFR 761.50(a)(2)].

Transformers

Transformers are used to raise or lower voltage depending on the user's needs. Large transformers can be the size of a trailer and contain hundreds or thousands of gallons of dielectric liquid. These large transformers are typically located in generating facilities or electrical substations. The vast majority of transformers, however, are considerably smaller.

Electrical transformers are often filled with a dielectric liquid that increases the resistance of the unit to arcing and acts as a heat-transfer medium, helping to cool the coils. Most transformers are filled with transformer oil, but some liquid-filled transformers are filled with a PCB-based chlorinated fire-resistant fluid.[1] Prior to 1979, PCB-based transformers contained 60 to 100 percent PCBs.

Transformers with less than 3 pounds of dielectric may assume that the PCB concentration is less than 50 ppm unless otherwise known. Transformers manufactured prior to July 2, 1979, which contain greater than 3 pounds of fluid is presumed to have a PCB concentration of 500 ppm or greater unless demonstrated otherwise (40 CFR 761.2). This demonstration *can* be accomplished by analysis, identification labels, or manufacturer's documentation (e.g., ransformer was filled with mineral oil, as discussed in the next section).

Mineral-Oil-Filled Transformers. Transformers with mineral oil as a dielectric manufactured before July 2, 1979 are presumed to be PCB-contaminated (50 to 499 ppm) unless otherwise known. All pole-top and pad-mounted distribution transformers manufactured before July 2, 1979, must be presumed to be mineral-oil filled. If testing the mineral oil demonstrates that the fluid has a PCB concentration of 500 ppm or greater, the item must be classified as a PCB item [40 CFR 761.2(a)(2)]. EPA has established a time frame for bringing "newly discovered" mineral-oil transformers into compliance with PCB transformer requirements [40 CFR 761.30(a)(1)(xv)]. A transformer owner has 7 days from testing to label and, if the transformer is located in or near a commercial building, 30 days to notify the appropriate fire-response personnel and building owner(s) of the discovery.

[1] Common trade names for PCB-based liquids include Askarel, Chlorextol, Asbestol, Pyranol, Saf-T-Kuhl, EEC-18, No-Flamol, Inerteen, Draclor, Chorinol, Adkerel, Eucarel, Nepolin, and Dykanol.

Authorized Uses of Transformers. PCB and PCB-contaminated transformers may be used for the remainder of their useful service lives, provided that they are intact and nonleaking, except as noted in the following "Prohibited Uses" discussion. (*Intact and nonleaking* means the transformer is structurally sound with all fluid intact and there are no PCBs on the external surface of the transformer.)[2] A PCB transformer that is not in use, but is intended for reuse, is considered an in-service transformer for the purposes of the rule and is authorized for the remainder of its useful service life.

PCB and PCB-contaminated transformers may be sold, provided that they were originally purchased or sold for purposes other than resale before July 1, 1979. In addition, the transformer must be intact, nonleaking, marked as a PCB transformer, and in working condition [40 CFR 761.20(c)(1)]. All owners of PCB transformers, including those in storage for reuse, were required to register them with EPA headquarters no later than December 28, 1998 [40 CFR 761.30(a)(vi)(A)]. Copies of the registration must be maintained with the annual records.

PCB transformers generally need to be serviced periodically and repaired to ensure proper operation. Authorized servicing under TSCA includes

- Sampling of the fluid to test its dielectric strength.
- "Topping off" with additional dielectric fluid.
- Replacement of gaskets, bushings, and insulators that may involve partial draining of the transformer.
- Removal and filtering of the PCB liquid and refilling of the unit.
- Removal of the PCB liquid and its replacement with non-PCB fluid.

Any PCB transformer being serviced or rebuilt may not have its core removed at any time; however, it may be "topped off" with dielectric fluid of any concentration. PCB-contaminated transformers may have their core removed but only dielectric fluid with less than 500 ppm PCB may be added. If dielectric fluid containing PCBs above 500 ppm is introduced to a PCB-contaminated or non-PCB transformer, the transformer must be classified as a PCB transformer regardless of the actual PCB concentration. PCBs removed from any servicing activity must be captured and either reused as dielectric fluid or disposed of in accordance with 40 CFR 761.60. Any dielectric fluid containing PCBs at concentrations of 50 ppm or greater and used for servicing transformers must be stored in accordance with the storage for disposal requirements of 40 CFR 761.65, addressed in the storage section of this chapter [40 CFR 761.30(a)(2)].

Prohibited Uses of Transformers. The use (and storage for reuse) of PCB transformers that pose an exposure risk to food or feed is prohibited. In addition, the installation and most uses of PCB transformers in or near commercial buildings has

[2] U.S.EPA, *PCB Q&A Manual*, 1994 Edition, pages 1–5.

been prohibited, except for those conditions outlined in this section [40 CFR 761.30(a)(1)].

All PCB transformers except lower and higher secondary-voltage radial transformers equipped with electrical protection to detect sustained high-current faults and provide for the complete deenergization of the transformer or the faulted phase of the transformer are prohibited from being used in or near commercial buildings [761.30(a)(1)(iv)(E)]. Pressure- and temperature-sensitive (or equally effective) devices must be used in these transformers to detect sustained low-current faults. Equipment must be disconnected to ensure complete deenergization in the event of a sensed abnormal condition (e.g., an overpressure or overtemperature condition in the transformer) caused by a sustained low-current fault. The disconnect equipment must be configured to operate automatically within 30 to 60 seconds of the receipt of a signal indicating an abnormal condition from a sustained low-current fault or to allow for manual deenergization from a *manned on-site control center* upon the receipt of an audio or visual signal indicating an abnormal condition caused by a sustained low-current fault. If automatic operation is selected and a circuit breaker is utilized for disconnection, it must also have the capacity to be manually opened if necessary. The detection and deenergization equipment must be properly installed and maintained and be set sensitive enough to detect abnormal operations [40 CFR 761.30(a)(1)(v)(b)].

Combustible materials including but not limited to paints, solvents, plastics, paper, and sawdust, must not be stored within a PCB transformer enclosure, within 5 meters of a PCB enclosure, or within 5 meters of a PCB transformer [40 CFR 761.30(a)(1)(viii)].

If a PCB transformer (or PCB voltage regulator) is involved in a fire-related incident, the owner or operator must immediately report the incident to the National Response Center (800-424-8802). A *fire-related incident* is defined as any incident involving a PCB transformer that involves the generation of sufficient heat and/or pressure (by any source) to result in the rupture (violent or nonviolent break) of a PCB transformer and the release of PCBs [40 CFR 761.30(a)(1)(xi)]. The owner or the operator must also take measures as soon as practical and safe to contain and control any potential releases of PCBs and incomplete combustion product into water [40 CFR 761.30(a)(1)(xi)]. These measures should include the following:

- The blocking of all floor drains in the vicinity of the transformer.
- The containment of water runoff.
- The control and treatment (prior to release) of any water used in subsequent cleanup operations.

Inspections. A visual inspection of certain PCB transformers in use, or stored for reuse, must be conducted at least once every three months, except as noted in the following list [40 CFR 761.30(a)(1)(ix)]. These inspections may take place at any time during the three-month time periods of January–March, April–June, July–

September, and October–December, provided there is a minimum of 30 days between inspections. The visual inspection must include observations for any leaks on or around the transformer. In-service PCB-contaminated transformers require no inspection.

- In-service PCB transformers with a PCB content of more than 60,000 ppm require a quarterly inspection.
- In-service PCB transformers with a PCB content of between 500 and 60,000 ppm require only an annual inspection.
- In-service PCB transformers with a PCB content of more than 60,000 ppm but with 100 percent impermeable secondary containment require only an annual inspection.

Records of a transformer's inspection and maintenance history must be prepared and must be kept as part of the annual records for at least three years after disposal of the transformer. These records must contain the following information:

- The name of the person conducting the inspection.
- The date of each visual inspection and the date when any leak was discovered.
- The transformer's location.
- The location of any leaks.
- An estimate of the amount of any dielectric fluid released from the leak.
- The date and description of any cleanup, containment, replacement, or repair performed on the transformer.
- The results of any containment and daily inspection required for uncorrected active leaks.

Leaks. If a PCB transformer is found to have a leak that results in *any* quantity of PCBs running off or about to run off its external surface, the transformer must be repaired or replaced to eliminate the source of the leak. PCB waste resulting from the cleanup of spills or leaks must be stored and disposed of in accordance with 40 CFR 761.60(a).

Capacitors

Capacitors are commonly used to improve the voltage and power factor. Virtually all capacitors manufactured before 1978 were filled with PCB dielectric and had a PCB concentration range of 75 to 100 percent. Capacitors are presumed to have a concentration of 500 ppm or greater unless demonstrated otherwise [40 CFR 761.2(a)(4)]. All non-PCB capacitors (except large, high-voltage capacitors) manufactured between July 1, 1978, and July 1, 1998, are required to be labeled with the statement NO PCBS [40 CFR 761.40(g)].

Under TSCA, there are three categories of PCB capacitors based on the amount of dielectric and voltage ratings:

- Large high-voltage capacitors (> 3 lbs. dielectric and rated at 2,000 volts)
- Large low-voltage capacitors (> 3 lbs. dielectric and rated at less than 2,000 volts)
- Small capacitors (< 3 lbs. dielectric)

PCBs at any concentrations may be used in capacitors for the remainder of their useful lives if the capacitors are intact and nonleaking, unless they present an exposure risk to food or feed, and only if they are in a restricted-access area.

Exposure Risk to Food or Feed. Large PCB capacitors used or stored for reuse that pose an "exposure risk to food or feed" are prohibited [40 CFR 761.30(l)(1)(i)]. In evaluating the *exposure risk* from a capacitor, it is best to consider a hypothetical situation in which PCBs are discharged in any way from the capacitor, such as through a leak or a rupture. Assuming that such a discharge occurs, thereby releasing all or a portion of the contained PCBs, and considering the capacitor's location and any relevant factors, it must be determined whether contact between PCBs and food or feed is possible. PCB items that are located directly adjacent to or above food or feed products are presumed to pose an exposure risk to food or feed unless there is some type of secondary containment or other physical structure that prevents discharges of PCBs from contaminating food or feed.

Restricted Access. Large PCB capacitors may be used only if they are located at a restricted-access electrical substation or in a restricted-access indoor location that has adequate roof, walls, and floor to contain any release of PCBs within the indoor location [40 CFR 761.30(l)(1)(ii)]. A *restricted-access* electrical substation is an outdoor, fenced or walled-in facility that restricts public access and is used in the transmission or distribution of electrical power (47 *FR* 37349, August 25, 1982).

Miscellaneous Equipment

All oil-filled switches, electromagnets, and voltage regulators are presumed to be PCB-contaminated (50 to 499 ppm) unless it is demonstrated otherwise. All oil-filled cables, reclosers, and circuit breakers are assumed to be non-PCB (< 50 ppm) unless verification demonstrates otherwise [40 CFR 761.2(a)(2)].

PCBs were widely used in hydraulic systems by steel manufacturing and die-casting plants to reduce fire hazards. Hydraulic systems normally leak several times their capacity each year because the fluid is pressurized to several thousand pounds per square inch and can leak at connections. PCBs may be used in hydraulic systems only at concentrations less than 50 ppm because these systems are not considered to be totally enclosed [40 CFR 761.30(e)].

PCB MARKING REQUIREMENTS

The following items *must* be marked with an appropriate PCB label. The label must be placed on the exterior of the PCB item or vehicle in a place that can be easily seen and read by anyone inspecting or servicing it (40 CFR 761.40). The following must be marked:

- All PCB transformers (PCB-contaminated transformers are not required to be marked).
- All large PCB capacitors.
- Equipment containing a PCB transformer or a large PCB high-voltage capacitor.
- All containers used to store PCB items.
- All accesses to PCB transformer locations in or near commercial buildings.
- PCB storage facilities.
- Any transport vehicle carrying one or more PCB transformers or 99.4 pounds of PCB liquid greater than 50 ppm (must be marked on all four sides).
- Electrical motors using PCB coolants.
- Hydraulic and heat transfer systems using PCBs.
- Voltage regulators with PCBs 500 ppm or greater and their locations.

Note: Although small PCB capacitors themselves are not required to be labeled with a PCB label as of January 1, 1979, all equipment containing small capacitors must be marked at the time of manufacture with the statement "This equipment contains PCB capacitors" [40 CFR 761.40(d)].

PCB STORAGE REQUIREMENTS

This section describes the storage requirements for PCBs and PCB items, including the following:

- Storage facility requirements.
- Temporary storage requirements.
- Inspections during storage.
- Bulk storage.
- Storage time limits.
- Laboratory storage.

Storage Facility Requirements

PCB items destined for disposal must be stored in a PCB storage facility that complies with 40 CFR 761.65(b). The requirements for the PCB storage facility include the following:

- The roof and walls must provide adequate protection from precipitation.
- The floor must have continuous curbing at a minimum height of 6 inches.
- The floor and curbing must provide containment for at least 25 percent of the total internal volume of all stored PCB items or two times the internal volume of the largest PCB item, whichever is greater.
- The PCB storage facility cannot have drain valves, floor drains, expansion joints, sewer lines, or other openings that could permit liquids to flow from the curbed area.
- The floor must be made of a smooth and impervious material (e.g., Portland cement, steel), which prevents or minimizes penetration of PCBs.
- The PCB storage facility cannot be located within a 100-year floodplain.

Additional Storage Options

There are two options for storing PCBs in areas other than a PCB storage facility that comply with 40 CFR 761.65(b). These options are commonly called *30-day temporary storage* and *pallet storage.*

Temporary Storage. Certain PCB items may be stored temporarily in an area that does not comply with the preceding PCB storage facility requirements for up to 30 days from the date of their removal from service. This practice is authorized only if a notation is attached to the PCB item indicating the date when the item was removed from service, and only if the storage area is marked with a PCB mark [40 CFR 761.65(c)(1)]. The only PCB items authorized for the 30-day temporary storage provision are the following:

- Nonleaking PCB articles and equipment.
- Leaking PCB articles and PCB equipment if the items are placed in a nonleaking PCB container that contains sufficient sorbent materials to absorb any liquid PCBs remaining in the PCB items.
- PCB containers holding nonliquid PCBs (e.g., contaminated soil, rags, and debris).
- PCB containers holding liquid PCBs at a concentration of 50 ppm or greater provided a Spill Prevention, Control, and Countermeasure Plan (SPCC) has been prepared for the temporary storage area in accordance with 40 CFR Part 112. In addition, all liquid PCBs must be packaged in accordance with HMTA (see Chapter 4).

Pallet Storage. The other storage option, known as pallet storage, is authorized for certain equipment on pallets provided that the equipment is stored next to a PCB storage facility that complies with 40 CFR 761.65(b). Furthermore, pallet storage is authorized only if the PCB storage facility has immediately available unfilled storage space equal to 10 percent of the volume of the equipment stored on the pallets outside of the PCB storage facility.

The only types of equipment authorized for pallet storage are nonleaking, structurally undamaged, large high-voltage PCB capacitors and PCB-contaminated electrical equipment not drained of free-flowing dielectric fluid. (PCB-contaminated electrical equipment drained of free-flowing dielectric fluid is no longer regulated under TSCA.) All equipment stored on pallets outside of the storage facility must be inspected for leaks at least weekly [40 CFR 761.65(c)(2)].

Inspections During Storage

All PCB items in a PCB storage facility must be checked for leaks at least once every 30 days [40 CFR 761.65(c)(5)]. Any leaking PCB items and their contents must be immediately transferred to properly marked nonleaking containers. Any spilled or leaked materials must be immediately cleaned up in accordance with the PCB Spill Cleanup Policy (see Chapter 18) and disposed of in accordance with 40 CFR 761.60(a)(4).

Bulk Storage

Large containers, such as storage tanks, may be used to store bulk PCB liquids for up to one year. These storage tanks must meet the design and construction standards adopted by the Occupational Safety and Health Administration (OSHA) contained in 29 CFR 1910.106 for flammable and combustible liquids. In addition, bulk storage facilities must have an SPCC plan similar to the plans required for oil-spill prevention [40 CFR 761.65(c)(7)].

For each batch of PCBs stored in bulk, records must be maintained that indicate the exact quantity of the batch and the date the batch was added to the container. The record must also include the date, quantity, and disposition of any PCBs removed from the container [40 CFR 761.65(c)(8)].

Storage Time Limits

In accordance with 40 CFR 761.65(a), all PCB items, articles, or containers must be disposed of within one year of the date they were placed into storage (40 CFR 761.65(a)(2) authorizes a conditional one-year extension). Because the one-year storage deadline is applicable to the generator *and* disposal facility, EPA has established a policy (48 *FR* 52304, November 17, 1983) concerning the appropriate storage time limits for the generator and disposal facility. This policy states that a generator has 275 days (from the date it went into storage) to remove the PCB item and deliver it to the disposal facility. The disposal facility has 90 days (from the date it received the PCB item) to dispose of the PCB item. If the one-year storage limit is violated, the party that exceeds his or her time limit (i.e., 275 days for the generator or 90 days for the disposer) is in violation. For example, a PCB transformer was put into storage on July 1, 1999, but was not disposed of until September 1, 2000. The generator, however, sent the transformer in a timely fashion, and it was received by the disposer on January 1, 2000. In this case, the disposer was in violation for having exceeded the 90-day limit, whereas the generator was in compliance with the 275-day limit.

When either party exceeds its respective time limit, a *one-year exception report* must be filed with the appropriate EPA Regional Administrator. The one-year exception report must include a legible copy of a manifest or other written communication relevant to the transfer and disposal of the affected PCBs or PCB item. In addition, a cover letter signed by the submitter must also be included, describing the following:

- The date(s) when the PCBs or the PCB item was removed from service for disposal.
- The date(s) when the affected PCBs or PCB item was received by the submitter of the report, if applicable.
- The date(s) when the affected PCBs or PCB item was transferred to a designated disposal facility.
- The identity of the transporter, commercial storer, or disposers known to be involved in the transaction.
- The reason, if known, for the delay in bringing about the disposal of the affected PCBs or PCB item within the one-year time limit.

Every PCB item destined for disposal must be marked with the date when the item was taken out of service. The PCB storage facility must be arranged and managed so that PCB items can be located by the date they entered storage [40 CFR 761.65(c)(8)].

An item being stored for reuse *may* be stored indefinitely if it is intact, nonleaking, and in working order. Any PCB item being stored for reuse is subject to all in-use requirements, such as quarterly inspections and marking.

Laboratory Storage

Laboratories storing samples are conditionally exempt from the notification and approval requirements for commercial storers provided they comply with 40 CFR 761.65(b)(1)(i) through (b)(1)(iv). A laboratory is defined in 40 CFR 761.3 as a facility that analyzes PCBs and is unaffiliated with any entity whose activities involve PCBs.

PCB DISPOSAL REQUIREMENTS

This section describes the disposal requirements for the following:[3]

- Transformers
- Capacitors
- PCB containers
- Liquids
- Soil and debris
- Alternative disposal methods

Transformers

There are two sets of requirements for the disposal of transformers, depending on the classification: PCB transformers or PCB-contaminated transformers.

PCB Transformers. There are two options for the disposal of PCB transformers (i.e., 500 ppm or greater PCB). The first option allows the transformer and the PCB dielectric fluid to be burned together in a TSCA-approved high-temperature incinerator. Under the second option, all free-flowing dielectric fluid must be removed from the transformer. The drained transformer carcass must then be filled with an appropriate solvent (e.g., xylene, toluene, kerosene) for 18 hours, then drained again. The original PCB dielectric fluid and the solvent used to decontaminate the carcass must be incinerated or decontaminated and the carcass must be disposed of in a TSCA-permitted landfill [40 CFR 761.60(b)(1)].

Note: PCB-containing dielectric fluids removed from electrical transformers, capacitors, and associated PCB-contaminated equipment that also exhibits the RCRA toxicity characteristic for organics (i.e., D004–D017) are excluded from RCRA and subject to TSCA (40 CFR 261.8 and 55 *FR* 11841, March 29, 1990). This means that certain solvents may be used to remove the dielectric and only TSCA need be followed, subject to state restrictions.

PCB-Contaminated Transformers. For PCB-contaminated transformers, all free-flowing PCB-contaminated dielectric fluid must be removed. This fluid must either be incinerated, burned in a high-efficiency boiler, disposed of in a landfill (provided that the dielectric is nonignitable), or disposed of in an alternative manner approved by EPA. The transformer carcass, provided that all free-flowing liquid is removed, must be sent to a permitted solid waste landfill or industrial furnace [40 CFR 761.60(b)(4)].

[3] All disposal facilities, except for high-efficiency boilers, must possess a permit issued under TSCA if they dispose of PCBs in concentrations at or above 50 ppm. The technical requirements for the disposal units are not addressed in this book because of their complexity and because they affect only a very small part of the regulated community. The requirements, however, can be found in 40 CFR 761.70 for incinerators, 40 CFR 761.60(a)(2)(iii) for high-efficiency boilers, and 40 CFR 761.75 for landfills.

Dielectric fluid removed only from mineral-oil-filled electrical equipment may be collected in a common container, provided that no other chemical substances are added to the container. This "common-container" option does not allow dilution of the collected oil. Dielectric fluid that is presumed or known to contain at least 50 ppm PCB must not be mixed with mineral oil known or assumed to contain less than 50 ppm PCB. For the purposes of complying with the marking and disposal requirements, representative samples may be taken from either the common containers or the individual electrical equipment articles to determine the PCB concentration. However, if PCBs at a concentration of 500 ppm or greater have been added to the container or equipment, the total contents must be considered to have a PCB concentration of 500 ppm or greater [40 CFR 761.60(g)(i)].

Capacitors

Large capacitors may be incinerated or disposed of by any other method that EPA has specifically permitted under TSCA [40 CFR 761.60(b)(2)]. Intact and nonleaking small capacitors may be disposed of in a municipal landfill (subject to state rules), except for those of PCB capacitor manufacturers or manufacturers of equipment containing capacitors, such as telephone booths and light fixtures [40 CFR 761.60(b)(2)(ii)]. Manufacturers must dispose of small capacitors in TSCA-permitted incinerators [40 CFR 761.60(b)(2)].[4]

Note: Regardless of the authorized disposal options under TSCA, §§104, 106, and 107 of Superfund (Part II of this book) establish strict liability for the release of hazardous substances (PCBs are a designated hazardous substance) and for any incurred cleanup costs and natural resource damages. Therefore, major consequences can arise from disposing of small capacitors in a unsecured manner.

Containers

PCB containers may be incinerated or decontaminated, after which they may disposed of or reused) [40 CFR 761.60(c)]. Any solvent may be used for decontamination if the solubility of the PCBs in the solvent is 5 percent or more by weight. Each rinse must use a volume of the solvent equal to approximately 10 percent of the container capacity. The solvent may be reused until it contains 50 ppm PCB. The solvent must be disposed of as a PCB liquid regardless of the PCB concentration [40 CFR 761.79(a)]. The solvents recommended for decontamination of PCB containers are xylene, toluene, and kerosene (44 *FR* 31546, May 31, 1979).

Under TSCA, containers used to contain only PCBs at a concentration less than 500 ppm may be disposed of as municipal solid waste, provided that the PCBs were in a liquid state, the container was first drained of all liquid, and the liquid was disposed of accordingly.

[4] EPA Memorandum from John Moore, Assistant Administrator, Office of Pesticide and Toxic Substances, to Gary O'Neal, Director of Air and Toxics, Region X, March 4, 1985.

Liquids

Liquids containing PCBs at a concentration of 50 ppm or greater must be incinerated [40 CFR 761.60(a)]. Liquids containing PCBs at concentrations less than 500 ppm may be incinerated, burned in a high-efficiency boiler, or disposed of in a TSCA-approved landfill, provided that the waste is nonignitable and that the waste is pretreated and/or stabilized to eliminate the presence of free liquids prior to final disposal [40 CFR 761.60(a)(2) for mineral oil and 40 CFR 761.60(a)(3) for non-mineral oil].

It is prohibited to process liquid PCBs into non-liquid forms to circumvent the high-temperature incineration requirements [40 CFR 761.50(a)(2)].

Containerized liquids with PCB concentrations of less than 500 ppm may be incinerated or disposed of in a TSCA-approved landfill. Containers that are landfilled must be surrounded by an amount of inert sorbent material capable of absorbing all of the liquid contents of the container, such as an overpack [40 CFR 761.75(b)(8)]. Again, however, caution should be exercised because of the strict liability provisions of Superfund.

Soil and Debris

Soil, rags, absorbents, dredge materials, municipal sewage treatment sludges, and debris that contains PCBs at concentrations of 50 ppm or greater may be disposed of in either a TSCA permitted landfill or incinerator. However, processing liquid PCBs into nonliquid forms to circumvent the incineration requirements is prohibited [40 CFR 761.60(a)(4) and (a)(5)].

Alternative Disposal Methods

Any alternative method of disposal may be approved by EPA, provided that the method has a destruction efficiency that is similar to the other methods, and it is protective of human health and the environment. Current examples of alternative disposal methods include solvent extraction, dechlorination, and biodegradation [40 CFR 761.60(e)].

EXPORTING AND IMPORTING PCBS FOR DISPOSAL

Historically, EPA has imposed a "closed border" policy prohibiting the export or import of PCBs and PCB items with a PCB concentration of 50 ppm or greater for the purposes of disposal. On March 18, 1996 (61 *FR* 11096), EPA reversed this policy to be consistent with the Basel Convention on the Control of Transboundary Movements of Hazardous Wastes and Their Disposal.

In response, the Sierra Club sued EPA. On July 7, 1997, the U.S. Court of Appeals for the Ninth Circuit overturned EPA's Import for Disposal Rule (see *Sierra Club v. EPA*, 118 F.3d 1324). As a result, EPA promulgated a final rule on June 29, 1998 (63 *FR* 35420) to close the border to exports and imports of PCBs ≥50 ppm for disposal.

RECORD KEEPING AND REPORTING

EPA has established comprehensive "cradle-to-grave" tracking requirements for shipments of PCB wastes nearly identical to the RCRA hazardous waste program. The requirements for tracking include the following:

- *Notification of PCB Waste Activity* (EPA Form 7710-53)
- EPA identification number
- Manifest system
- Annual documents
- Certificate of disposal
- Annual records

Notification of PCB Waste Activity

Nonexempt generators, commercial storers, transporters, and disposers of PCB waste must notify EPA and obtain an EPA identification number before handling any PCB waste. *PCB waste* means those PCBs and PCB items that are subject to the disposal requirements of 40 CFR 761.60. That is, PCBs and PCB items become subject to the disposal requirements when it has been determined they no longer serve their intended purpose and are to be disposed of or are involved in a release. Items unregulated for disposal, such as intact, nonleaking small capacitors and drained PCB-contaminated equipment, are not subject to the disposal requirements of 40 CFR 761.60 and therefore are not included in the definition of PCB waste.

Generators of PCB waste must not process, store, offer for transport, transport, or dispose of PCBs without having an EPA identification number.[5] In addition, transporters, commercial storers, and disposers must not accept PCB waste without an EPA identification number. An EPA identification number is obtained by submitting to EPA Form 7710-53, *Notification of PCB Waste Activity*, as described above (40 CFR 761.202). After notification and verification of the information, persons will be issued a unique 12-digit EPA identification number.

Facilities that have previously been issued an EPA identification number for hazardous waste activities under RCRA may indicate on Item 11 of EPA Form 7710-53 the facility's RCRA identification number. After verification, EPA will use the existing RCRA number for TSCA purposes. Generators exempt from TSCA notification,

[5] Only generators who maintain PCB storage facilities subject to the 40 CFR 761.65(b) and (c)(7) storage facility standards are required to notify EPA of PCB waste activities. Generators who do not maintain such storage facilities are exempt from the requirements to notify EPA and obtain EPA identification numbers. Thus, persons who store PCB waste temporarily per 40 CFR 761.65(c)(1) are exempt from notification. Because they are not, however, excluded from manifest requirements, these generators must use the generic identification number "40 CFR Part 761" on their manifests where it calls for the shipper's EPA identification number [40 CFR 761.205(c)(1)] unless they have an EPA identification number.

however, may elect to use their previously issued RCRA identification number without having to notify [40 CFR 761.205(c)(1)].

Note: Even though a facility may already have an EPA identification number, a person engaged in PCB waste activities must still give notification of PCB activities [40 CFR 761.205(b)].

Manifest System

A generator who transports or offers PCB waste for transportation to an off-site commercial storage or disposal must use a uniform hazardous waste manifest (EPA Form 8700-22) [40 CFR 761.207(a)].

Any person shipping PCB waste to a state (the consignment state) that supplies and requires use of its manifest must use that state's manifest. If the consignment state does not require use of its own manifest, and the state in which the PCB waste was generated (the generator state) supplies and requires the use of its manifest, the person shipping the PCB waste must use the generator state's manifest. If both states require use of their manifests, the manifest of the consignment state must be used. If neither the consignment state nor the generator state requires use of its own manifest, the shipper *may* use a manifest from any source [40 CFR 761.207(b)-(f)].

The generator must do the following:

- Sign the manifest certification by hand.
- Obtain the handwritten signature of the initial transporter and the date of acceptance on the manifest.
- Retain one copy in its records.
- Give to the transporter the remaining copies of the manifest, which will accompany the shipment of PCB waste.

Generators who use independent transporters to ship waste to a storage or disposal facility are required to confirm, by telephone or by other means of confirmation agreed to by both parties, that the commercial storer or disposer actually received the manifested waste.

Samples. A sample is conditionally exempt from the manifesting requirements when it is being transported either to the laboratory for testing, or from the laboratory after it has been tested [40 CFR 761.65(i)]. A sample is also conditionally exempt from manifesting when it is being stored by the sample collector before being transported to a laboratory, stored in a laboratory before testing, stored by a laboratory after testing but before being returned to the sample collector, or stored because of a court case or enforcement action. To qualify for such an exemption, the sample handlers must ensure that the following information accompanies the shipment while being transported to or from the laboratory:

- The sample collector's name, mailing address, and telephone number.

- The laboratory's name, address, and telephone number.
- The quantity, date of shipment, and description of the sample.

The sample must be packaged so that it does not leak, spill, or vaporize from its packaging. Depending on the mode of shipment (e.g., by mail or surface transport), the requirements of DOT or the U.S. Postal Service must be followed.

When the concentration of the PCB sample has been determined, and its use terminated, the sample must be properly disposed of in accordance with its classification [40 CFR 761.65(i)(4)].

Manifest Exception Report. If the generator has not received the hand-signed manifest within 35 days after the initial transporter accepted the PCB waste, the generator must contact the disposer or commercial storer to determine whether the PCB waste has been received [40 CFR 761.215(a)]. If the generator has not received a hand-signed manifest from an EPA-approved facility within 45 days after the waste was originally shipped, the generator must submit an exception report to the EPA Regional Administrator for the region in which the generator is located.

The exception report must contain both a legible copy of the manifest and a cover letter signed by the generator describing the efforts taken to locate the missing shipment [40 CFR 761.215(b)].

Manifest Discrepancy Report. If a significant discrepancy is discovered between the quantity or type of PCB waste designated on the manifest and that which is actually received by the designated facility, the owner or operator of the commercial storer or disposal facility must attempt to reconcile the discrepancy verbally. (A *significant discrepancy* means a variation of 10 percent or more in weight, any variation in piece count of containers, a difference in classification levels of PCB concentration, or a different physical state.) If the discrepancy is not reconciled within 15 days of receiving the waste, a manifest discrepancy report must be filed with the EPA Regional Administrator. A manifest discrepancy report consists of a letter outlining attempts to reconcile the discrepancy and a copy of the applicable manifest [40 CFR 761.210(b)].

Unmanifested Waste Report. If a commercial storer or disposer received any shipment from off-site that has no manifest or shipping paper, the owner or operator must contact the shipper to obtain the paperwork or return the shipment. If the shipper cannot be found, the EPA Regional Administrator must be notified. Within 15 days of receiving the unmanifested shipment, the owner or operator of the commercial facility must submit an unmanifested waste report (using either EPA Form 8700-13B or a written letter) to the EPA Regional Administrator 40 CFR 761.211). The report must contain the following information:

- EPA identification number, name, and address of the commercial facility.
- Date the unmanifested waste was received.

- If available, the EPA identification number, name, and address of the shipper.
- Description of the type and quantity in the shipment.
- Brief description as to why the shipment was unmanifested (if known).
- Disposition of the waste.

Certificate of Disposal

The owner or operator of a disposal facility that accepts and disposes of a PCB waste shipment must prepare a certificate of disposal for that shipment [40 CFR 761.218(a)]. The certificate must include the following:

- The identity of the disposal facility by name, address, and EPA identification number.
- The identity of the PCB waste, including reference to the manifest number.
- A statement certifying that disposal has occurred, including the date of disposal and the process employed.
- The following certification statement:

> Under civil and criminal penalties of law for the making or submission of false or fraudulent statements or representations (18 U.S.C. 1001 and 15 U.S.C. 2615), I certify that the information contained in or accompanying this document is true, accurate, and complete. As to the identified section(s) of this document for which I cannot personally verify truth and accuracy, I certify as the company official having supervisory responsibility for the persons who, acting under my direct instructions, made the verification that this information is true, accurate, and complete.

The certificate of disposal must be sent to the generator within 30 days of the date of disposal [40 CFR 761.218(b)]. The disposer, the commercial storer, and the generator must keep a copy of each certificate as part of their facility records [40 CFR 761.218(c) and (d)].

Annual Records

Annual records must be maintained if an owner or an operator of a facility uses or stores at any time during the previous calendar year the following [40 CFR 761.180(a)]:[6]

- 45 kg (99.4 lbs.) of PCBs in containers,
- One or more PCB transformers, or
- 50 or more large, high- or low-voltage PCB capacitors.

[6] Commercial storers and disposal facilities must comply with the record-keeping requirements contained in 40 CFR 761.180(b) and (c) and submit annual reports.

Annual records are yearly summaries documenting the handling and disposition of PCBs managed during the previous year (i.e., January 1 to December 31), including all signed manifests and all certificates of disposal received.

And by July 1 of each year, an *annual document log* containing the following information must be prepared for the previous year's activities:

- The name, address, and EPA identification number of the facility and the calendar year being reported.
- The unique manifest number of every manifest generated by the facility during the calendar year.
- For *bulk PCB waste* (e.g., tanker or truck), its weight (in kg), the first date it was removed from service for disposal, the date it was placed into transport for off-site storage or disposal, and the date of disposal.
- The serial number or other means of identifying each *PCB article.*
- The weight (in kg) of the PCB waste in each transformer or capacitor, the date it was removed from service for disposal, the date it was placed in transport for off-site storage or disposal, and the date of disposal.
- A unique number identifying each *PCB container;* a description of the contents of each PCB container (e.g., liquid, soil, and cleanup debris), including the total weight of the material (in kg) in each PCB container; the first date the material was placed in each PCB container for disposal; the date each container was placed in transport for off-site storage or disposal; and the date of disposal.
- A unique number identifying each *PCB article container;* a description of the contents of each PCB article container (e.g., capacitors, electric motors, pumps), including the total weight (in kgs) of the contents of each PCB article container; the initial date a PCB article was placed in each PCB article container for disposal; the total weight of the PCB articles (in kgs) in each PCB article container; the date the PCB article container was placed in transport for off-site storage or disposal; and the date of disposal.
- A record of each telephone call, or other means of verification agreed upon by both parties, made to each designated commercial storer or designated disposer to confirm receipt of PCB waste transported by an independent transporter.
- A copy of all required PCB Transformer registration records.
- The total quantities (in kg) and number of PCB items (e.g., transformers, capacitors) and PCBs remaining in service at the end of each calendar year.

The annual records and document log have to be *maintained,* but they do not have to be submitted unless requested by EPA. An owner or operator of multiple facilities may maintain the records and document logs at one designated facility that is normally occupied for at least eight hours a day, provided that the identity of the designated facility is available at each facility using or storing PCBs and PCB items.

Weight Calculations. The most accurate way to calculate the weight of the PCBs in PCB items is to have the dielectric analyzed by a laboratory. However, if the exact weight of the dielectric fluid is unknown, assumptions about its weight may be used. As a general rule, Aroclor fluids weigh approximately 12 pounds per gallon, mineral oil weighs approximately eight pounds per gallon, and dielectric fluid from capacitors weighs approximately 14 pounds per gallon.[7]

Record Retention. The annual document log must be maintained for at least three years after the facility ceases using or storing PCBs and PCB items in the prescribed quantities. Annual records, manifests, and certificates of disposal must also be maintained for three years [40 CFR 761.180(a)].

In addition to these record-keeping requirements, records of inspections and maintenance history must be prepared for PCB transformers and maintained for at least 3 years after the disposal of the transformer [40 CFR 761.30(a)(xii)].

TSCA ENFORCEMENT PROVISIONS

Under the authority of TSCA, EPA may inspect any establishment where *chemical substances* are manufactured, processed, imported into, stored, or held before or after their distribution in commerce. The term *chemical substance,* under TSCA, includes PCBs. No inspection may include financial, sales, pricing, personnel, or research data, unless specified in an inspection notice. EPA may, however, subpoena witnesses, documents, and other information as necessary to carry out a TSCA inspection.

Civil actions concerning violations of or lack of compliance with TSCA may be brought to a U.S. district court to restrain or compel the taking of an action. Any chemical substance or mixture, including PCBs that were manufactured, processed, or distributed in commerce in violation of TSCA, may be subject to seizure.

A specific enforcement policy for implementing TSCA regulations has been developed by EPA. The policy seeks to identify and rank possible violations of a particular regulation, identify the tools available for compliance monitoring and how they will be used, provide a formula for determining the application of inspection resources, and establish policy for determining civil penalties under TSCA. This enforcement policy is known as the *TSCA Civil Penalty Policy.* In addition, EPA issues periodic updates to its *PCB Penalty Policy*, which is used to determine penalties for violations of the PCB rules in conjunction with the *TSCA Civil Penalty Policy.*

[7] EPA regulatory interpretation letter from Glen Kuntz, Office of Toxic Substances, Washington, DC, to Moorehead Electrical Machinery Company, 1981 (date unspecified).

Citizen Suits

Under §21 of TSCA, any person may bring a civil suit to restrain a violation of TSCA by any party or to compel EPA to perform any nondiscretionary duty required by this law. EPA has 90 days to respond to a petition. If no action is taken or a petition is denied, the party has the opportunity for judicial review in a U.S. district court. In both civil suits and citizens' petitions, the court may award reasonable legal costs and attorneys' fees, if appropriate.

Civil and Criminal Penalties

Any person who fails or refuses to comply with any requirement under TSCA may be subject to a civil penalty of up to $25,000 per day per violation. Persons who knowingly or willfully violate the law, in addition to any civil penalties, may be fined up to $25,000 for each day of violation, imprisoned for up to one year, or both.

Imminent Hazard

Although it has not been used to date, EPA may issue an Imminent Hazard Order under §7 of TSCA. An imminent hazard exists when a chemical substance or mixture *presents an imminent and unreasonable risk of serious or widespread injury to health or the environment.* When such a condition exists, EPA is authorized to initiate an action in a U.S. district court. Remedies to an imminent hazard may include seizure, recall, replacement, and notification of the affected population.

PCB COMPLIANCE CHECKLIST[8]

The following is intended only to be an overview of the major requirements and is not intended to be all-inclusive.

Inventory

Does facility have in service, stored for reuse, or for disposal?

- Large high- or low-voltage PCB capacitors Yes___ No___
- PCB transformers Yes___ No___
- PCB-contaminated transformers Yes___ No___
- Electromagnets with PCBs Yes___ No___
- Hydraulic systems with PCBs Yes___ No___
- Switches with PCBs Yes___ No___
- Reclosers with PCBs Yes___ No___

[8]Because there is no provision for state authority under TSCA, no state operates the PCB program in lieu of the federal government. However, some states classify PCBs as a hazardous waste and some states assist EPA by enforcing the TSCA rules; thus, states may have some different and/or additional requirements.

Reclassification

Has the dielectric of the PCB transformer been drained and replaced with non-PCB fluid? Yes___ No___

Has the refilled transformer been subjected to a minimum three-month, in-service test? Yes___ No___

Has the dielectric been tested using laboratory analysis? Yes___ No___

Inspections

Are quarterly visual inspections being conducted for PCB transformers with a PCB concentration > 60,000 ppm? Yes___ No___

Are annual visual inspections being conducted for PCB transformers with a PCB concentration of < 60,000 ppm or those with >60,000, but with impermeable secondary containment? Yes___ No___

Are PCB items stored on pallets inspected weekly? Yes___ No___

Has it been verified that no PCB transformer or capacitor poses an exposure risk to food or feed? Yes___ No___

Has it been verified that only authorized PCB transformers are located in or near a commercial building? Yes___ No___

Are records maintained on the results of all inspections? Yes___ No___

Marking

Are the following items marked with a PCB label?

- PCB transformer Yes___ No___
- Large PCB capacitor Yes___ No___
- Equipment containing PCB transformers or large capacitors Yes___ No___
- PCB container Yes___ No___
- PCB storage facility Yes___ No___
- Access to a PCB transformer in or near a commercial building Yes___ No___
- Transport vehicle carrying a PCB transformer or 99.4 lbs. of PCB liquid Yes___ No___
- Electrical motors using PCB coolants Yes___ No___
- Hydraulic systems using PCB hydraulic fluid Yes___ No___
- Heat-transfer systems using PCBs Yes___ No___

Storage within Storage Facility

Is there an on-site PCB storage facility? Yes___ No___

Does the storage facility meet the following requirements:

- Adequate protection from rainfall? Yes___ No___
- Impervious flooring with six-inch continuous curbing? Yes___ No___
- No open floor drains or grates? Yes___ No___
- Adequate containment volume? Yes___ No___
- Facility located above the 100-year floodplain? Yes___ No___

Are all PCB items located within the storage facility dated? Yes___ No___

Has an SPCC plan been prepared and implemented? Yes___ No___

Are all PCB items arranged so they can be located by date? Yes___ No___

Are PCB items stored and handled in a manner that protects them from accidental breakage or damage? Yes___ No___

Storage in Other Storage Areas

Are authorized intact PCB items stored in compliance with the 30-day requirements? Yes___ No___

Are authorized intact PCB items stored in compliance with the pallet storage requirements? Yes___ No___

Disposal

Are PCB transformers or other pieces of equipment containing PCBs drained or decontaminated prior to disposal? Yes___ No___

Is the drainage and solvent-filling site adequate to protect against spills and leaks and consequent contamination of surrounding areas and waterways? Yes___ No___

Do solvents used for decontamination contain less than 50 ppm PCB? Yes___ No___

Was the rinse volume of the solvent approximately equal to 10% of the container's total volume? Yes___ No___

Are PCB transformers completely filled with solvent and allowed to stand for at least 18 hours before being drained? Yes___ No___

Is the PCB transformer carcass sent to a permitted chemical waste landfill? Yes___ No___

Are the drained PCB fluids properly stored and or disposed of? Yes___ No___

Are liquid PCBs sent off-site for incineration? Yes___ No___

Are PCB-contaminated transformers drained, with the liquid incinerated or burned? Yes___ No___

Are PCB containers incinerated or properly decontaminated? Yes___ No___

Are PCB capacitors incinerated or disposed of in an approved manner? Yes___ No___

Tracking

Has EPA been notified of PCB waste activities (if required)? Yes___ No___

Has an EPA identification number been obtained? Yes___ No___

Are uniform hazardous waste manifests used for PCB waste shipments? Yes___ No___

If required, has a manifest exception report been filed with EPA? Yes___ No___

Have certificates of disposal been received for all items as required? Yes___ No___

Have all PCB transformers been registered with EPA? Yes___ No___

Record Keeping

Do records indicate the following dates?

- When the PCBs were removed from service? Yes___ No___
- When PCBs were placed in storage for disposal? Yes___ No___
- When PCBs were placed in transport for disposal? Yes___ No___

Do records indicate the following quantities for items related to the preceding list:

- The weight of PCBs and PCB items in PCB containers? Yes___ No___
- The identification of contents of PCB containers? Yes___ No___
- The number of PCB transformers? Yes___ No___
- The weight of PCBs in PCB transformers? Yes___ No___
- The number of large, high-, and low-voltage PCB capacitors? Yes___ No___

Do records indicate the quantities of PCBs remaining in service, broken down as follows?

- The weight of PCBs and PCB items in PCB containers Yes___ No___
- The identification of contents of PCB containers Yes___ No___
- The number of PCB transformers Yes___ No___
- The weight of PCBs in PCB transformers Yes___ No___
- The number of large, high-, and low-voltage PCB capacitors Yes___ No___

Are copies of manifests maintained? Yes___ No___

Are copies of PCB transformer registrations maintained? Yes___ No___

Is all of the information listed in the record keeping section contained in annual records? Yes___ No___

Is all of the information listed in the record keeping section compiled in an annual document log? Yes___ No___

18

PCB SPILL CLEANUP REQUIREMENTS

In 1987 EPA established its National PCB Spill Cleanup Policy, applicable to spills of PCBs at concentrations of 50 ppm or greater. This chapter describes the policy as well as spill-reporting requirements for PCBs, in addition to the requirements addressed in Chapter 12.

INTRODUCTION

On April 2, 1987, EPA promulgated (52 *FR* 10688) its National PCB Spill Cleanup Policy. This policy establishes reporting requirements in addition to those under Superfund for spills involving materials containing PCBs at concentrations of 50 ppm or greater. The policy also establishes cleanup standards for such spills (40 CFR 761.120).

Before the promulgation of the National PCB Spill Cleanup Policy, persons responsible for a spill were required to contact the appropriate EPA Regional PCB Coordinator for a determination of adequate cleanup levels and procedures. The result was inconsistent cleanup policies, depending on the region. EPA issued the national policy to establish a nationally consistent PCB spill cleanup policy based on the most common types of spills involving electrical equipment. However, non-typical spills or spills onto high exposure areas such as water, sewers, grazing land, or gardens are not subject to this policy and must be reported directly to the EPA Regional PCB Coordinator. The cleanup of these spills is handled on a case-by-case basis.

Because PCBs must be disposed of in accordance with TSCA regulations, EPA classifies spills, leaks, and other uncontrolled discharges of PCBs at concentrations of 50 ppm or greater as an illegal disposal of PCBs [40 CFR 761.50(a)(4) and 761.135(a)]. The term *spill* includes spills, leaks, or other uncontrolled discharges of PCBs in which the release results in any quantity of PCBs running off, or about to run off, the surface of the equipment or other PCB source, as well as the contamination resulting from those releases (40 CFR 761.123). When PCBs are improperly disposed of as a result of a spill of material containing 50 ppm or greater, EPA has the authority under §17 of TSCA, as well as under §§104, 106, and 107 of Superfund (because PCBs are a hazardous substance), to compel persons to take actions to remediate damages and/or clean up contamination resulting from the spill. Section 107 of Superfund also authorizes the remediation of natural resource damages.

Previously, it was EPA policy not to charge a party with a disposal violation if the responsible party could show that the spill, leak, or uncontrolled discharge occurred during the authorized use of electrical equipment, and that adequate cleanup measures were initiated within 48 hours. The National PCB Spill Cleanup Policy has not changed the first part of this policy with regard to citing a facility for a disposal violation. However, EPA has changed the action-initiation time frame to 24 hours (or 48 hours involving a PCB transformer) for high-concentration spills and low-concentration spills involving more than 1 lb. of PCBs [40 CFR 761.125(c)(1)].

Compliance with the National PCB Spill Cleanup Policy

Although a spill containing PCB concentrations of 50 ppm or greater is considered improper disposal, the National PCB Spill Cleanup Policy establishes requirements that EPA considers to ensure adequate cleanup of spilled PCBs. Cleanup in accordance with the national policy means compliance with its procedural as well as numerical requirements (40 CFR 761.135); and also creates a presumption against enforcement action for penalties and the need for further cleanup under TSCA. EPA does reserve the right, however, to initiate appropriate action to complete cleanup if, upon review of the records of cleanup or sampling following cleanup, EPA finds that the decontamination levels in the policy have not been achieved. EPA also reserves the right to seek penalties if it is believed that the responsible party has not made a good-faith effort to comply with all provisions of the policy, such as prompt notification of a spill when required [40 CFR 761.120(b)].

Applicability. The National PCB Spill Cleanup Policy is applicable to any spill of material containing PCBs of 50 ppm or greater concentration.[1] Spills that are not subject to the national policy, and are instead addressed by EPA Regional Offices, include spills directly into surface water, drinking water, sewers, grazing lands, and vegetable (residential and commercial) gardens. EPA retains the authority to require additional, more stringent, less stringent, or alternative decontamination procedures

[1] The 50-ppm requirement is also applicable to materials assumed to be 50 ppm or greater (i.e., untested mineral oil dielectric) unless demonstrated otherwise.

for any spill that warrants such action (40 CFR 761.135). These additional procedures must be based on a finding by the EPA Regional Administrator that further cleanup is necessary to prevent unreasonable risk, or if there is a finding that the cleanup standards are unwarranted because of risk-mitigating factors, impracticable compliance, or cost-prohibitive factors.

PCB SPILL REPORTING REQUIREMENTS

As depicted in Figure 18-1, any spill of a material containing a PCB concentration of 50 ppm or greater into surface waters, drinking-water supplies, sewers, grazing land, or vegetable gardens must be reported immediately to the designated EPA Regional PCB Coordinator. The report must be made immediately after discovery, and in no case later than 24 hours after discovery [40 CFR 761.125(a)(1)(i)]. In addition, any spill involving 10 lbs. or more of a material that contains PCBs at 50 ppm released to a location not addressed in the preceding sentences must be reported to the designated EPA Regional PCB Coordinator.

Regardless of the PCB spill requirements under TSCA, a release of 1 lb. or more of "pure" PCBs into the environment, must be reported to the National Response Center. (A "pure" release of PCB-containing material at a concentration of 500 ppm would require a release of 200 lbs., or about 20 gallons.) This is required because PCBs are a Superfund-listed hazardous substance (see Chapter 12).

Any other spill of material containing a PCB concentration of 50 ppm or greater, although not necessarily reportable, must still be cleaned up in accordance with National PCB Spill Cleanup Policy [40 CFR 761.125(a)(1)(iv)]. No matter what the quantity of PCB-containing material spilled, the spill is covered under the liability provisions of §107 of Superfund regardless of the quantity or concentration of PCBs released (see Chapter 13).

Also, if a PCB transformer is involved in a fire-related incident, the owner or the operator of the transformer must immediately report the incident to the NRC (800-424-8802). A *fire-related incident* is defined as any incident involving a PCB transformer (or PCB voltage regulator) that involves the generation of sufficient heat and/or pressure (by any source) to result in the violent or nonviolent rupture of a PCB transformer or voltage regulator and the release of PCBs.

PCB SPILL CLEANUP REQUIREMENTS

As depicted in Figure 18-2, there are three types of spills addressed by EPA's National PCB Spill Cleanup Policy:

- Small low-concentration spills
- Large low-concentration spills
- High concentration spills

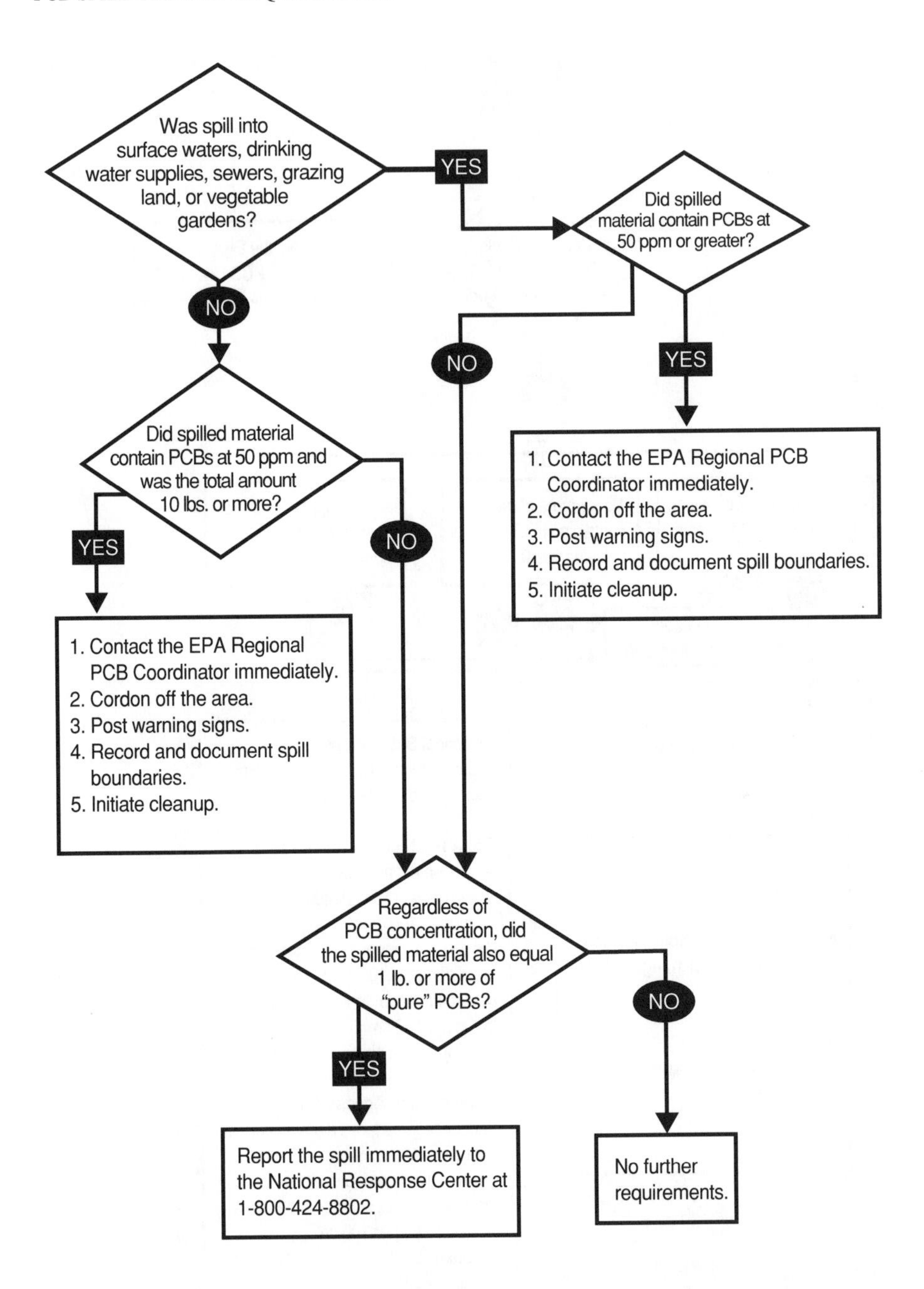

**Note:* Regardless of the quantity or amount of PCB material spilled, the spill is covered under the liability provisions of §107 of Superfund for the cleanup and natural resource damages resulting from the spill.

Figure 18-1. PCB Spill Reporting Requirements

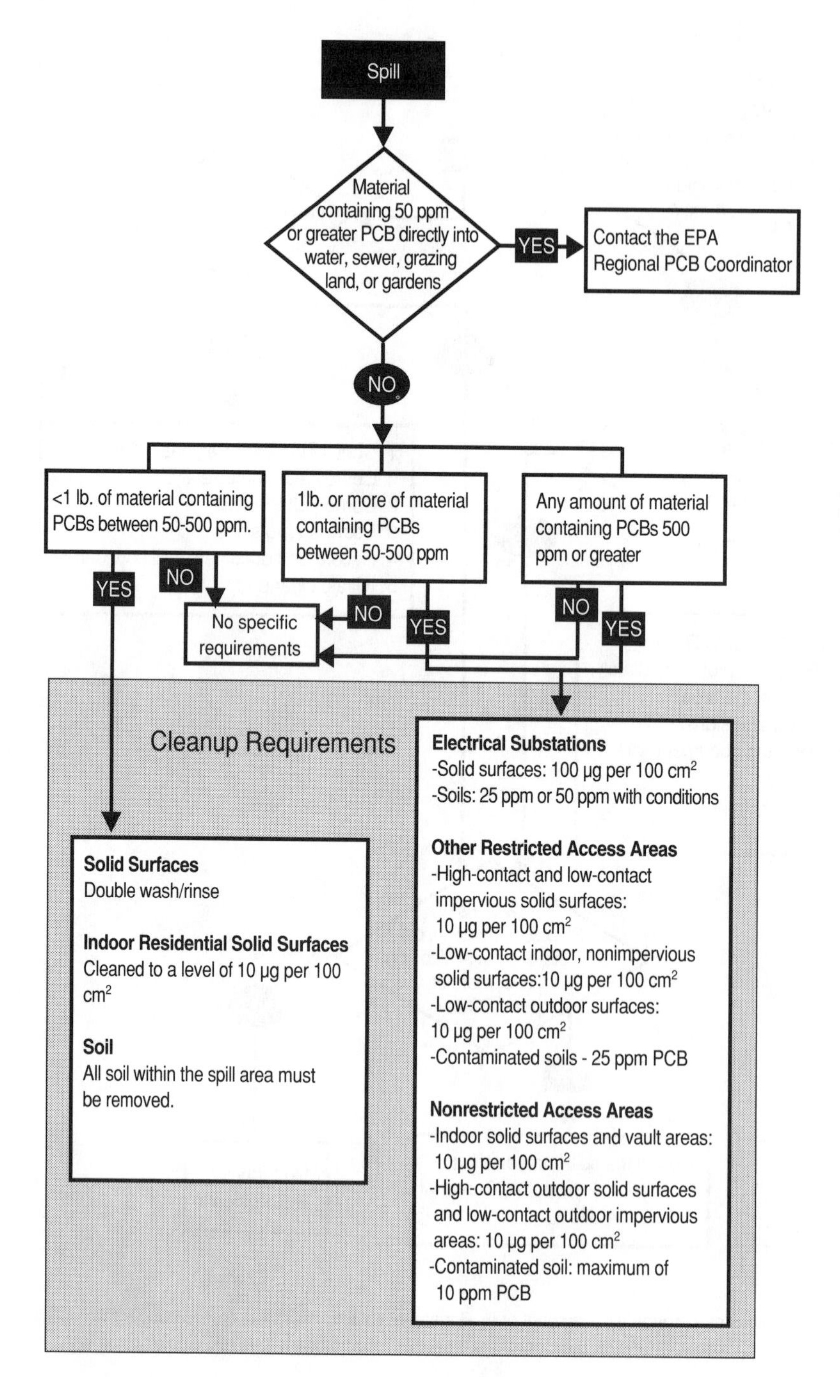

Figure 18-2. Summary of PCB Spill Cleanup Requirements

Note: Currently, large low-concentration and high-concentration spills are handled the same way.

The cleanup requirements also depend on the predetermined exposure potential based on the area receiving the spill (i.e., electrical substations, other restricted access areas, and nonrestricted access areas).

Under this policy, all contaminated soils, solvents, rags, and other materials resulting from the cleanup of PCBs must be stored, labeled, and disposed of in accordance with the applicable PCB disposal requirements (see Chapter 17) [40 CFR 761.125(a)(2)]. For spills where there are insufficient visible traces, yet evidence of a leak or spill, the boundaries of the spill are to be determined by testing the area for PCBs using a statistically based sampling scheme [40 CFR 761.125(a)(3)].

Small Low-Concentration Spills

Small low-concentration spills are spills of materials with PCB concentrations between 50 and 500 ppm that involve less than 1 lb. of PCBs by weight (less than 270 gallons of untested mineral oil). *By weight* means the actual amount of PCB molecules in a material and does not refer to other substances contained in the mixture. For spills involving untested mineral oil, it is presumed that the mineral oil is PCB-contaminated (between 50 and 500 ppm) unless demonstrated otherwise (40 CFR 761.2(a)(2)). Any spill in this category must be cleaned up in accordance with the appropriate requirements within 48 hours of its discovery [40 CFR 761.125(b)]. The specific cleanup requirements depend on the whether the spill occurred on solid surfaces or soil.

Solid Surfaces. Solid surfaces, except for indoor residential areas, must be double washed/rinsed. *Double wash/rinse* means to cleanse the solid surface (both impervious and nonimpervious) a minimum of two times with an appropriate solvent or other material in which PCBs are at least 5 percent soluble (by weight). A volume of PCB-free fluid sufficient to cover the contaminated surface must be completely used in each wash/rinse. The wash/rinse requirement does not mean the mere spreading of solvent or other fluid over the surface, nor does the requirement mean a once-over wipe with a soaked cloth. Precautions must be taken to contain any resultant runoff and to dispose properly of wastes generated during the cleansing [40 CFR 761.125(b)].

Indoor, residential solid surfaces must be cleaned up to 10 micrograms (μg) per 100 square centimeters (cm^2) by using a standard wipe test, which means, for spills on solid surfaces, a cleanup to numerical surface standards and sampling by a standard wipe test to verify that the numerical standards have been met. This definition constitutes the minimum requirements for an appropriate wipe testing protocol [40 CFR 761.125(b)(i)]. A standard-size template (10 cm × 10 cm) should be used to delineate the area of cleanup; the wiping medium should be a gauze pad or glass wool of known size that has been saturated with hexane. It is important that the wipe be performed very quickly after the hexane is exposed to air. The gauze (or glass wool) should be prepared with hexane in the laboratory, and the wiping medium should be stored in sealed glass vials until it is used for the wipe test.

Soils. All soil within a spill area (including a minimum buffer of 1 lateral foot beyond and 10 inches below the visible spill boundary) must be excavated. The ground must be restored to its original configuration by backfilling with clean, less than 1-ppm-PCB soil [40 CFR 761.125(b)(ii)].

High-Concentration and Large Low-Concentration Spills

The following criteria are applicable to any high-concentration spill (500 ppm or greater), and all low-concentration spills involving 1 lb. or more of PCBs by weight (including 270 lbs. or more of untested mineral oil) [40 CFR 761.125(c)].

The following five actions must be taken within 24 hours (or 48 hours for PCB transformers) of spill discovery [40 CFR 761.125(c)(1)]:

1. Notify, if applicable, the appropriate EPA regional office (Regional PCB Coordinator), and, if required by §103 of Superfund, the National Response Center (800-424-8802).
2. Cordon off the immediate spill area plus a three-foot buffer zone.
3. Clearly visible warning signs, requiring avoidance of the area, are to be placed adjacent to the cordoned area.
4. Record and document the area of visible contamination to establish the spill boundaries. If visible observations are not adequate to delineate the boundaries, appropriate statistical sampling must be used.
5. The cleanup of all visible traces of fluid on hard surfaces and all visible traces of fluid on soil, gravel, sand, oyster shells, and other similar media must be initiated.

It is important to note that there is no time limit for the completion of the cleanup of spills in this category. However, EPA will consider the promptness of the completion of cleanup in determining whether a response party made good-faith efforts to clean up in accordance with this policy (52 *FR* 10693, April 2, 1987).

Decontamination Requirements for Electrical Substations. Solid surfaces within electrical substations, which are considered restricted-access areas, must be cleaned up to a level of 100 μg PCB per 100 cm^2 as determined by a standard wipe test [40 CFR 761.125(c)(2)]. Contaminated soils must be cleaned up to a level of 25 ppm *or* 50 ppm PCB. The 50-ppm soil cleanup option requires the prominent display of a sign that, at a minimum, includes information concerning the date of the spill, the quantity of spilled material, the concentration of PCBs remaining at the spill site, and the recommendation to use protective clothing [40 CFR 761.125(c)(2)(ii)].

Decontamination Requirements for Other Restricted-Access Areas. Spills that occur in restricted-access areas other than electrical substations must be cleaned up as follows [40 CFR 761.125(c)(3)]:

- High-contact solid surfaces and low-contact, indoor, impervious solid surfaces must be cleaned up to 10 μg per 100 cm^2.
- Low-contact, indoor, nonimpervious surfaces can be cleaned up either to 10 μg per 100 cm^2, or to 100 μg per 100 cm^2 if the spill is also encapsulated.
- Low-contact outdoor surfaces must be cleaned up to 100 μg per 100 cm^2.
- Contaminated soil must be cleaned up to 25 ppm PCBs by weight.

Decontamination Requirements for Nonrestricted-Access Areas. Spills that occur in nonrestricted-access areas must be cleaned up in accordance with the following criteria [40 CFR 761.125(c)(4)]:

- Furnishings, toys, and other easily replaceable household items must be disposed of in accordance with 40 CFR 761.60 and replaced by the responsible party.
- Indoor solid surfaces, indoor vault areas, high-contact, outdoor, solid surfaces, and low-contact, outdoor, impervious solid areas must be cleaned up to 10 μg per 100 cm^2.
- Contaminated soil must be cleaned up to a minimum depth of 10 inches and to a minimum PCB concentration of 10 ppm. However, if less than 1 ppm of PCBs is found in the soil before the 10-inch minimum is reached, no further excavation is required. The resultant hole then must be filled in with clean soil containing less than 1 ppm of PCBs (53 *FR* 40883, October 19, 1988).

Post-Cleanup Sampling Requirements

Post-cleanup sampling is required to verify that the required cleanup levels under the National PCB Spill Cleanup Policy have been attained (40 CFR 761.130). The responsible party may use any statistically valid, reproducible sampling scheme (either random or grid samples), provided that the following two requirements are met:

1. The sampling area is equal to the greater of an area equal to the area cleaned plus an additional one-foot boundary or an area 20 percent larger than the original area of contamination.
2. The sampling scheme ensures a 95 percent confidence level against false-positives.

The number of samples must be sufficient to ensure that areas of contamination of a radius of two feet or more within the sampling area will be detected; the minimum number of samples is 3 and the maximum number of samples is 40. The sampling scheme must account for expected variability due to analytical error [40 CFR 761.130(c)].

EPA recommends the use of the EPA guidance manuals *Verification of PCB Spill Cleanup by Sampling and Analysis* and *Field Manual for Grid Sampling of PCB Spill Sites to Verify Cleanup* for the purposes of sampling for PCBs under this policy. (Contact EPA's TSCA Assistance hotline at 202-554-1404 for copies.) The major advantage of using the recommended sampling schemes is that they are designed to characterize the degree of contamination within the entire sampling area with a high degree of confidence while requiring fewer samples than other grid or random sampling schemes. These recommended schemes allow some sites to be characterized on the basis of using composite samples instead of grab samples [40 CFR 761.130(e)].

Record-Keeping Requirements

The responsible party must document the cleanup with written records [40 CFR 761.125(c)(5)]. These records must be maintained for at least five years, and must contain the following information:

- The source of the spill.
- The date and the time of the spill.
- The date and the time when cleanup was completed or terminated.
- A brief description of the spill location and the materials contaminated, including the location and type (e.g., outdoor electrical substation).
- Precleanup sampling data used to establish the spill boundaries, if required.
- A description of the decontamination procedures for solid surfaces and a listing of those solid surfaces.
- The approximate depth and amount of soil excavated.
- Post-cleanup verification sampling data, sampling methodology, and analytical techniques used.

As part of the record-keeping requirements, a certification statement signed by the responsible party stating that the cleanup requirements have been met, and that the information contained in the record is true to the best of his or her knowledge, is required [40 CFR 761.125(c)(5)].

APPENDIX A

LIST OF RCRA HAZARDOUS WASTES, CERCLA REPORTABLE QUANTITIES, AND EPCRA REPORTABLE QUANTITIES

EPA Hazardous Waste No.	Description		Hazard Code*	CERCLA RQ†	EPCRA RQ†
Characteristic Hazardous Wastes (D Codes)					
D001	Unlisted ignitable hazardous wastes		I	100	
D002	Unlisted corrosive hazardous wastes		C	100	
D003	Unlisted reactive hazardous wastes		R	100	
D004-43	Unlisted toxic hazardous wastes:		T		
	D004	Arsenic		1	
	D005	Barium		1,000	
	D006	Cadmium		10	
	D007	Chromium		10	
	D008	Lead		1	
	D009	Mercury		1	
	D010	Selenium		1	

*Hazard codes are as follows: I = ignitable, C = corrosive, R = reactive, T = toxic, and H = acutely hazardous.

†This RQ represents the lowest reportable amount for a mixture or if the individual components are unknown. If the specific chemical identity is known, that designated reportable quantity may be used.

Note: The reportable quantities listed in this category should be used as guides only. For example, there may be other components in the waste that also have a reportable quantity. See Chapter 12 for more information on the requirements for mixtures and reporting under both of these statutes.

EPA Hazardous Waste No.	Description	Hazard Code*	CERCLA RQ†	EPCRA RQ†
D011	Silver		1	
D012	Endrin		1	
D013	Lindane		1	
D014	Methoxychlor		1	
D015	Toxaphene		1	
D016	2,4-D		100	
D017	2,4,5-TP Silvex		100	
D018	Benzene		10	
D019	Carbon tetrachloride		10	
D020	Chlordane		1	
D021	Chlorobenzene		100	
D022	Chloroform		10	
D023	*o*-Cresol		100	
D024	*m*-Cresol		100	
D025	*p*-Cresol		100	
D026	Cresol		100	
D027	1,4-Dichlorobenzene		100	
D028	1,2-Dichloroethane		100	
D029	1,1-Dichloroethylene		100	
D030	2,4-Dinitrotoluene		10	
D031	Heptachlor (and epoxide)		1	
D032	Hexachlorobenzene		10	
D033	Hexachlorobutadiene		1	
D034	Hexachloroethane		100	
D035	Methyl ethyl ketone		5,000	
D036	Nitrobenzene		1,000	
D037	Pentachlorophenol		10	
D038	Pyridine		1,000	
D039	Tetrachloroethylene		100	
D040	Trichloroethylene		100	
D041	2,4,5-Trichlorophenol		10	
D042	2,4,6-Trichlorophenol		10	
D043	Vinyl Chloride		100	
Hazardous Wastes from Nonspecific Sources (F Codes)				
F001	The following spent halogenated solvents used in degreasing:	T	1‡	
	tetrachloroethylene,		1	
	trichloroethylene,		1000	
	methylene chloride,		1	

EPA Hazardous Waste No.	Description	Hazard Code*	CERCLA RQ†	EPCRA RQ†
	1,1,1-trichloroethane,		1	
	carbon tetrachloride, and		5000	
	chlorinated fluorocarbons			
	All spent solvent mixtures/blends used in degreasing containing, before use, a total of 10 percent **or** more (by volume) of one **or** more of the above halogenated solvents **or** those solvents listed in F002, F004, and F005; and still bottoms from the recovery of these spent solvents and spent solvent mixtures.			
F002	The following spent halogenated solvents:	T	1‡	
	tetrachloroethylene		1	
	methylene chloride		1	
	trichloroethylene		1000	
	1,1,1-trichloroethane		1	
	chlorobenzene		100	
	1,1,2-trichloro-1,2,2-trifluoroethane		100	
	o-dichlorobenzene		1	
	trichlorofluoromethane		1	
	1,1,2-trichloroethane		1	
	All spent solvent mixtures/blends containing, before use, a total of 10 percent **or** more of one **or** more of the above halogenated solvents **or** those listed in F001, F004, **or** F005; and still bottoms from the recovery of these spent solvents and spent solvent mixtures.			
F003	The following spent nonhalogenated solvents:	I	100†	
	xylene		1,000	
	acetone		5,000	
	ethyl acetate		5,000	
	ethyl benzene		1,000	
	ethyl ether		100	
	methyl isobutyl ketone		5,000	
	n-butyl alcohol		5,000	
	cyclohexanone		5,000	
	methanol		5,000	
	All spent solvent mixtures/blends containing, before use, only the above spent nonhalogenated solvents; and all spent solvent mixtures/blends containing, before use, one **or** more of the above nonhalogenated solvents, and, a total of 10 percent **or** more (by volume) of one **or** more of those solvents listed in F001, F002, F004, and F005; and still bottoms from the recovery of these spent solvents and spent solvent mixtures.			
F004	The following spent nonhalogenated solvents:	T	1,000†	

EPA Hazardous Waste No.	Description	Hazard Code*	CERCLA RQ†	EPCRA RQ†
	cresols		1,000	
	cresylic acid		1,000	
	nitrobenzene		1,000	
	All spent solvent mixtures/blends containing, before use, a total of 10 percent **or** more (by volume) of one **or** more of the above nonhalogenated solvents **or** those solvents listed in F001, F002, and F005; and still bottoms from the recovery of these spent solvents and spent solvent mixtures.			
F005	The following spent nonhalogenated solvents:	I, T	100‡	
	toluene		1,000	
	methyl ethyl ketone		5,000	
	carbon disulfide		100	
	isobutanol		5,000	
	pyridine		1,000	
	benzene		10	
	2-ethoxyethanol		10	
	2-nitropropane		10	
	All spent solvent mixtures/blends containing, before use, a total of 10 percent **or** more (by volume) of one or more of the above nonhalogenated solvents **or** those solvents listed in F001, F002, **or** F004; and still bottoms from the recovery of these spent solvents and spent solvent mixtures.			
F006	Wastewater treatment sludges from electroplating operations except from the following processes: (1) Sulfuric acid anodizing of aluminum; (2) tin plating on carbon steel; (3) zinc plating (segregated basis) on carbon steel; (4) aluminum or zinc-aluminum plating on carbon steel; (5) cleaning/stripping associated with tin, zinc, and aluminum plating on carbon steel; and (6) chemical etching and milling of aluminum.	T	10	
F007	Spent cyanide plating bath solutions from electroplating operations	R,T	10	
F008	Plating bath residues from the bottom of plating baths from electroplating operations where cyanides are used in the process	R,T	10	
F009	Spent stripping and cleaning bath solutions from electroplating operations where cyanides are used in the process	R,T	10	

EPA Hazardous Waste No.	Description	Hazard Code*	CERCLA RQ†	EPCRA RQ†
F010	Quenching bath residues from oil baths from metal heat treating operations where cyanides are used in the process	R,T	10	
F011	Spent cyanide solutions from salt bath pot clearing from metal heat treating operations	R, T	10	
F012	Quenching wastewater treatment sludges from metal heat treating operations where cyanides are used in the process	T	10	
F019	Wastewater treatment sludges from the chemical conversion coating of aluminum	T	10	
F020	Wastes (except wastewater and spent carbon from hydrogen chloride purification) from the production or manufacturing use (as a reactant, chemical intermediate, or component in a formulating process) of tri- or tetrachlorophenol, or of intermediates used to produce their pesticide derivatives. (This listing does not include wastes from the production of hexachlorophene from highly purified 2,4,5-trichlorophenol.)	H	1	
F021	Wastes (except wastewater and spent carbon from hydrogen chloride purification) from the production or manufacturing use (as a reactant, chemical intermediate, or component in a formulating process) of pentachlorophenol, or of intermediates used to produce its derivatives	H	1	
F022	Wastes (except wastewater and spent carbon from hydrogen chloride purification) from the manufacturing use (as a reactant, chemical intermediate, or component in a formulating process) of tetra-, penta-, or hexachlorobenzenes under alkaline conditions	H	1	
F023	Wastes (except wastewater and spent carbon from hydrogen chloride purification) from the production of materials on equipment previously used for the production or manufacturing use (as a reactant, chemical intermediate, or component in a formulating process) of tri- and	H	1	

EPA Hazardous Waste No.	Description	Hazard Code*	CERCLA RQ†	EPCRA RQ†
	tetrachlorophenols. (This listing does not include wastes from equipment used only for the production or use of hexachlorophene from highly purified 2,4,5-trichlorophenol.)			
F024	Wastes, including but not limited to distillation residues, heavy ends, tars, and reactor cleanout wastes from the production of chlorinated aliphatic hydrocarbons, having carbon content from one to five, utilizing free radical catalyzed processes. (This listing does not include light ends, spent filters and filter aids, spent desiccants, wastewater, wastewater treatment sludges, spent catalysts, and wastes listed in §261.32.)	T	1	
F025	Condensed light ends, spent filters and filter aids, and spent desiccant wastes from the production of certain chlorinated aliphatic hydrocarbons, by free-radical catalyzed processes. These chlorinated aliphatic hydrocarbons are those having carbon chain lengths ranging from one to and including five, with varying amounts and positions of chlorine substitution	T	1	
F026	Wastes (except wastewater and spent carbon from hydrogen chloride purification) from the production of materials on equipment previously used for the manufacturing use (as a reactant, chemical intermediate, or component in a formulating process) of tetra-, penta-, or hexachlorobenzene under alkaline conditions	H	1	
F027	Discarded unused formulations containing tri-, tetra-, or pentachlorophenol or discarded unused formulations containing compounds derived from these chlorophenols. (This listing does not include formulations containing hexachlorophene synthesized from prepurified 2,4,5-trichlorophenol as the sole component.)	H	1	

EPA Hazardous Waste No.	Description	Hazard Code*	CERCLA RQ†	EPCRA RQ†
F028	Residues resulting from the incineration or thermal treatment of soil contaminated with EPA Hazardous Waste Nos. F020, F021, F022, F023, F026, and F027	T	1	
F032	Wastewaters, process residuals, preservative drippage, and spent formulations from wood preserving processes generated at plants that currently use or have previously used chlorophenolic formulations (except potentially cross-contaminated wastes that have had the F032 waste code deleted in accordance with §261.35 of this chapter and where the generator does not resume or initiate use of chlorophenolic formulations). This listing does not include K001 bottom sediment sludge from the treatment of wastewater from wood preserving processes that use creosote and/or pentachlorophenol.	T	1	

Note: The listing of wastewaters that have not come into contact with process contaminants is stayed administratively. The listing for plants that have previously used chlorophenolic formulations is administratively stayed whenever these wastes are covered by the F034 or F035 listings. These stays will remain in effect until further administrative action is taken.

EPA Hazardous Waste No.	Description	Hazard Code*	CERCLA RQ†	EPCRA RQ†
F034	Wastewaters, process residuals, preservative drippage, and spent formulations from wood preserving processes generated at plants that use creosote formulations. This listing does not include K001 bottom sediment sludge from the treatment of wastewater from wood preserving processes that use creosote and/or pentachlorophenol.	T	1	

Note: The listing of wastewaters that have not come into contact with process contaminants is stayed administratively. The stay will remain in effect until further administrative action is taken.

EPA Hazardous Waste No.	Description	Hazard Code*	CERCLA RQ†	EPCRA RQ†
F035	Wastewaters, process residuals, preservative drippage, and spent formulations from wood preserving processes generated at plants that use inorganic preservatives containing arsenic or chromium. This listing does not include K001 bottom sediment sludge from the treatment of wastewater from wood preserving processes that use creosote and/or pentachlorophenol.	T	1	

Note: The listing of wastewaters that have not come into contact with process contaminants is stayed administratively. The stay will remain in effect until further administrative action is taken.

EPA Hazardous Waste No.	Description	Hazard Code*	CERCLA RQ†	EPCRA RQ†
F037	Petroleum refinery primary oil/water/solids separation sludge—any sludge generated from the gravitational separation of oil/water/solids during the storage or treatment of process wastewaters and oily cooling wastewaters from petroleum refineries. Such sludges include but are not limited to those generated in: oil/water/solids separators; tanks and impoundments; ditches and other conveyances; sumps; and stormwater units receiving dry weather flow. Sludge generated in stormwater units that do not receive dry weather flow, sludges generated from noncontact once through cooling waters segregated for treatment from other process or oily cooling waters, sludges generated in aggressive biological treatment units as defined in §261.31(b)(2) (including sludges generated in one or more additional units after wastewaters have been treated in aggressive biological treatment units), and K051 wastes are not included in this listing.	T	1	
F038	Petroleum refinery secondary (emulsified) oil/water/solids separation sludge—any sludge and/or float generated from the physical and/or chemical separation of oil/water/solids in process wastewaters and oily cooling wastewaters from petroleum refineries. Such wastes include but are not limited to all sludges and floats generated in induced air flotation (IAF) units, tanks, and impoundments, and all sludges generated in DAF units. Sludges generated in stormwater units that do not receive dry weather flow, sludges generated from noncontact once-through cooling waters segregated for treatment from other process or oily cooling waters, sludges and floats generated in aggressive biological treatment units as defined in §261.31(b)(2) (including sludges and floats generated in one or more additional units after wastewaters have been treated in aggressive biological treatment units), and F037, K048, and K051 wastes are not included in this listing.	T	1	

EPA Hazardous Waste No.	Description	Hazard Code*	CERCLA RQ†	EPCRA RQ†
F039	Leachate (liquids that have percolated through land disposal wastes) resulting from the disposal of more than one restricted waste classified as hazardous under Subpart D of this part. (Leachate resulting from the disposal of one **or** more of the following EPA hazardous wastes and no other hazardous wastes retains its EPA hazardous waste number(s): F020, F021, F022, F026, F027, and/or F028.)	T	1	
Hazardous Wastes from Specific Sources (K Codes)				
K001	Bottom sediment sludge from the treatment of wastewaters from wood preserving processes that use creosote and/or pentachlorophenol	T	1	
K002	Wastewater treatment sludge from the production of chrome yellow and orange pigments	T	10	
K003	Wastewater treatment sludge from the production of molybdate orange pigments	T	10	
K004	Wastewater treatment sludge from the production of zinc yellow pigments	T	10	
K005	Wastewater treatment sludge from the production of chrome green pigments	T	10	
K006	Wastewater treatment sludge from the production of chrome oxide green pigments (anhydrous and hydrated)	T	10	
K007	Wastewater treatment sludge from the production of iron blue pigments	T	10	
K008	Oven residue from the production of chrome oxide green pigments	T	10	
K009	Distillation bottoms from the production of acetaldehyde from ethylene	T	10	
K010	Distillation side cuts from the production of acetaldehyde from ethylene	T	10	
K011	Bottom stream from the wastewater stripper in the production of acrylonitrile	R,T	10	
K013	Bottom stream from the acetonitrile column in the production of acrylonitrile	R,T	10	
K014	Bottoms from the acetonitrile purification column in the production of acrylonitrile	T	5,000	

EPA Hazardous Waste No.	Description	Hazard Code*	CERCLA RQ†	EPCRA RQ†
K015	Still bottoms from the distillation of benzyl chloride	T	10	
K016	Heavy ends or distillation residues from the production of carbon tetrachloride	T	1	
K017	Heavy ends (still bottoms) from the purification column in the production of epichlorohydrin	T	10	
K018	Heavy ends from the fractionation column in ethyl chloride production	T	1	
K019	Heavy ends from the distillation of ethylene dichloride in ethylene dichloride production	T	1	
K020	Heavy ends from the distillation of vinyl chloride in vinyl chloride monomer production	T	1	
K021	Aqueous spent antimony catalyst waste from fluoromethanes production	T	10	
K022	Distillation bottom tars from the production of phenol/acetone from cumene	T	1	
K023	Distillation light ends from the production of phthalic anhydride from naphthalene	T	5,000	
K024	Distillation bottoms from the production of phthalic anhydride from naphthalene	T	5,000	
K025	Distillation bottoms from the production of nitrobenzene by the nitration of benzene	T	10	
K026	Stripping still tails from the production of methyl ethyl pyridines	T	1,000	
K027	Centrifuge and distillation residues from toluene diisocyanate production	R,T	10	
K028	Spent catalyst from the hydrochlorinator reactor in the production of 1,1,1-trichloroethane	T	1	
K029	Waste from the product steam stripper in the production of 1,1,1-trichloroethane	T	1	
K030	Column bottoms or heavy ends from the combined production of trichloroethylene and perchloroethylene	T	1	
K031	By-product salts generated in the production of MSMA and cacodylic acid	T	1	

EPA Hazardous Waste No.	Description	Hazard Code*	CERCLA RQ†	EPCRA RQ†
K032	Wastewater treatment sludge from the production of chlordane	T	10	
K033	Wastewater and scrub water from the chlorination of cyclopentadiene in the production of chlordane	T	10	
K034	Filter solids from the filtration of hexachlorocyclopentadiene in the production of chlordane	T	10	
K035	Wastewater treatment sludges generated in the production of creosote	T	1	
K036	Still bottoms from toluene reclamation distillation in the production of disulfoton	T	1	
K037	Wastewater treatment sludges from the production of disulfoton	T	1	
K038	Wastewater from the washing and stripping of phorate production	T	10	
K039	Filter cake from the filtration of diethylphosphorodithioic acid in the production of phorate	T	10	
K040	Wastewater treatment sludge from the production of phorate	T	10	
K041	Wastewater treatment sludge from the production of toxaphene	T	1	
K042	Heavy ends **or** distillation residues from the distillation of tetrachlorobenzene in the production of 2,4,5-T	T	10	
K043	2,6-Dichlorophenol waste from the production of 2,4-D	T	10	
K044	Wastewater treatment sludges from the manufacturing and processing of explosives	R	10	
K045	Spent carbon from the treatment of wastewater containing explosives	R	10	
K046	Wastewater treatment sludges from the manufacturing, formulation and loading of lead-based initiating compounds	T	10	
K047	Pink/red water from TNT operations	R	10	
K048	Dissolved air flotation (DAF) float from the petroleum refining industry	T	10	

EPA Hazardous Waste No.	Description	Hazard Code*	CERCLA RQ†	EPCRA RQ†
K049	Slop oil emulsion solids from the petroleum refining industry	T	10	
K050	Heat exchanger bundle cleaning sludge from the petroleum refining industry	T	10	
K051	API separator sludge from the petroleum refining industry	T	10	
K052	Tank bottoms (leaded) from the petroleum refining industry	T	10	
K060	Ammonia still lime sludge from coking operations	T	1	
K061	Emission control dust/sludge from the primary production of steel in electric furnaces	T	10	
K062	Spent pickle liquor generated by steel finishing operations of facilities within the iron and steel industry (SIC Codes 331 and 332)	C,T	10	
K064	Acid plant blowdown slurry/sludge resulting from the thickening of blowdown slurry from primary copper production	T	10	
K065	Surface impoundment solids contained in and dredged from surface impoundments at primary lead smelting facilities	T	10	
K066	Sludge from treatment of process wastewater and/or acid plant blowdown from primary zinc production	T	10	
K069	Emission control dust/sludge from secondary lead smelting	T	10	
K071	Brine purification muds from the mercury cell process in chlorine production, where separately prepurified brine is not used	T	1	
K073	Chlorinated hydrocarbon waste from the purification step of the diaphragm cell process using graphite anodes in chlorine production	T	10	
K083	Distillation bottoms from aniline production	T	100	
K084	Wastewater treatment sludges generated during the production of veterinary pharmaceuticals from arsenic or organo-arsenic compounds	T	1	

EPA Hazardous Waste No.	Description	Hazard Code*	CERCLA RQ†	EPCRA RQ†
K085	Distillation or fractionation column bottoms from the production of chlorobenzenes	T	10	
K086	Solvent washes and sludges, caustic washes and sludges, or water washes and sludges from cleaning tubs and equipment used in the formulation of ink from pigments, driers, soaps, and stabilizers containing chromium and lead	T	10	
K087	Decanter tank tar sludge from coking operations	T	100	
K088	Spent potliners from primary aluminum	T	10	
K090	Emission control dust or sludge from ferrochromiumsilicon production	T	10	
K091	Emission control dust or sludge from ferrochromium production	T	10	
K093	Distillation light ends from the production of phthalic anhydride from *o*-xylene	T	5,000	
K094	Distillation bottoms from the production of phthalic anhydride from *o*-xylene	T	5,000	
K095	Distillation bottoms from the production of 1,1,1-trichloroethane	T	100	
K096	Heavy ends from the heavy ends column from the production of 1,1,1-trichloroethane	T	100	
K097	Vacuum stripper discharge from the chlordane chlorinator in the production of chlordane	T	1	
K098	Untreated process wastewater from the production of toxaphene	T	1	
K099	Untreated wastewater from the production of 2,4-D	T	10	
K100	Waste leaching solution from acid leaching of emission control dust/sludge from secondary lead smelting	T	10	
K101	Distillation tar residues from the distillation of aniline-based compounds in the production of veterinary pharmaceuticals from arsenic or organo-arsenic compounds	T	1	
K102	Residue from the use of activated carbon for decolorization in the production of veterinary pharmaceuticals from arsenic or organo-arsenic compounds	T	1	

EPA Hazardous Waste No.	Description	Hazard Code*	CERCLA RQ†	EPCRA RQ†
K103	Process residues from aniline extraction from the production of aniline	T	100	
K104	Combined wastewater streams generated from nitrobenzene/aniline production	T	10	
K105	Separated aqueous stream from the reactor product washing step in the production of chlorobenzenes	T	10	
K106	Wastewater treatment sludge from the mercury cell process in chlorine production	T	1	
K107	Column bottoms from product separation from the production of 1,1-dimethylhydrazine (UDMH) from carboxylic acid hydrazines	C,T	10	
K108	Condensed column overheads from product separation and condensed reactor vent gases from the production of 1,1-dimethylhydrazine (UDMH) from carboxylic acid hydrazides	I,T	10	
K109	Spent filter cartridges from product purification from the production of 1,1-dimethylhydrazine (UDMH) from carboxylic acid hydrazides	T	10	
K110	Condensed column overheads from intermediate separation from the production of 1,1-dimethylhydrazine (UDMH) from carboxylic acid hydrazines	T	10	
K111	Product washwaters from the production of dinitrotoluene via nitration of toluene	C, T	10	
K112	Reaction by-product water from the drying column in the production of toluenediamine via hydrogenation of dinitrotoluene	T	10	
K113	Condensed liquid light ends from the purification of toluenediamine in the production of toluenediamine via hydrogenation of dinitrotoluene	T	10	
K114	Vicinals from the purification of toluenediamine in the production of toluenediamine via hydrogenation of dinitrotoluene	T	10	

EPA Hazardous Waste No.	Description	Hazard Code*	CERCLA RQ†	EPCRA RQ†
K115	Heavy ends from the purification of toluenediamine in the production of toluenediamine via hydrogenation of dinitrotoluene	T	10	
K116	Organic condensate from the solvent recovery column in the production of toluene diisocyanate via phosgenation of toluenediamine	T	10	
K117	Wastewater from the reactor vent gas scrubber in the production of ethylene dibromide via bromination of ethene	T	1	
K118	Spent adsorbent solids from purification of ethylene dibromide in the production of ethylene dibromide via bromination of ethene	T	1	
K123	Process wastewater (including supernates, filtrates, and washwaters) from the production of ethylenebisdithiocarbamic acid and its salt	T	10	
K124	Reactor vent scrubber water from the production of ethylenebisdithiocarbamic acid and its salts	C, T	10	
K125	Filtration, evaporation, and centrifugation solids from the production ethylenebisdithiocarbamic acid and its salts	T	10	
K126	Bag house dust and floor sweepings in milling and packaging operations from the production or formulation of ethylenebisdithiocarbamic acid and its salts	T	10	
K131	Wastewater from the reactor and spent sulfuric acid from the acid dryer from the production of methyl bromide	C, T	100	
K132	Spent absorbent and wastewater treatment separator solids from the production of methyl bromide	T	1,000	
K136	Still bottoms from the purification of ethylene dibromide in the production of ethylene dibromide via bromination of ethene	T	1	

EPA Hazardous Waste No.	Description	Hazard Code*	CERCLA RQ†	EPCRA RQ†
K141	Process residues from the recovery of coal tar, including but not limited to collecting sump residues from the production of coke from coal or the recovery of coke by-products produced from coal. This listing does not include K087 (decanter tank tar from coking operations).	T	1	
K142	Tar storage tank residues from the production of coke from coal or from the recovery of coke by-products produced from coal.	T	1	
K143	Process residues from the recovery of light oil, including but not limited to those generated in stills, decanters, and wash oil recovery units from the recovery of coke by-products produced from coal.	T	1	
K144	Wastewater sump residues from light oil refining, including but not limited to intercepting or contamination sump sludges from the recovery of coke by-products produced from coal.	T	1	
K145	Residues from naphthalene collection and recovery operations from the recovery of coke by-products produced from coal	T	1	
K147	Tar storage tank residues from coal tar refining	T	1	
K148	Residues from coal tar distillation, including but not limited to still bottoms	T	1	
K149	Distillation bottoms from the production of α- (or methyl-) chlorinated toluenes, ring-chlorinated toluenes, benzoyl chlorides, and compounds with mixtures of these functional groups. (This waste does not include still bottoms from the distillation of benzyl chloride.)	T	10	
K150	Organic residuals, excluding spent carbon adsorbent, from the spent chlorine gas and hydrochloric acid recovery processes associated with the production of α- (or methyl-) chlorinated toluenes, ring-chlorinated toluenes, benzoyl chlorides, and compounds with mixtures of these functional groups	T	10	

EPA Hazardous Waste No.	Description	Hazard Code*	CERCLA RQ†	EPCRA RQ†
K151	Wastewater treatment sludges, excluding neutralization and biological sludges, generated during the treatment of wastewaters from the production of α- (or methyl-) chlorinated toluenes, ring-chlorinated toluenes, benzoyl chlorides, and compounds with mixtures of these functional groups	T	10	
K156	Organic waste (including heavy ends, still bottoms, light ends, spent solvents, filtrates, and decantates) from the production of carbamates and carbamoyl oximes	T	1	
K157	Wastewaters (including scrubber waters, condenser waters, washwaters, and separation waters) from the production of carbamates and carbamoyl oximes	T	1	
K158	Bag house dusts and filter/separation solids from the production of carbamates and carbamoyl oximes	T	1	
K159	Organics from the treatment of thiocarbamate wastes	T	1	
K160	Solids (including filter wastes, separation solids, and spent catalysts) from the production of thiocarbamates and solids from the treatment of thiocarbamate wastes	T	1	
K161	Purification solids (including filtration, evaporation, and centrifugation solids), bag house dust, and floor sweepings from the production of dithiocarbamate acids and their salts. (This listing does not include K125 or K126.)	R,T	1	

Discarded Acutely Toxic Commercial Chemical Products, Off-Specification Species, Container Residues and Spills (P Codes)

EPA Hazardous Waste No.	Description	Hazard Code*	CERCLA RQ†	EPCRA RQ†
P001	2H-1-Benzopyran-2-one, 4-hydroxy-3-(3-oxo-l-phenylbutyl)-, & salts, when present at concentrations greater than 0.3% **or** Warfarin, & salts, when present at concentrations greater than 0.3%	H	100	100
P002	1 -Acetyl-2-thiourea **or** Acetamide, N-(aminothioxomethyl)-	H	1,000	

EPA Hazardous Waste No.	Description	Hazard Code*	CERCLA RQ†	EPCRA RQ†
P003	Acrolein **or** 2-Propenal	H	1	1
P004	Aldrin **or** 1,4,5,8-Dimethanonaphthalene, 1,2,3,4,10,10-hexa-chloro-1,4,4a,5,8,8a,-hexahydro (1alpha,4alpha,4abeta,5alpha, 8alpha,8abeta)-	H	1	1
P005	Allyl alcohol **or** 2-Propen-1-ol	H	100	100
P006	Aluminum phosphide	R,T,H	100	100
P007	5-(Aminomethyl)-3-isoxazolol **or** 3(2H)-Isoxazolone, 5-(aminomethyl)-	H	1,000	1,000
P008	4-Aminopyridine **or** 4-Pyridinamine	H	1,000	1,000
P009	Ammonium picrate **or** Phenol, 2,4,6-trinitro-, ammonium salt	R,H	10	
P010	Arsenic acid H_3AsO_4	T,H	1	
P011	Arsenic oxide As_2O_5 (arsenic pentoxide)	H	1	1
P012	Arsenic oxide As_2O_3 (arsenic trioxide)	H	1	1
P013	Barium cyanide	H	10	
P014	Benzenethiol **or** Thiophenol	H	100	100
P015	Beryllium	H	10	
P016	Dichloromethyl ether **or** Methane, oxybis[chloro-	H	10	10
P017	Bromoacetone **or** 2-Propanone, 1-bromo-	H	1,000	
P018	Brucine (strychnidin-10-one, 2,3-dimethoxy-)	H	100	
P020	Dinoseb **or** Phenol, 2-(I-methylpropyl)-4,6-dinitro-	H	1,000	1,000
P021	Calcium cyanide (Calcium cyanide $Ca(CN)_2$)	H	10	
P022	Carbon disulfide	H	100	100
P023	Chloroacetaldehyde (Acetaldehyde, chloro)	H	1,000	
P024	*p*-Chloroaniline **or** Benzenamine, 4-chloro-	H	1,000	
P026	1-(*o*-Chlorophenyl)thiourea **or** Thiourea, (2-chlorophenyl)-	H	100	100
P027	3-Chloropropionitrile **or** Propanenitrile, 3-chloro-	H	1,000	1,000
P028	Benzyl chloride **or** Benzene, (chloromethyl)-	H	100	100
P029	Copper cyanide	H	10	
P030	Cyanides (soluble cyanide salts), not otherwise specified	H	10	

EPA Hazardous Waste No.	Description	Hazard Code*	CERCLA RQ†	EPCRA RQ†
P031	Cyanogen **or** Ethanedinitrile	H	100	
P033	Cyanogen chloride	H	10	
P034	2-Cyclohexyl-4,6-dinitrophenol **or** Phenol, 2-cyclohexyl-4,6-dinitro-	H	100	
P036	Arsonous dichloride, phenyl- **or** Dichlorophenylarsine	H	1	1
P037	Dieldrin [2,7:3,6-Dimethanonaphth[2,3-bloxirene,3,4,5,6,9,9-hexachloro-1a,2,2a,3,6,6a,7,7a-octahydro-, (1aalpha,2beta,2aalpha,3beta,6beta,6aalpha, 7beta, 7aalpha)- & metabolites]	H	1	
P038	Diethylarsine **or** Arsine, diethyl-	H	1	
P039	Disulfoton **or** Phosphorodithioic acid, O,O-diethyl S-[2-(ethylthio)ethyl] ester	H	1	1
P040	O,O-Diethyl *o*-pyrazinyl phosphorothioate **or** Phosphorothioic acid, O,O-diethyl O-pyrazinyl ester	H	100	100
P041	Diethyl-*p*-nitrophenyl phosphate **or** Phosphoric acid, diethyl 4-nitrophenyl ester	H	100	
P042	Epinephrine **or** 1,2-Benzenediol, 4-[1-hydroxy-2-(methylamino)ethyl]-, (R)-	H	1,000	
P043	Diisopropyl fluorophosphate (DFP) **or** Phosphorofluoridic acid, bis(1-methylethyl) ester	H	100	100
P044	Dimethoate **or** Phosphorodithioic acid, O,O-dimethyl S-[2-(methylamino) -2-oxoethyl] ester	H	10	10
P045	Thiofanox **or** 2-Butanone, 3,3-dimethyl-1-(methylthio)-, *o*-methylamino) carbonyl]oxime	H	100	100
P046	α,α-Dimethylphenethylamine **or** Benzeneethanamine, α,α-dimethyl-	H	5,000	
P047	4,6-Dinitro-*o*-cresol & salts **or** Phenol, 2-methyl-4,6-dinitro-, & salts	H	10	10
P048	2,4-Dinitrophenol **or** Phenol, 2,4-dinitro-	H	10	
P049	Dithiobiuret **or** Thioimidodicarbonic diamide $[(H_2N)C(S)]_2NH$	H	100	100

EPA Hazardous Waste No.	Description	Hazard Code*	CERCLA RQ†	EPCRA RQ†
P050	Endosulfan **or** 6,9-Methano-2,4,3-benzodioxathiepin, 6,7,8,9,10,10-hexachloro-1,5,5a,6,9,9a-hexahydro-, 3-oxide	H	1	1
P051	Endrin & metabolites **or** [2,7:3,6-Dimethanonaphth [2,3-b]oxirene,3,4,5,6,9,9-hexachloro-1a,2,2a,3,6,6a,7,7a-octahydro(1aalpha,2beta,2abeta,3alpha, 6alpha,6abeta,7beta,7aalpha)- & metabolites	H	1	1
P054	Ethyleneimine **or** Aziridine	H	1	1
P056	Fluorine	H	10	10
P057	Fluoroacetamide **or** Acetamide, 2-fluoro-	H	100	100
P058	Fluoroacetic acid, sodium salt, **or** Acetic acid, fluoro-, sodium salt	H	10	10
P059	Heptachlor **or** 4,7-Methano-1H-indene, 1,4,5,6,7,8,8-heptachloro-3a,4,7,7a-tetrahydro	H	1	
P060	Isodrin **or** 1,4,5,8-Dimethanonaphthalene, 1,2,3,4,10,10-hexa-chloro-1,4,4a,5,8,8a-hexahydro-1alpha,4alpha,4abeta, 5beta,8beta,8abeta)-	H	1	1
P062	Hexaethyl tetraphosphate **or** Tetraphosphoric acid, hexaethyl ester	H	100	
P063	Hydrocyanic acid **or** Hydrogen cyanide	H	10	10
P064	Methane, isocyanato- **or** Methyl isocyanate	H	10	10
P065	Fulminic acid, mercury(2+) salt **or** Mercury fulminate	R,T,H	10	
P066	Methomyl **or** Ethanimidothioic acid, N-[[(methylamino)carbonyl]oxy]-, methyl ester	H	100	100
P067	Aziridine, 2-methyl- **or** 1,2-Propylenimine	H	1	1
P068	Methyl hydrazine **or** Hydrazine, methyl-	H	10	10
P069	2-Methyllactonitrile **or** Propanenitrile, 2-hydroxy-2-methyl-	H	10	10
P070	Aldicarb **or** Propanal, 2-methyl-2-(methylthio)-, *o*-[(methylamino) carbonyl]oxime	H	1	1

EPA Hazardous Waste No.	Description	Hazard Code*	CERCLA RQ†	EPCRA RQ†
P071	Methyl parathion **or** Phosphorothioic acid, 0,0,-dimethyl O-(4-nitrophenyl) ester	H	100	100
P072	alpha-Naphthylthiourea **or** Thiourea, 1-naphthalenyl-	H	100	100
P073	Nickel carbonyl **or** Nickel carbonyl $Ni(CO)_4$, (T-4)-	H	10	10
P074	Nickel cyanide $Ni(CN)_2$	H	10	
P075	Nicotine & salts **or** Pyridine, 3-(I-methyl-2-pyrrolidinyl)-, (S)-, & salts	H	100	100
P076	Nitric oxide	H	10	10
P077	*p*-Nitroaniline **or** Benzenamine, 4-nitro-	H	5,000	
P078	Nitrogen dioxide **or** Nitrogen oxide NO_2	H	10	10
P081	Nitroglycerine **or** 1,2,3-Propanetriol, trinitrate	R,H	10	
P082	N-Nitrosodimethylamine **or** Methanamine, N-methyl-N-nitroso-	H	10	10
P084	N-Nitrosomethylvinylamine **or** Vinylamine, N-methyl-N-nitroso-	H	10	
P085	Octamethylpyrophosphoramide **or** Diphosphoramide, octamethyl-	H	100	100
P087	Osmium oxide OsO_4,(T-4)- **or** Osmium tetroxide	H	1,000	
P088	7-Oxabicyclo[2.2.1]heptane-2,3-dicarboxylic acid **or** Endothall	H	1,000	
P089	Parathion **or** Phosphorothloic acid, O,O-diethyl o-(4-nitrophenyl) ester	H	10	10
P092	Mercury, (acetato-O)phenyl- **or** Phenylmercury acetate	H	100	100
P093	Phenylthiourea **or** Thiourea, phenyl-	H	100	100
P094	Phorate **or** Phosphorodithioic acid, O,O-diethyl S-[(ethylthio)methyl] ester	H	10	10
P095	Phosgene **or** Carbonic dichloride	H	10	10
P096	Phosphine **or** Hydrogen phosphide	H	100	100
P097	Famphur **or** Phosphorothioic acid, O-14-[(dimethylamino)sulfonyl]phenyl] O,O-dimethyl ester	H	1,000	
P098	Potassium cyanide	H	10	10

EPA Hazardous Waste No.	Description	Hazard Code*	CERCLA RQ†	EPCRA RQ†
P099	Potassium silver cyanide **or** Argentate(1-), bis(cyano-C)-, potassium	H	1	1
P101	Propanenitrile **or** Ethyl cyanide	H	10	10
P102	2-Propyn-1-ol **or** Propargyl alcohol	H	1,000	
P103	Selenourea	H	1,000	
P104	Silver cyanide	H	1	
P105	Sodium azide	H	1,000	1,000
P106	Sodium cyanide	H	10	10
P108	Strychnine & salts **or** Strychnidin-10-one & salts	H	10	10
P109	Tetraethyldlthiopyrophosphate	H	100	100
P110	Tetraethyl lead **or** Thiodiphosphoric acid, tetraethyl ester **or** Plumbane, tetraethyl-	H	10	10
P111	Tetraethyl pyrophosphate **or** Diphosphoric acid, tetraethyl ester	H	10	
P112	Tetranitromethane **or** Methane, tetranitro	R,H	10	10
P113	Thallic oxide	H	100	
P114	Thallium (l) selenite **or** Selenious acid, dithallium (l+) salt	H	1,000	
P115	Thallium (l) sulfate **or** Sulfuric acid, dithallium(1+) salt	H	100	100
P116	Thiosemicarbazide **or** Hydrazlnecarbothioamide	H	100	100
P118	Methanethiol, trichloro **or** Trichloromethanethiol	H	100	
P119	Vanadic acid, ammonium salt, **or** Ammonium vanadate	H	1,000	
P120	Vanadium oxide V_2O_1 **or** Vanadium pentoxide	H	1,000	1,000
P121	Zinc cyanide	H	100	
P122	Zinc phosphide Zn_3P_2, when present at concentrations greater than 10%	R,T,H	100	100
P123	Toxaphene	H	1	1
P127	Carbofuran **or** 7-Benzofuranol, 2,3-dihydro-2,2- dimethyl-, methylcarbamate	H	10	10
P128	Mexacarbate **or** Phenol, 4-(dimethylamino)-3,5-dimethyl-, methylcarbamate (ester)	H	1,000	1,000

EPA Hazardous Waste No.	Description	Hazard Code*	CERCLA RQ†	EPCRA RQ†
P185	Tirpate **or** 1,3-Dithiolane-2-carboxaldehyde, 2,4- dimethyl-, O- [(methylamino)-carbonyl]oxime	H	1	1
P188	Physostigmine salicylate **or** Benzoic acid, 2-hydroxy-, compd. with (3aS-cis)-1,2,3,3a,8,8a-hexahydro-1,3a,8-trimethylpyrrolo[2,3-b]indol-5- yl methylcarbamate ester 1:1	H	1	1
P189	Carbosulfan **or** Carbamic acid, [(dibutylamino)- thio]methyl-, 2,3-dihydro-2,2-dimethyl- 7-benzofuranyl ester	H	1	
P190	Metolcarb **or** Carbamic acid, methyl-, 3-methylphenyl ester	H	1	1
P191	Dimetilan **or** Carbamic acid, dimethyl-, 1-[(dimethyl- amino)carbonyl]- 5-methyl-1H- pyrazol-3-yl ester	H	1	1
P192	Isolan **or** Carbamic acid, dimethyl-, 3-methyl-1- (1-methylethyl)-1H- pyrazol-5-yl ester	H	1	1
P194	Oxamyl **or** Ethanimidothioc acid, 2-(dimethylamino)-N-[[(methylamino) carbonyl]oxy]-2-oxo-, methyl ester	H	1	1
P196	Manganese dimethyldithiocarbamate **or** Manganese, bis(dimethylcarbamodithioato-S,S′)-	H	1	
P197	Formparanate **or** Methanimidamide, N,N-dimethyl-N′-[2- methyl-4-[methylamino)carbonyl] oxy]phenyl]-	H	1	1
P198	Formetanate hydrochloride **or** Methanimidamide, N,N-dimethyl-N′-[3-[[(methylamino)-carbonyl]oxy]phenyl]-, monohydrochloride	H	1	1
P199	Methiocarb **or** Phenol, (3,5-dimethyl-4-(methylthio)-, methylcarbamate	H	10	10
P201	Promecarb **or** Phenol, 3-methyl-5-(1-methylethyl)-, methyl carbamate	H	1	1
P202	m-Cumenyl methylcarbamate **or** 3-Isopropylphenyl N-methylcarbamate **or** Phenol, 3-(1-methylethyl)-, methyl carbamate	H	1	1

EPA Hazardous Waste No.	Description	Hazard Code*	CERCLA RQ†	EPCRA RQ†
P203	Aldicarb sulfone **or** Propanal, 2-methyl-2-(methyl-sulfonyl)- , O-[(methylamino) carbonyl]oxime	H	1	
P204	Physostigmine **or** Pyrrolo[2,3-b]indol-5-ol1,2,3,3a,8,8a-hexahydro-1,3a,8-, trimethyl methylcarbamate (ester), (3aS-cis)-	H	1	1
P205	Ziram **or** Zinc, bis(dimethylcarbamodithioato-S,S′)-	H	1	

Discarded Acutely Toxic Commercial Chemical Products, Off-Specification Species, Container Residues and Spills (U Codes)

EPA Hazardous Waste No.	Description	Hazard Code*	CERCLA RQ†	EPCRA RQ†
U001	Acetaldehyde **or** Ethanal	I	1,000	
U002	Acetone **or** 2-Propanone	I	5,000	
U003	Acetonitrile	I,T	5,000	
U004	Acetophenone or Ethanone, 1-phenyl-	T	5,000	
U005	2-Acetylaminofluorene **or** Acetamide, N-9H-fluoren-2-yl-	T	1	
U006	Acetyl chloride	C,R,T	5,000	
U007	Acrylamide **or** 2-Propenamide	T	5,000	5,000
U008	Acrylic acid **or** 2-Propenoic acid (1)	I	5,000	
U009	Acrylonitrile **or** 2-Propenenitrile	T	100	100
U010	Mitomycin C **or** Azirino[2′,3′:3,4]pyrrolo[1,2-a]indole-4,7-dione, 6-amino-8-[[(aminocarbonyl)oxy]methyl], 1a,2,8,8a,8bhexal,ydro-8a-methoxy-5-methyl-, [1aS-(1aalpha, 8beta, 8aalpha,8balpha)]-	T	10	10
U011	Amitrole **or** 1H-1,2,4-Triazol-3-amine	T	10	
U012	Aniline **or** Benzenamine	I,T	5,000	5,000
U014	Auramine **or** Benzenamine, 4,4′-carbonimidoylbis[N,N-dimethyl-	T	100	
U015	Azaserine **or** L-Serine, diazoacetate (ester)	T	10	
U016	Benz[c]acridine	T	10	
U017	Benzal chloride **or** Benzene, (dichloromethyl)-	T	5,000	5,000
U018	Benz[a]anthracene	T	10	

EPA Hazardous Waste No.	Description	Hazard Code*	CERCLA RQ†	EPCRA RQ†
U019	Benzene	I,T	10	
U020	Benzenesulfonic acid chloride **or** Benzenesulfonyl chloride	C,R	100	
U021	Benzidine **or** [1,1′-Biphenyll -4,4′-diamine	T	1	
U022	Benzo[a]pyrene	T	1	
U023	Benzotrichloride **or** Benzene, (trichloromethyl)-	C,R,T	10	10
U024	Dichloromethoxy ethane **or** Ethane,1,1′-[methylenebis(oxy)]bis[2-chloro-	T	1,000	
U025	Dichloroethyl ether **or** Ethane, 1,1′-oxybis[2-chloro-	T	10	10
U026	Naphthalenamine, N,N′-bis(2-chloroethyl)-	T	100	
U027	Dichloroisopropyl ether **or** Propane, 2,2′-oxybis[2-chloro-	T	1,000	
U028	Diethylhexyl phthalate **or** 1,2-Benzenedicarboxylic acid, bis(2-ethylhexyl) ester	T	100	
U029	Methyl bromide **or** Methane, bromo-	T	1,000	
U030	4-Bromophenyl phenyl ether **or** Benzene, 1-bromo-4-phenoxy-	T	100	
U031	Butanol **or** *n*-Butyl alcohol	I	5,000	
U032	Chromic acid H_2CrO_4, calcium salt, **or** Calcium chromate	T	10	
U033	Carbon oxyfluoride **or** Carbonic difluoride	R,T	1,000	1,000
U034	Chloral **or** Acetaldehyde, trichloro-	T	5,000	
U035	Chlorambucil **or** Benzenebutanoic acid, 4-[bis(2-chloroethyl)amino]-	T	10	
U036	Chlordane, alpha & gamma isomers **or** 4,7-Methano-1H-indene, 1,2,4,5,6,7,8,8-octachloro-2,3,3a,4,7,7a-hexahydro-	T	1	1
U037	Chlorobenzene **or** Benzene, chloro-	T	100	
U038	Chlorobenzilate **or** Benzeneaceticacid,4-chloro-alpha- (4-chlorophenyl)-alpha-hydroxy-,ethyl ester	T	10	
U039	*p*-Chloro-*m*-cresol **or** Phenol, 4-chloro-3-methyl-	T	5,000	
U041	Epichlorohydrin **or** Oxirane, (chloromethyl)-	T	100	100

EPA Hazardous Waste No.	Description	Hazard Code*	CERCLA RQ†	EPCRA RQ†
U042	2-Chloroethyl vinyl ether **or** Ethene, (2-chloroethoxy)-	T	1,000	
U043	Vinyl chloride **or** Ethene, chloro-	T	1	
U044	Chloroform **or** Methane, trichloro-	T	10	10
U045	Methyl chloride **or** Methane, chloro-	I,T	100	
U046	Chloromethyl methyl ether **or** Methane, chloromethoxy-	T	10	10
U047	Naphthalene, 2-chloro-or beta-Chloronaphthalene	T	5,000	
U048	*o*-Chlorophenol **or** Phenol, 2-chloro-	T	100	
U049	Benzenamine, 4-chloro-2-methyl-, hydrochloride **or** 4-Chloro-*o*-toluidine, hydrochloride	T	100	
U050	Chrysene	T	100	
U051	Creosote	T	1	
U052	Cresol (Cresylic acid) **or** Phenol, methyl-	T	100	100
U053	Crotonaldehyde **or** 2-Butenal	T	100	100
U055	Cumene **or** Benzene, (1-methylethyl)-	I	5,000	
U056	Cyclohexane **or** Benzene, hexahydro-	I	1,000	
U057	Cyclohexanone	I	5,000	
U058	Cyclophosphamide **or** 2H-1, 3,2-Oxazaphosphorin-2-amine, N,N-bis(2-chloroethyl)tetrahydro-, 2-oxide	T	10	
U059	Daunomycin **or** 5,12-Naphthacenedione, 8-acetyl-10-[(3-amino-2,3,6-trideoxy) -alpha-L-Iyxo-hexopyranosyl)oxy] -7,8,9, 10-tetrahydro-6,8, 1 1-trihydroxy-1-methoxy-, (8S-cis)-	T	10	
U060	DDD **or** Benzene, 1,1′-(2,2-dichloroethylidene)bis[4-chloro	T	1	
U061	DDT **or** Benzene, 1,1′-(2,2,2-trichloroethylidene)bis[4-chloro-	T	1	
U062	Diallate **or** Carbamothioic acid, bis(1-methylethyl)-, S-(2,3-dichloro-2-propenyl) ester	T	100	
U063	Dibenz[a,h]anthracene	T	1	
U064	Dibenzo[a,i]pyrene **or** Benzo[rst]pentaphene	T	10	

EPA Hazardous Waste No.	Description	Hazard Code*	CERCLA RQ†	EPCRA RQ†
U066	1,2-Dibromo-3-chloropropane **or** Propane, 1,2-dibromo-3-chloro-	T	1	
U067	Ethylene dibromide **or** Ethane, 1,2-dibromo-	T	1	
U068	Methylene bromide **or** Methane, dibromo-	T	1,000	
U069	Dibutyl phthalate **or** 1,2-Benzenedicarboxylic acid, dibutyl ester	T	10	
U070	*o*-Dichlorobenzene **or** Benzene, 1,2-dichloro-	T	100	
U071	*m*-Dichlorobenzene **or** Benzene, 1,3-dichloro-	T	100	
U072	*p*-Dichlorobenzene **or** Benzene, 1,4-dichloro-	T	100	
U073	[1,1′-Biphenyll-4,4′-diamine, 3,3′-dichloro- **or** 3,3′-Dichlorobenzidine	T	1	
U074	1,4-Dichloro-2-butene **or** 2-Butene, 1,4-dichloro-	I,T	1	
U075	Dichlorodifluoromethane **or** Methane, dichlorodifluoro	T	5,000	
U076	Ethylidene dichloride **or** Ethane, 1,1-dichloro-	T	1,000	
U077	Ethylene dichloride **or** Ethane, 1,2-dichloro-	T	100	
U078	1,1 -Dichloroethylene **or** Ethene, 1,1-dichloro-	T	100	
U079	1,2-Dichloroethylene **or** Ethene, 1,2-dichloro-,(E)-	T	1,000	
U080	Methylene chloride **or** Methane, dichloro-	T	1,000	
U081	2,4-Dichlorophenol	T	100	
U082	2,6-Dichlorophenol	T	100	
U083	Propylene dichloride **or** Propane, 1,2-dichloro	T	1,000	
U084	1,3-Dichloropropene	T	100	
U085	1,2:3,4-Dieooxybutane **or** 2,2′-Bioxirane	I,T	10	10
U086	Hydrazine, 1,2-diethyl- **or** N,N′-Diethylhydrazine	T	10	
U087	O,O-Diethyl S-methyl dithiophosphate **or** Phosphorodithioic acid, O,O-diethyl S-methyl ester	T	5,000	
U088	Diethyl phthalate **or** 1,2-Benzenedicarboxylic acid, diethyl ester	T	1,000	
U089	Diethylstilbesterol **or** Phenol, 4,4′-(1,2-diethyl-1,2-ethenediyl)bis-, (E)-	T	1	

EPA Hazardous Waste No.	Description	Hazard Code*	CERCLA RQ†	EPCRA RQ†
U090	Dihydrosafrole **or** 1,3-Benzodioxole, 5-propyl-	T	10	
U091	[1,1′-Biphenyl]-4,4′-diamine, 3,3′-dimethoxy- **or** 3,3′-Dimethoxybenzidine	T	100	
U092	Dimethylamine **or** Methanamine, N-methyl-	I	1,000	
U093	*p*-Dimethylaminoazobenzene **or** Benzenamine,N-dimethyl-4-(phenylazo)-	T	10	
U094	Benz[a]anthracene, 7,12-dimethyl- **or** 7,12-Dimethylbenz[a] anthracene	T	1	
U095	3,3′-Dimethylbenzidine **or** [1,1′-Biphenyl]-4,4′-diamine, 3,3′-dimethyl	T	10	
U096	Hydroperoxide, 1-methyl-l-phenylethyl- **or** alpha, alpha-Dimethylbenzyl-hydroperoxide	R	10	
U097	Dimethylcarbamoyl chloride **or** Carbamic chloride, dimethyl-	T	1	
U098	1,1-Dimethylhydrazine **or** Hydrazine, 1,1-dimethyl-	T	10	
U099	1,2-Dimethylhydrazine **or** Hydrazine, 1,2-dimethyl-	T	1	
U101	2,4-Dimethylphenol **or** Phenol, 2,4-dimethyl-	T	100	
U102	Dimethyl phthalate **or** 1,2-Benzenedicarboxylic acid, dimethyl ester	T	5,000	
U103	Dimethyl sulfate **or** Sulfuric acid, dimethyl ester	T	100	100
U105	2,4-Dinitrotoluene **or** Benzene, 1-methyl-2,4-dinitro-	T	10	
U106	2,6-Dinitrotoluene **or** Benzene, 2-methyl-1,3-dinitro-	T	100	
U107	Di-*n*-octyl phthalate **or** 1,2-Benzenedicarboxylic acid, dioctyl ester	T	5,000	
U108	1,4-Diethyleneoxide **or** 1,4-Dioxane	T	100	
U109	1,2-Diphenylhydrazine **or** Hydrazine, 1,2-diphenyl-	T	10	
U110	Dipropylamine **or** 1-Propanamine, N-propyl-	I	5,000	
U111	Di-*n*-propylnitrosamine **or** 1-Propanamine, N-nitroso-N-propyl-	T	10	

EPA Hazardous Waste No.	Description	Hazard Code*	CERCLA RQ†	EPCRA RQ†
U112	Ethyl acetate **or** Acetic acid ethyl ester	I	5,000	
U113	Ethyl acrylate **or** 2-Propenoic acid, ethyl ester	T	1,000	
U114	Carbamodithioic acid, 1,2-ethanediylbis-, salts & esters **or** Ethylenebisdithiocarbamic acid, salts, & esters	T	5,000	
U115	Ethylene oxide **or** Oxirane	I,T	10	10
U116	Ethylenethiourea **or** 2-Imidazolidinethione	T	10	
U117	Ethane, 1,1′-oxybis- **or** Ethyl ether	I	100	
U118	Ethyl methacrylate **or** 2-Propenoic acid, 2-methyl-, ethyl ester	T	1,000	
U119	Ethyl methanesulfonate **or** Methanesulfonic acid, ethyl ester	T	1	
U120	Fluoranthene	T	100	
U121	Trichloromonofluoromethane **or** Methane, trichlorofluoro-	T	5,000	
U122	Formaldehyde	T	100	100
U123	Formic acid	C,T	5,000	
U124	Furan **or** Furfuran	I	100	100
U125	Furfural **or** 2-Furancarboxaldehyde	I	5,000	
U126	Glycidylaldehyde **or** Oxiranecarboxyaldehyde	T	10	
U127	Hexachlorobenzene	T	10	
U128	1,3-Butadiene, 1,1,2,3,4,4-hexachloro-	T	1	
U129	Lindane **or** Cyclohexane, 1,2,3,4,5,6-hexachloro-,1alpha,2alpha,3beta, 4alpha,5alpha,6beta)-	T	1	1
U130	Hexachlorocyclopentadiene **or** 1,3-Cyclopentadiene, 1,2,3,4,5,5-hexachloro-	T	10	10
U131	Hexachloroethane	T	100	
U132	Hexachlorophene **or** Phenol, 2′-methylenebis[3,4,6-trichloro-	T	100	
U133	Hydrazine	R,T	1	1
U134	Hydrofluoric acid **or** Hydrogen fluoride	C,T	100	100
U135	Hydrogen sulfide	T	100	100
U136	Cacodylic acid **or** Arsinic acid, dimethyl-	T	1	
U137	Indeno[1,2,3-cdlpyrene	T	100	

EPA Hazardous Waste No.	Description	Hazard Code*	CERCLA RQ†	EPCRA RQ†
U138	Methyl iodide	T	100	
U140	1-Propanol, 2-methyl- **or** Isobutyl alcohol	I,T	5,000	
U141	Isosafrole **or** 1,3-Benzodioxole, 5-(1-propenyl)-	T	100	
U142	Kepone **or** 1,3,4-Metheno-2H-cyclobuta[cd]pentalen-2-one,1,1a,3,3a,4,5,5,5a,5b,6-decachlorooctahydro-	T	1	
U143	Lasiocaipine **or** 2-Butenoic acid, 2-methyl-, 7-[[2,3-dihydroxy-2-(1-methoxyethyl)-3-methyl-1-oxobutoxy]methyl]-2,3,5,7a-tetrahydro-1H-pyrrolizin-1-yl ester,[1S-[1alpha(Z),7(2S8*,3R*),7aalpha]]-	T	10	
U144	Lead acetate **or** Acetic acid, lead(2 +) salt	T	10	
U145	Lead phosphate **or** Phosphoric acid, lead(2 +) salt (2:3)	T	10	
U146	Lead subacetate **or** Lead, bis(acetato-O)tetrahydroxytri	T	10	
U147	Maleic anhydride **or** 2,5-Furandione	T	5,000	
U148	Maleic hydrazide **or** 3,6-Pyridazinedione, 1,2-dihydro-	T	5,000	
U149	Malononitrile **or** Propanedinitrile	T	1,000	1,000
U150	Melphalan **or** L-Phenylalanine, 4-[bis(2-chloroethyl)amino]-	T	1	
U151	Mercury	T	1	
U152	Methacrylonitrile **or** 2-Propenenitrile, 2-methyl-	I,T	1,000	1,000
U153	Methanethiol **or** Thiomethanol	I,T	100	100
U154	Methanol	I	5,000	
U155	Methapyrilene **or** 1,2-Ethanediamine, N,N-dimethyl-N P-2-pyridinyl-N P-(2-thienylmethyl)-	T	5,000	
U156	Methyl chlorocarbonate **or** Carbonochloridic acid, methyl ester	I,T	1,000	1,000
U157	3-Methylcholanthrene **or** Benz[i]aceanthrylene, 1,2-dihydro-3-methyl-	T	10	

EPA Hazardous Waste No.	Description	Hazard Code*	CERCLA RQ†	EPCRA RQ†
U158	4,4′-Methylenebis(2-chloroaniline) **or** Benzenamine, 4,4′-methylenebis[2-chloroaniline]	T	10	
U159	Methyl ethyl ketone (MEK) **or** 2-Butanone	I,T	5,000	
U160	Methyl ethyl ketone peroxide **or** 2-Butanone, peroxide	R,T	10	
U161	Methyl isobutyl ketone **or** Pentanol, 4-methyl- **or** 4-Methyl-2-pentanone	I	5,000	
U162	Methyl methacrylate **or** 2-Propenoic acid, 2-methyl-, methyl ester	I,T	1,000	1,000
U163	MNNG **or** Guanidine, N-methyl-N P-nitro-N-nitroso-	T	10	
U164	Methylthiouracil **or** 4(1H)-Pyrimidinone, 2,3-dihydro-6-methyl-2-thioxo-	T	10	
U165	Naphthalene	T	100	
U166	1,4-Naphthalenedione **or** 1,4-Naphthoquinone	T	5,000	
U167	1-Naphthalenamine **or** α-Naphthylamine	T	100	
U168	2-Naphthalenamine **or** β-Naphthylamine	T	10	
U169	Nitrobenzene	I,T	1,000	1,000
U170	*p*-Nitrophenol **or** Phenol,-nitro-	T	100	
U171	2-Nitropropane **or** Propane, 2-nitro-	I,T	10	
U172	N-Nitrosodi-*n*-butylamine **or** 1-Butanamine, N-butyl-N-nitroso-	T	10	
U173	Ethanol, 2,2′-(nitrosoimino)bis- **or** N-Nitrosodiethanolamine	T	1	
U174	N-Nitrosodiethylamine **or** Ethanamine, N-ethyl-N-nitroso-	T	1	
U176	Urea, N-ethyl-N-nitroso- **or** N-Nitroso-N-ethylurea	T	1	
U177	Urea, N-methyl-N-nitroso- **or** N-Nitroso-N-methylurea	T	1	
U178	Carbamic acid, methylnitroso-, ethyl ester **or** N-Nitroso-N-methylurethane	T	1	
U179	N-Nitrosopiperidine **or** Piperidine, -nitroso-	T	10	
U180	N-Nitrosopyrrolidine **or** Pyrrolidine, 1-nitroso-	T	1	

EPA Hazardous Waste No.	Description	Hazard Code*	CERCLA RQ†	EPCRA RQ†
U181	5-Nitro-*o*-toluidine **or** Benzenamine, 2-methyl-5-nitro-	T	100	
U182	Paraldehyde **or** 1,3,5-Trioxane, 2,4,6-trimethyl-	T	1,000	
U183	Pentachlorobenzene **or** Benzene, pentachloro-	T	10	
U184	Pentachloroethane **or** Ethane, pentachloro-	T	10	
U185	Pentachloronitrobenzene (PCNB) **or** Benzene, pentachloronitro-	T	100	
U186	1,3-Pentadiene **or** 1-Methylbutadiene	I	100	
U187	Phenacetin **or** Acetamide, N-(4-ethoxyphenyl)-	T	100	
U188	Phenol	T	1,000	1,000
U189	Phosphorus sulfide **or** Sulfur phosphide	R	100	
U190	Phthalic anhydride **or** 1,3-Isobenzofurandione	T	5,000	
U191	2-Picoline **or** Pyridine, 2-methyl-	T	5,000	
U192	Pronamide **or** Benzamide, 3,5-dichloro-N-(1,1-dimethyl-2-propynyl)-	T	5,000	
U193	1,2-Oxathiolane, 2,2-dioxide **or** 1,3-Propane sultone	T	10	
U194	1-Propanamine **or** n-Propylamine	I,T	5,000	
U196	Pyridine	T	1,000	
U197	p-Benzoquinone **or** 2,5-Cyclohexadiene-1,4-dione	T	10	
U200	Reserpine **or** Yohimban-16-carboxylic acid,11,17-dimethoxy-18-[(3,4,5-)oxy]-, methyl, (3beta, 16beta,17alpha, 18beta,20alpha)-	T	5,000	
U201	Resorcinol **or** 1,3-Benzenediol	T	5,000	
U202	Saccharin & salts **or** 1,2-Benzisothiazol-3(2H)-one, 1,1-dioxide & salts	T	100	
U203	Safrole **or** 1,3-Benzodioxole, 5-(2-propenyl)-	T	100	
U204	Selenious acid **or** Selenium dioxide	T	10	10
U205	Selenium sulfide	R,T	10	
U206	Streptozotocin **or** Glucopyranose, 2-deoxy-2-(3-methyl-3-nitrosoureido)-, D- **or** D-Glucose, 2-deoxy-2-[[(methylnitrosoamino)-carbonyl]amino]-	T	1	

EPA Hazardous Waste No.	Description	Hazard Code*	CERCLA RQ†	EPCRA RQ†
U207	1, 2,4,5 -Tetrachlorobenzene **or** Benzene, 1,2,4,5-tetrachloro-	T	5,000	
U208	1,1,1,2-Tetrachloroethane **or** Ethane, 1,1,1,2-tetrachloro-	T	100	
U209	1,1, 2,2 -Tetrachloroethane **or** Ethane, 1,1,2,2-tetrachloro-	T	100	
U210	Tetrachloroethylene **or** Ethene, tetrachloro-	T	100	
U211	Carbon tetrachloride **or** Methane, tetrachloro-	T	10	
U213	Tetrahydrofuran **or** Furan, tetrahydro-(	I	1,000	
U214	Thallium (l) acetate **or** Acetic acid, thallium (1+) salt	T	100	
U215	Thallium (1) carbonate **or** Carbonic acid, dithallium (1+) salt	T	100	100
U216	Thallium chloride TlCl	T	100	100
U217	Nitric acid, thallium (1+) salt **or** Thallium (1) nitrate	T	100	
U218	Thioacetamide **or** Ethanethioamide	T	10	
U219	Thiourea	T	10	
U220	Toluene **or** Benzene, methyl-	T	1,000	
U221	Toluenediamine **or** Benzenediamine, ar-methyl-	T	10	
U222	*o*-Toluidine hydrochloride **or** Benzenamine, 2-methyl-, hydrochloride	T	100	
U223	Toluene diisocyanate **or** Benzene, 1,3-diisocyanatomethyl-	R,T	100	
U225	Bromoform **or** Methane, tribromo-	T	100	
U226	Methyl chloroform **or** Ethane, 1,1,1-trichloro-	T	1,000	
U227	1,1,2-Trichloroethane **or** Ethane, 1,1,2-trichloro-	T	100	
U228	Trichloroethylene **or** Ethene, trichloro-	T	100	
U234	1,3,5-Trinitrobenzene **or** Benzene, 1,3,5-trinitro-	R,T	10	
U235	1-Propanol, 2,3-dibromo-, phosphate (3:1) **or** Tris(2,3-dibromopropyl) phosphate	T	10	
U236	Trypan blue **or** 2,7-Naphthalenedisulfonic acid,3,3′-[(3,3′-dimethyl[1,1′-biphenyl]-4,4′-diyl) bis(azo)bis[5-aniino-4-hydroxy]-, tetrasodium salt	T	10	

EPA Hazardous Waste No.	Description	Hazard Code*	CERCLA RQ†	EPCRA RQ†
U237	Uracil mustard **or** 2,4-(1H,3H)-Pyrimidinedione, 5-[bis(2-chloroethyl)amino]-	T	10	
U238	Ethyl carbamate (urethane) **or** Carbamic acid, ethyl ester	T	100	
U239	Xylene **or** Benzene, dimethyl-	I,T	100	
U240	2,4-D, salts & esters **or** Acetic acid, (2,4-dichlorophenoxy)-, salts & esters	T	100	
U243	Hexachloropropene **or** 1-Propene, 1,1,2,3,3,3-hexachloro-	T	1,000	
U244	Thiram **or** Thioperoxydicarbonic diamide $[(H_2N)C(S)]_2S_2$, tetramethyl-	T	10	
U246	Cyanogen bromide (CN)BR	T	1,000	1,000
U247	Methoxychlor **or** Benzene, 1,1′-(2,2,2-trichloroethylidene)bis[4-methoxy-	T	1	
U248	Warfarin & salts, when present at concentrations of 0.3% **or** less **or** 2H-1-Benzopyran-2-one, 4-hydroxy-3-(3-oxo-1-phenyl-butyl)-, & salts, when present at concentrations of 0.3% **or** less	T		100
U249	Zinc phosphide Zn_3P_2, when present at concentrations of 10% **or** less	T	100	100
U271	Benomyl **or** Carbamic acid, [1-[(butylamino)carbonyl]-1H benzimidazol-2-yl]-, methyl ester-	T	1	
U277	Sulfallate **or** Carbamodithioic acid, diethyl-, 2- chloro-2-propenyl ester-	T	1	
U278	Bendiocarb **or** 1,3-Benzodioxol-4-ol, 2,2-dimethyl-, methyl carbamate	T	1	
U279	Carbaryl **or** 1-Naphthalenol, methylcarbamate	T	100	
U280	Barban **or** Carbamic acid, (3-chlorophenyl)-, 4- chloro-2-butynyl ester	T	1	
U328	*o*-Toluidine **or** Benzenamine, 2-methyl-	T	100	
U353	*p*-Toluidine **or** Benzenamine, 4-methyl-	T	100	
U359	Ethylene glycol monoethyl ether **or** Ethanol, 2-ethoxy-	T	1,000	
U364	Bendiocarb phenol **or** 1,3-Benzodioxol-4-ol, 2,2-dimethyl-,	T	1	

EPA Hazardous Waste No.	Description	Hazard Code*	CERCLA RQ†	EPCRA RQ†
U365	Molinate **or** H-Azepine-1-carbothioic acid hexahydro-, S-ethyl ester	T	1	
U366	Dazomet **or** 2H-1,3,5-Thiadiazine-2thione, tetrahydro-3,5-dimethyl-	T	1	
U367	Carbofuran phenol **or** Benzofuranol, 2,3-dihydro-2,2- dimethyl-	T	1	
U372	Carbendazim **or** Carbamic acid, 1H-benzimidazol-2-yl	T	1	
U373	Propham **or** Carbamodithioic acid, methyl,-monopotassium salt	T	1	
U375	Carbamic acid, butyl-, 3-iodo-2- propynyl ester **or** 3-Iodo-2-propynyl *n*-butylcarbamate	T	1	
U376	Selenium, tetrakis(dimethyldithio-carbamate) **or** Carbamodithioic acid, dimethyl-, tetraanhydrosulfide with orthothioselenious acid	T	1	
U377	Potassium *n*-methyldithiocarbamate	T	1	
U378	Potassium *n*-hydroxymethyl- *n*-methyldi-thiocarbamate **or** Carbamodithioic acid, (hydroxymethyl)methyl-, monopotassium salt	T	1	
U379	Sodium dibutyldithiocarbamate **or** Carbamodithioic acid, dibutyl, sodium salt	T	1	
U381	Sodium diethyldithiocarbamate **or** Carbamodithioic acid, diethyl-, sodium salt	T	1	
U382	Sodium dimethyldithiocarbamate **or** Carbamodithioic acid, dimethyl-, sodium salt	T	1	
U383	Potassium dimethyldithiocarbamate **or** Carbamodithioic acid, dimethyl, potassium salt	T	1	
U384	Metam Sodium **or** Carbamodithioic acid, methyl-, monosodium salt	T	1	
U385	Vernolate **or** Carbamothioic acid, dipropyl-, S-propyl ester	T	1	
U386	Cycloate **or** Carbamothioic acid, cyclohexylethyl-, S-ethyl ester	T	1	

EPA Hazardous Waste No.	Description	Hazard Code*	CERCLA RQ†	EPCRA RQ†
U387	Prosulfocarb **or** Carbamothioic acid, dipropyl-, S- (phenylmethyl) ester	T	1	
U389	Triallate **or** Carbamothioic acid, bis(1-methylethyl)- , S-(2,3,3-trichloro-2-propenyl) ester	T	1	
U390	EPTC **or** Carbamothioic acid, dipropyl-, S-ethyl ester	T	1	
U391	Pebulate **or** Carbamothioic acid, butylethyl-, S- propyl ester	T	1	
U392	Butylate **or** Carbamothioic acid, bis(2-methylpropyl)-, S-ethyl ester	T	1	
U393	Copper, bis(dimethylcarbamodithioato-S,S′ **or** Copper dimethyldithiocarbamate)-	T	1	
U394	A2213 **or** Ethanimidothioic acid, 2-(dimethylamino)-N-hydroxy-2-oxo-, methyl ester	T	1	
U395	Diethylene glycol, dicarbamate **or** Ethanol, 2,2′-oxybis-, dicarbamate	T	1	
U396	Ferbam **or** Iron, tris(dimethylcarbamodithioato- S,S′)-	T	1	
U400	Piperdine, 1,1′(tetrathiocarbonothioyl)-bis- **or** Bis(pentamethylene)thiuram tetrasulfide	T	1	
U401	Tetramethyliuram monosulfide **or** Bis(dimethylthiocarbamoyl)	T	1	
U402	Thioperokydicarbonic diamide, tetrabutyl **or** Tetrabutylthiurium disulfide	T	1	
U403	Disulfiram **or** Thioperokydicarbonic diamide, tetraethyl	T	1[1]	
U404	Triethylamine **or** Ethanamine, N,N-diethyl-	T	5,000	
U407	Zinc, bis(diethylcarbamodithioato-S,S′)- **or** Ethyl Ziram	T	1	
U409	Thiophanate-methyl **or** Carbamic acid, [1,2-phenylenebis (iminocarbonothioyl)]bis-, dimethyl ester	T	1	
U410	Thiodicarb **or** Ethanimidothioic acid, N,N′-[thiobis[(methylimino)-carbonyloxy]]bi s-, dimethyl ester	T	1	
U411	Propoxur **or** Phenol, 2-(1-methylethoxy)-, methylcarbamate	T	100	

APPENDIX B

HAZARDOUS CONSTITUENTS (THE APPENDIX VIII CONSTITUENTS)

A2213
Acetonitrile
Acetophenone
2-Acetylaminefluarone
Acetyl chloride
1-Acetyl-2-thiourea
Acrolein
Acrylamide
Acrylonitrile
Aflatoxins
Aldicarb
Aldicarb sulfone
Aldrin
Allyl alcohol
Allyl chloride
Aluminum phosphide
4-Aminobiphenyl
5-(Aminomethyl)-3-isoxazolol
4-Aminopyridine
Amitrole
Ammonium vanadate
Aniline
Antimony
Antimony and compounds, N.O.S.
Aramite
Arsenic
Arsenic and compounds, N.O.S.
Arsenic acid
Arsenic pentoxide
Arsenic trioxide
Auramine
Azaserine

Barium
Barium and compounds, N.O.S.
Barium cyanide
Bendiocarb
Bendiocarb phenol
Benomyl
Benz[c]acridine
Benz[a]anthracene
Benzal chloride
Benzene
Benzenearsonic acid
Benzidine
Benzo[b]fluoranthene
Benzo[i]fluoranthene
Benzo[k]fluoranthene

Benzo[a]pyrene
p-Benzoquinone
Benzotrichloride
Benzyl chloride
Beryllium powder
Beryllium and compounds, N.O.S.
Bis (pentamethylene)-thiuram tetrasulfide
Bromoacetone
Bromoform
4-Bromophenyl phenyl ether
Brucine
Butyl benzyl phthalate
Butylate

Cacodylic acid
Cadmium
Cadmium and compounds, N.O.S.
Calcium chromate
Calcium cyanide
Carbaryl
Carbendazim
Carbofuran
Carbofuran phenol
Carbon disulfide
Carbon oxyfluoride
Carbon tetrachloride
Chloral
Chlorambucil
Chlordane (alpha and gamma isomers)
Chlorinated benzenes, N.O.S.
Chlorinated ethane, N.O.S.
Chlorinated fluorocarbons, N.O.S.
Chlorinated naphthalene, N.O.S.
Chlorinated phenol, N.O.S.
Chlornaphazin
Chloroacetaldehyde
Chloroalkyl ethers, N.O.S.
p-Chloroaniline
Chlorobenzene
Chlorobenzilate
p-Chloro-*m*-cresol
2-Chloro vinyl ether
Chloroform
Chloromethyl methyl ether
beta-Chloronaphthalene
o-Chlorophenol
1-(*o*-Chlorophenyl)thiourea
Chloroprene
3-Chloropropionitrile
Chromium and compounds, N.O.S.
Chrysene
Citrus red No. 2
Coal tar creosote
Copper cyanide
Copper dimethyldithiocarbamate
Creosote
Cresol (Cresylic acid)
Crotonaldehyde
Cyanides (soluble salts and complexes), N.O.S.
Cyanogen
Cyanogen bromide
Cyanogen chloride
Cycasin
Cycloate
2-Cyclohexyl-4,6-dinitrophenol
Cyclophosphamide

2,4-D, salts and esters
Daunomycin
Dazomet
DDD
DDE
DDT
Diallate
Dibenz[a,h]acridine
Dibenz[a, j]acridine
Dibenz[a,h]anthracene
7H-Dibenzo[c,g]carbazole
Dibenzo[a,e]pyrene
Dibenzo[a,h]pyrene
Dibenzo[a,i]pyrene
1,2-Dibromo-3-chloro-propane
Dibutyl phthalate
o-Dichlorobenzene
m-Dichlorobenzene
p-Dichlorobenzene
Dichlorobenzene, N.O.S.
3,3′-Dichlorobenzidine

1,4-Dichloro-2-butene
Dichlorodifluoromethane
Dichloroethylene, N.O.S.
1,1-Dichloroethylene
1,2-Dichloroethylene
Dichloroethyl ether
Dichloroisopropyl ether
Dichloromethoxy ethane
Dichloromethyl ether
2,4-Dichlorophenol
2,6-Dichlorophenol
Dichlorophenylarsine
Dichloropropane, N.O.S.
Dichloropropanol, N.O.S.
Dichloropropene, N.O.S.
1,3-Dichloropropene
Dieldrin
1,2:3,4-Diepoxybutane
Diethylarsine
Diethyleneglycol, dicarbamate
1,4-Diethyleneoxide
Diethylhexyl phthalate
N,N′-Diethylhydrazine
O,O-Diethyl S-methyl dithiophosphate
Diethyl-*p*-nitrophenyl phosphate
Diethyl phthalate
O,O-Diethyl O-pyrazinyl phosphorothioate
Diethylstilbesterol
Dihydrosafrole
Diisopropylfluorophosphate (DFP)
Dimethoate
3,3′-Dimethoxybenzidine
p-Dimethylaminoazobenzene
7,12-Dimethylbenz[a]anthracene
3,3′-Dimethylbenzidine
Dimethylcarbamoylchloride
1,1-Dimethylhydrazine
1,2-Dimethylhydrazine
alpha,alpha-Dimethylphenethylamine
2,4-Dimethylphenol
Dimethyl phthalate
Dimethyl sulfate
Dimetilan
Dinitrobenzene, N.O.S.
4,6-Dinitr-*o*-cresol and salts
2,4-Dinitrophenol
2,4-Dinitrotoluene
2,6-Dinitrotoluene
Dinoseb
Di-*n*-octyl phthalate
Diphenylamine
1,2-Diphenylhydrazine
Di-*p*-propylnitrosamine
Disulfoton
Dithiobiuret

Endosulfan
Endothall
Endrin and metabolites
Epichlorhydrin
Epinephrine
EPTC
Ethyl carbamate (urethane)
Ethyl cyanide
Ethylenebisdithiocarbamic acid, salts and esters
Ethylene dibromide
Ethylene dichloride
Ethylene glycol monoethyl ether
Ethyleneimine
Ethylene oxide
Ethylenethiourea
Ethylidene dichloride
Ethyl methacrylate
Ethyl methanesulfonate
Ethyl ziram

Famphur
Ferbam
Fluoranthene
Fluorine
Fluoroacetamide
Fluoroacetic acid, sodium salt
Formaldehyde
Formaldehyde hydrochloride
Formic acid
Formparante

Glycidylaldehyde

Halomethanes, N.O.S.
Heptachlor
Heptachlor epoxide (alpha, beta, and gamma isomers)
Heptachlorodibenzofurans
Heptachlorodibenzo-*p*-dioxins
Hexachlorobenzene
Hexachlorobutadiene
Hexachlorocyclopentadiene
Hexachloroethane
Hexachlorophene
Hexachloropropene
Hexaethyl tetraphosphate
Hydrazine
Hydrogen cyanide
Hydrogen fluoride
Hydrogen sulfide

3-Iodo-2-propynyl-n-butylcarbamate
Indeno[1,2,3-cd]pyrene
Isobutyl alcohol
Isodrin
Isolan

Ketone

Lasiocarpine
Lead and compounds, N.O.S.
Lead acetate
Lead phosphate
Lead subacetate
Lindane

Maleic anhydride
Maleic hydrazide
Malononitrile
Managanese dimethyldithiocarbamate
Melphalan
Mercury and compounds, N.O.S.
Mercury fulminate
Metam Sodium
Methacrylonitrile
Methiocarb
Methapyrilene
Methomyl
Methoxychlor
Methyl bromide
Methyl chloride
Methyl chlorocarbonate
Methyl chloroform
3-Methylcholanthrene
4,4′-Methylenebis(2-chloroaniline)
Methylene bromide
Methylene chloride
Methyl ethyl ketone [MEK]
Methyl ethyl ketone peroxide
Methyl hydrazine
Methyl iodide
Methyl isocyanate
2-Methyllactonitrile
Methyl methacrylate
Methyl methanesulfonate
Methyl parathion
Methylthiouracil
Mexacarbate
Mitomycin C
MNNG
Molinate
Mustard gas

Naphthalene
1,4-Naphthoquinone
alpha-Naphthylamine
beta-Naphthylamine
alpha-Naphthylthiourea
Nickel and compounds, N.O.S.
Nickel carbonyl
Nickel cyanide
Nicotine and salts
Nitric oxide
p-Nitroaniline
Nitrobenzene
Nitrogen dioxide
Nitrogen mustard
Nitrogen mustard, hydrochloride salt
Nitrogen mustard N-oxide
Nitrogen mustard N-oxide, hydrochloride salt
Nitroglycerin
p-Nitrophenol

2-Nitropropane
Nitrorosamines, N.O.S.
N-Nitrosodi-*n*-butylamine
N-Nitrosodiethanolamine
N-Nitrosodimethylamine
N-Nitroso-N-ethylurea
N-Nitrosomethylethylamine
N-Nitroso-N-methylurea
N-Nitroso-N-methylurethane
N-Nitrosomethylvinylamine
N-Nitrosomorpholine
N-Nitrosonornicotine
N-Nitrosopiperidine
N-Nitrosopyrrolidine
N-Nitrososarcosine
5-Nitro-*o*-toluidine

Octamethylpyrophosphoramide
Osmium tetroxide
Oxamyl

Paraldehyde
Parathion
Pebulate
Pentachlorobenzene
Pentachlorodibenzo-*p*-dioxins
Pentachlorodibenzofurans
Pentachloroethane
Pentachloronitrobenzene (PCNB)
Pentachlorophenol
Phenacetin
Phenol
Phenylenediamine
Phenylmercury acetate
Phenylthiourea
Phorate
Phosgene
Phosphine
Phthalic acid esters, N.O.S.
Phthalic anhydride
Physostigmine
Physostigmine salicylate
2-Picoline
Polychlorinated biphenyls, N.O.S.
Potassium cyanide
Potassium dimethyldithiocarbamate
Potassium hydroxymethyl- *n*-methyl-dithiocarbamate
Potassium *n*-methyldithiocarbamate
Potassium pentachlorophenate
Potassium silver cyanide
Promecarb
Pronamide
1,3-Propane sultone
Propargyl alcohol
Propham
Proxpur
Propylene dichloride
1,2-Propylenimine
Propylthiouracil
Prosulcarb
Pyridine

Reserpine

Saccharin and salts
Safrole
Selenium and compounds, N.O.S.
Selenium dioxide
Selenium sulfide
Selenium, tetrakis (dimethyl-dithiocarbamate)
Selenourea
Silver and compounds, N.O.S.
Silver cyanide
Silvex (2,4,5-TP)
Sodium cyanide
Sodium dibutyldithiocarbamate
Sodium diethtyldithiocarbamate
Sodium dimethyldithiocarbamate
Sodium pentachlorophenate
Streptozotocin
Strychnine and salts
Sulfallate

TCDD
Tetrabutylthiuram disulfide
Tetrabutylthiuram monosulfide
1,2,4,5-Tetrachlorobenzene
Tetrachlorodibenzo-*p*-dioxins

Tetrachlorodibenzofurans
Tetrachloroethane, N.O.S.
1,1,1,2-Tetrachloroethane
1,1,2,2-Tetrachloroethane
Tetrachloroethylene
2,3,4,6-Tetrachlorophenol
2,3,4,6-Tetrachlorophenol, potassium salt
2,3,4,6-Tetrachlorophenol, sodium salt
Tetraethyldithiopyrophosphate
Tetraethyl lead
Tetraethyl pyrophosphate
Tetranitromethane
Thallium and compounds, N.O.S.
Thallic oxide
Thallium (I) acetate
Thallium (I) carbonate
Thallium (I) chloride
Thallium (I) nitrate
Thallium selenite
Thallium (I) sulfate
Thioacetamide
Thiodicarb
Thiofanox
Thiomethanol
Thiophanate-methyl
Thiophenol
Thiosemicarbazide
Thiourea
Thiram
Tirpate
Toluene
Toluenediamine
Toluene-2,4-diamine
Toluene-2,6-diamine
Toluene-3,4-diamine
Toluene diisocyanate
o-Toluidine
o-Toluidine hydrochloride
p-Toluidine
Toxaphene
Triallate
1,2,4-Trichlorobenzene
1,1,2-Trichloroethane
Trichloroethylene
Trichloromethanethiol
Trichloromonofluoromethane
2,4,5-Trichlorophenol
2,4,6-Trichlorophenol
2,4,5-T
Trichloropropane, N.O.S.
1,2,3-Trichloropropane
Triethylamine
O,O,O-Triethyl phosphorothioate
1,3,5-Trinitrobenzene
Trs(1-aziridinyl)phosphine sulfide
Tris(2,3-dibromopropyl)phosphate
Trypan blue

Uracil mustard

Vanadium pentoxide
Vermolate
Vinyl chloride

Warfarin and salts

Zinc cyanide
Zinc phosphide
Ziram

APPENDIX C

GROUNDWATER MONITORING CONSTITUENTS (THE APPENDIX IX CONSTITUENTS)

Acenaphthene
Acenaphthylene
Acetone
Acetonitrile (Methyl cyanide)
Acetophenone
2-Acetylaminofluorene (2-AAF)
Acrolein
Acrylonitrile
Aldrin
Allyl chloride
4-Aminobiphenyl
Aniline
Anthracene
Antimony
Aramite
Arsenic

Barium
Benzene
Benzo(a)anthracene
Benzo(b)fluoranthene
Benzo(k)fluoranthene
Benzo(ghi)perylene
Benzo(a)pyrene
Benzyl alcohol
Beryllium
alpha-BHC
beta-BHC
delta-BHC
gamma-BHC; Lindane
Bis(2-chloroethoxy)methane
Bis(2-chloroethyl)ether
Bis(2-chloro-1-methyl-ethyl) ether; 2,2′-Dichlorodiisopropyl ether
Bis(2-ethylhexyl) phthalate
Bromodichloromethane
Bromoform (Tribromomethane)
4-Bromophenyl phenyl ether
Butyl benzyl phthalate (Benzl butyl phthalate)

Cadmium
Carbon disulfide
Carbon tetrachloride
Chlordane
p-Chloroaniline
Chlorobenzene
Chlorobenzilate

p-Chloro-*m*-cresol
Chloroethane (Ethyl chloride)
Chloroform
2-Chloronaphthalene
2-Chlorophenol
4-Chlorophenyl phenyl ether
Chloroprene
Chromium
Chrysene
Cobalt
Copper
m-Cresol
o-Cresol
p-Cresol
Cyanide

2,4-D (2,4-Dichlorophenoxyacetic acid)
4,4′-DDD
4,4′-DDE
4,4′-DDT
Diallate
Dibenz(a,h)anthracene
Dibenzofuran
Dibromochloromethane
1,2-Dibromo-3-chloropropane (DBCP)
1,2-Dibromoethane (Ethylene dibromide)
Di-*n*-butyl phthalate
m-Dichlorobenzene
o-Dichlorobenzene
p-Dichlorobenzene
3,3′-Dichlorobenzidine
trans-1,4-Dichloro-2-butene
Dichlorodifluoromethane
1,1-Dichloroethane
1,2-Dichloroethane
1,1-Dichloroethylene (Vinylidene chloride)
trans-1,2-Dichloroethylene
2,4-Dichlorophenol
2,6-Dichlorophenol
1,2-Dichloropropane
cis-1,3-Dichloropropene
trans-1,3-Dichloropropene
Dieldrin
Diethyl phthalate
O,O-Diethyl 0-2-pyrazinyl phosphorothioate (Thionazin)
Dimethoate
p-(Dimethylamino)azobenzene
7,12-Dimethylbenz(a)anthracene
3,3′-Dimethylbenzidine
alpha,alpha-Dimethylphenethylamine
2,4-Dimethylphenol
Dimethyl phthalate
m-Dinitrobenzene
4,6-Dinitro-*o*-cresol
2,4-Dinitrophenol
2,4-Dinitrotoluene
2,6-Dinitrotoluene
Di-*n*-octyl phthalate
Dinoseb (DNBP or 2-*sec*-Butyl-4,6-dinitrophenol)
1,4-Dioxane
Diphenylamine
Disulfoton

Endosulfan I
Endosulfan II
Endosulfan sulfate
Endrin
Endrin aldehyde
Ethylbenzene
Ethyl methacrylate
Ethyl methanesulfonate

Famphur
Fluoranthene
Fluorene

Heptachlor
Heptachlor epoxide
Hexachlorobenzene
Hexachlorobutadiene
Hexachlorocyclopentadiene
Hexachloroethane
Hexachlorophene
Hexachloropropene
2-Hexanone

Indeno(1,2,3-cd)pyrene
Isobutyl alcohol
Isodrin
Isophorone
Isosafrole

Ketone

Lead

Mercury
Methacrylonitrile
Methapyrilene
Methoxychlor
Methyl bromide (Bromomethane)
Methyl chloride (Chloromethane)
3-Methylcholanthrene
Methylene bromide (Dibromomethane)
Methylene chloride (Dichloromethane)
Methyl ethyl ketone (MEK)
Methyl iodide (Iodomethane)
Methyl methacrylate
Methyl methanesulfonate
2-Methylnaphthalene
Methyl parathion
4-Methyl-2-pentanone (Methyl isobutyl ketone)

Naphthalene
1,4-Naphthoquinone
1-Naphthylamine
2-Naphthylamine
Nickel
m-Nitroaniline
o-Nitroaniline
p-Nitroaniline
Nitrobenzene
5-Nitro-o-toluidine
p-Nitrophenol
4-Nitroquinoline 1-oxide
N-Nitrosodi-*n*-butylamine
N-Nitrosodiethylamine
N-Nitrosodimethylamine
N-Nitrosodiphenylamine
N-Nitrosodipropylamine; Di-*n*-propylnitrosamine
N-Nitrosomethylethylamine
N-Nitrosomorpholine
N-Nitrosopiperidine
N-Nitrosopyrrolidine

Parathion
Pentachlorethane
Pentachlorobenzene
Pentachloronitrobenzene
Pentachlorophenol
Phenacetin
Phenanthrene
Phenol
p-Phenylenediamine
Phorate
2-Picoline
Polychlorinated biphenyls (PCBs)
Polychlorinated dibenzofurans (PCDFs)
Polychlorinated di-benzo-*p*-dioxins (PCDDs)
Pronamide
Propionitrile (Ethyl cyanide)
Pyrene
Pyridine

Safrole
Selenium
Silver
Silvex (2,4,5-TP)
Styrene
Sulfide

2,4,5-T (2,4,5-Trichlorophenoxyacetic acid)
2,3,7,8-TCDD (2,3,7,8-Tetrachlorodibenzo *p*-dioxin)
1,2,4,5-Tetrachlorobenzene
1,1,1,2-Tetrachloroethane
1,1,2,2-Tetrachloroethane
Tetrachloroethene (Perchloroethylene)
Tetrachloroethylene
2,3,4,6-Tetrachlorophenol

Tetraethyl dithiopyrophosphate (Sulfotone)
Thallium
Tin
Toluene
o-Toluidine
Toxaphene
1,2,4-Trichlorobenzene
1,1,1-Trichloroethane (Methylchloroform)
1,1,2-Trichloroethane
Trichloroethylene (Trichloroethene)
Trichlorofluoromethane
2,4,5-Trichlorophenol
2,4,6-Trichlorophenol
1,2,3-Trichloropropane
O,O,O-Triethyl phosphorothioate
sym-Trinitrobenzene

Vanadium
Vinyl acetate
Vinyl chloride

Xylene (total)

Zinc

APPENDIX D

CLASSIFICATION OF RCRA PERMIT MODIFICATIONS

Modifications	Class
A. General Permit Provisions	
1. Administrative and informational changes	1
2. Correction of typographical errors	1
3. Equipment replacement or upgrading with functionally equivalent components (e.g., pipes, valves, pumps)	1
4. Changes in the frequency of or procedures for monitoring, reporting, sampling, or maintenance activities by the permittee:	
a. To provide for more frequent monitoring, reporting, sampling, or maintenance	1
b. Other changes	2
5. Schedule of compliance:	
a. Changes in interim compliance dates, with prior approval of the Director	*1
b. Extension of final compliance date	3
6. Changes in expiration date of permit to allow earlier permit termination, with prior approval of the Director	*1
7. Changes in ownership or operational control of a facility, provided the procedures of 40 CFR 270.40(b) are followed	*1

*Permit modifications requiring prior EPA/state approval

Modifications	Class
B. General Facility Standards	
1. Changes to waste sampling or analysis methods:	
a. To conform with agency guidance or regulations	1
b. To incorporate changes associated with F039 (multi-source leachate) sampling or analysis methods	1
c. To incorporate changes associated with underlying hazardous constituents in ignitable or corrosive wastes	*1
d. Other changes	2
2. Changes to analytical quality assurance/control plan:	
a. To conform with agency guidance or regulations	1
b. Other changes	2
3. Changes in procedures for maintaining the operating record	1
4. Changes in frequency or content of inspection schedules	2
5. Changes in the training plan:	
a. That affect the type or decrease the amount of training given to employees	2
b. Other changes	1
6. Contingency plan:	
a. Changes in emergency procedures (i.e., spill or release response procedures)	2
b. Replacement with functionally equivalent equipment, upgrade, or relocation of emergency equipment listed	1
c. Removal of equipment from emergency equipment list	2
d. Changes in name, address, or phone number of coordinators or other persons or agencies identified in the plan	1
7. Construction quality assurance plan:	
a. Changes that the CQA officer certifies in the operating record will provide equivalent or better certainty that the unit components meet the design specifications	1
b. Other changes	2

Note: When a permit modification (such as introduction of a new unit) requires a change in facility plans or other general facility standards, that change shall be reviewed under the same procedures as the permit modification.

Modifications	Class
C. Groundwater Protection	
1. Changes to wells:	
a. Changes in the number, location, depth, or design of upgradient or downgradient wells of permitted groundwater monitoring system	2

Modifications	Class
b. Replacement of an existing well that has been damaged or rendered inoperable, without change to location, design, or depth of the well	1
2. Changes in groundwater sampling or analysis procedures or monitoring schedule, with prior approval of the Director	*1
3. Changes in statistical procedure for determining whether a statistically significant change in groundwater quality between upgradient and downgradient wells has occurred, with prior approval of the Director	*1
4. Changes in point of compliance	*2
5. Changes in indicator parameters, hazardous constituents, or concentration limits (including ACLs):	
a. As specified in the groundwater protection standard	3
b. As specified in the detection monitoring program	2
6. Changes to a detection monitoring program as required by 40 CFR 264.98(j), unless otherwise specified in this appendix	2
7. Compliance monitoring program:	
a. Addition of compliance monitoring program as required by 40 CFR 264.98(h)(4) and 264.99	3
b. Changes to a compliance monitoring program as required by 40 CFR 264.99(k), unless otherwise specified in this appendix	2
8. Corrective action program:	
a. Addition of a corrective action program as required by 40 CFR 264.99(i)(2) and 264.100	3
b. Changes to a corrective action program as required by 40 CFR 264.100(h), unless otherwise specified in this appendix	2
D. Closure	
1. Changes to the closure plan:	
a. Changes in estimate of maximum extent of operations or maximum inventory of waste on-site at any time during the active life of the facility, with prior approval of the Director	*1
b. Changes in the closure schedule for any unit, changes in the final closure schedule for the facility, or extension of the closure period, with prior approval of the Director	*1
c. Changes in the expected year of final closure, where other permit conditions are not changed, with prior approval of the Director	*1
d. Changes in procedures for decontamination of facility equipment or structures, with prior approval of the Director	*1

Modifications	Class
e. Changes in approved closure plan resulting from unexpected events occurring during partial or final closure, unless otherwise specified in this appendix	2
f. Extension of the closure period to allow a landfill, surface impoundment, or land treatment unit to receive nonhazardous wastes after final receipt of hazardous wastes under 40 CFR 264.113(d) and (e)	2
2. Creation of a new landfill unit as part of closure	3
3. Addition of the following new units to be used temporarily for closure activities:	
a. Surface impoundments	3
b. Incinerators	3
c. Waste piles that do not comply with 40 CFR 264.250(c)	3
d. Waste piles that comply with 40 CFR 264.250(c)	2
e. Tanks or containers (other than specified below)	2
f. Tanks used for neutralization, dewatering, phase separation, or component separation, with prior approval of the Director	*1
E. Post-Closure	
1. Changes in name, address, or phone number of contact in post-closure plan	1
2. Extension of post-closure care period	2
3. Reduction in the post-closure care period	3
4. Changes to the expected year of final closure, where other permit conditions are not changed	1
5. Changes in post-closure plan necessitated by events occurring during the active life of the facility, including partial and final closure	2
F. Containers	
1. Modification or addition of container units:	
a. Resulting in greater than 25% increase in the facility's container storage capacity, except as provided in F(1)(c) and F(4)(a)	3
b. Resulting in up to 25% increase in the facility's container storage capacity, except as provided in F(1)(c) and F(4)(a)	2
c. Or treatment processes necessary to treat wastes that are restricted from land disposal to meet some or all of the applicable treatment standards or to treat wastes to satisfy (in whole or in part) the standard of "use of practically available technology that yields the greatest environmental benefit" contained in 40	*1

Modifications	Class
CFR 268.8(a)(2)(ii), with prior approval of the Director. This modification may also involve addition of new waste codes or narrative descriptions of wastes. It is not applicable to dioxin-containing wastes (F020, F021, F022, F023, F026, F027, and F028).	
2. a. Modification of a container unit without increasing the capacity of the unit	2
b. Addition of a roof to a container unit without alteration of the containment system	1
3. Storage of different wastes in containers, except as provided in (F)(4):	
a. That require additional or different management practices from those authorized in the permit	3
b. That do not require additional or different management practices from those authorized in the permit	2
Note: See 40 CFR 270.42(g) for modification procedures to be used for the management of newly listed or identified wastes.	
4. Storage of treatment of different wastes in containers:	
a. That require addition of units or change in treatment process or management standards, provided that the wastes are restricted from land disposal and are to be treated to meet some or all of the applicable treatment standards, or that are to be treated to satisfy (in whole or in part) the standard of "use of practically available technology that yields the greatest environmental benefit" contained in 40 CFR 268.8(a)(2)(ii). This modification is not applicable to dioxin-containing wastes (F020, F021, F022, F023, F026, F027, and F028).	1
b. That do not require the addition of units or a change in the treatment process or management standards, and provided that the units have previously received wastes of the same type (e.g., incinerator scrubber water). This modification is not applicable to dioxin-containing wastes (F020, F021, F022, F023, F026, F027, and F028).	*1
G. Tanks	
1. a. Modification or addition of tank units resulting in greater than 25% increase in the facility's tank capacity, except as provided in G(1)(c), G(1)(d), and G(1)(e)	3
b. Modification or addition of tank units resulting in up to 25% increase in the facility's tank capacity, except as provided in G(1)(d) and G(1)(e)	2

Modifications	Class
c. Addition of a new tank that will operate for more than 90 days using any of the following physical or chemical treatment technologies: neutralization, dewatering, phase separation, or component separation	2
d. After prior approval of the Director, addition of a new tank that will operate for up to 90 days using any of the following physical or chemical treatment technologies: neutralization, dewatering, phase separation, or component separation	*1
e. Modification or addition of tank units or treatment processes necessary to treat wastes that are restricted from land disposal to meet some or all of the applicable treatment standards or to treat wastes to satisfy (in whole or in part) the standard of "use of practically available technology that yields the greatest environmental benefit" contained in 40 CFR 268.8(a)(2)(ii), with prior approval of the Director. This modification may also involve addition of new waste codes. It is not applicable to dioxin-containing wastes (F020, F021, F022, F023, F026, F027, and F028).	*1
2. Modification of a tank unit or secondary containment system without increasing the capacity of the unit	2
3. Replacement of a tank with a tank that meets the same design standards and has a capacity within ±10% of the replaced tank provided —The capacity difference is no more than 1500 gallons, —The facility's permitted tank capacity is not increased, and —The replacement tank meets the same conditions in the permit.	1
4. Modification of a tank management practice	2
5. Management of different wastes in tanks:	
a. That require additional or different management practices, a different tank design, different fire protection specifications, or a significantly different tank treatment process from that authorized in the permit, except as provided in (G)(5)(c)	3
b. That do not require additional or different management practices, a different tank design, different fire protection specifications, or a significantly different tank treatment process than authorized in the permit, except as provided in (G)(5)(d)	2

Modifications	Class
c. That require addition of units or a change in treatment processes or management standards, provided that the wastes are restricted from land disposal and are to be treated to meet some or all of the applicable treatment standards or that are to be treated to satisfy (in whole or in part) the standard of "use of practically available technology that yields the greatest environmental benefit" contained in 40 CFR 268.8(a)(2)(ii). The modification is not applicable to dioxin-containing wastes (F020, F021, F022, F023, F026, F027, and F028).	*1
d. That do not require the addition of units or a change in treatment processes or management standards, and provided that the units have previously received wastes of the same type (e.g., incinerator scrubber water). This modification is not applicable to dioxin-containing wastes (F020, F021, F022, F023, F026, F027, and F028).	1
Note: See 40 CFR 270.42(g) for modification procedures to be used for the management of newly listed or identified wastes.	
H. Surface Impoundments	
1. Modification or addition of surface impoundment units that result in increasing the facility's surface impoundment storage or treatment capacity	3
2. Replacement of a surface impoundment unit	3
3. Modification of a surface impoundment unit without increasing the facility's surface impoundment storage or treatment capacity and without modifying the unit's liner, leak detection system, or leachate collection system	2
4. Modification of a surface impoundment management practice	2
5. Treatment, storage, or disposal of different wastes in surface impoundments:	
a. That require additional or different management practices or different designs of the liner or leak detection system than authorized in the permit	3
b. That do not require additional or different management practices or different designs of the liner or leak detection system than authorized in the permit	2
c. That are wastes restricted from land disposal that meet applicable treatment standards or that are treated to satisfy the standard of "use of practically available technology that yields the greatest environmental benefit'' contained in 40 CFR 269.8(a)(2)(ii), and provided that the unit meets the minimum technological requirements stated in 40 CFR 268.5(h)(2). This	1

Modifications	Class
modification is not applicable to dioxin-containing wastes (F020, F021, F022, F023, F026, F027, and F028).	
d. That are residues from wastewater treatment or incineration, provided that disposal occurs in a unit that meets the minimum technological requirements stated in 40 CFR 268.5(h)(2), and provided further that the surface impoundment has previously received wastes of the same type (for example, incinerator scrubber water). This modification is not applicable to dioxin-containing wastes (F020, F021, F022, F023, F026, F027, and F028).	1
6. Modifications of unconstructed units to comply with 40 CFR 264.221(c), 264.222, 264.223, and 264.226(d)	*1
7. Changes in response action plan:	
a. Increase in action leakage rate	3
b. Change in a specific response reducing its frequency or effectiveness	3
c. Other changes	2

Note: See 40 CFR 270.42(g) for modification procedures to be used for the management of newly listed or identified wastes.

Modifications	Class
I. Enclosed Waste Piles	
For all waste piles except those complying with 40 CFR 264.250(c), modifications are treated the same as for a landfill. The following modifications are applicable only to waste piles complying with 40 CFR 264.250(c).	
1. Modification or addition of waste pile units:	
a. Resulting in greater than 25% increase in the facility's waste pile storage or treatment capacity	3
b. Resulting in up to 25% increase in the facility's waste pile storage or treatment capacity	2
2. Modification of waste pile unit without increasing the capacity of the unit	2
3. Replacement of a waste pile unit with another waste pile unit of the same design and capacity and meeting all waste pile conditions in the permit	1
4. Modification of a waste pile management practice	2
5. Storage or treatment of different wastes in waste piles:	
a. That require additional or different management practices or different design of the unit	3
b. That do not require additional or different management practices or different design of the unit	2

Modifications	Class
6. Conversion of an enclosed waste pile to a containment building unit	2

Note: See 40 CFR 270.42(g) for modification procedures to be used for the management of newly listed or identified wastes.

Modifications	Class
J. Landfills and Unenclosed Waste Piles	
1. Modification or addition of landfill units that result in increasing the facility's disposal capacity	3
2. Replacement of a landfill	3
3. Addition or modification of a liner, leachate collection system, leachate detection system, runoff control, or final cover system	3
4. Modification of a landfill unit without changing a liner, leachate collection system, leachate detection system, runoff control, or final cover system	2
5. Modification of a landfill management practice	2
6. Landfill different wastes:	
a. That require additional or different management practices, or a different design of the liner, leachate collection system, or leachate detection system	3
b. That do not require additional or different management practices, or a different design of the liner, leachate collection system, or leachate detection system	2
c. That are wastes restricted from land disposal that meet the applicable treatment standards or that are treated to satisfy the standard of "use of practically available technology that yields the greatest environmental benefit" contained in 40 CFR 268.8(a)(2)(ii), and provided that the landfill unit meets the minimum technological requirements stated in 40 CFR 268.5(h)(2). This modification is not applicable to dioxin-containing wastes (F020, F021, F022, F023, F026, F027, and F028).	1
d. That are residues from wastewater treatment or incineration, provided that disposal occurs in a landfill unit that meets the minimum technological requirements stated in 40 CFR 268.5(h)(2), and provided further that the landfill has previously received wastes of the same type (for example, incinerator ash). This modification is not applicable to dioxin-containing wastes (F020, F021, F022, F023, F026, F027, and F028).	1
7. Modifications of unconstructed units to comply with 40 CFR 264.251(c), 264.252, 264.253, 264.254(c), 264.301(c), 264.302, 264.303(c), and 264.304	*1

Modifications	Class
8. Changes in response action plan:	
a. Increase in action leakage rate	3
b. Change in a specific response reducing its frequency or effectiveness	3
c. Other changes	2

Note: See 40 CFR 270.42(g) for modification procedures to be used for the management of newly listed or identified wastes.

Modifications	Class
K. Land Treatment	
1. Lateral expansion of or other modification of a land treatment unit to increase areal extent	3
2. Modification of runon control system	2
3. Modify runoff control system	3
4. Other modifications of land treatment unit component specifications or standards required in permit	2
5. Management of different wastes in land treatment units:	
a. That require a change in permit operating conditions or unit design specifications	3
b. That do not require a change in permit operating conditions or unit design specifications	2

Note: See 40 CFR 270.42(g) for modification procedures to be used for the management of newly listed or identified wastes.

Modifications	Class
6. Modification of a land treatment unit management practice to:	
a. Increase rate or change method of waste application	3
b. Decrease rate of waste application	1
7. Modification of a land treatment unit management practice to change measures of pH or moisture content, or to enhance microbial or chemical reactions	2
8. Modification of a land treatment unit management practice to grow food-chain crops, to add to or replace existing permitted crops with different food-chain crops, or to modify operating plans for distribution of animal feeds resulting from such crops	3
9. Modification of operating practice due to detection of releases from the land treatment unit pursuant to 40 CFR 264.278(g)(2)	3
10. Changes in the unsaturated zone monitoring system, resulting in a change to the location, depth, or number of sampling points, or that replace unsaturated zone monitoring devices or components of devices with devices or components that have specifications different from permit requirements	3

Modifications	Class
11. Changes in the unsaturated zone monitoring system that do not result in a change to the location, depth, or number of sampling points, or that replace unsaturated zone monitoring devices or components of devices with devices or components having specifications different from permit requirements	2
12. Changes in background values for hazardous constituents in soil and soil-pore liquid	2
13. Changes in sampling, analysis, or statistical procedure	2
14. Changes in land treatment demonstration program prior to or during the demonstration	2
15. Changes in any condition specified in the permit for a land treatment unit to reflect results of the land treatment demonstration, provided performance standards are met, and the Director's prior approval has been received	*1
16. Changes to allow a second land treatment demonstration to be conducted when the results of the first demonstration have not shown the conditions under which the wastes can be treated completely, provided the conditions for the second demonstration are substantially the same as the conditions for the first demonstration and have received the prior approval of the Director	*1
17. Changes to allow a second land treatment demonstration to be conducted when the results of the first demonstration have not shown the conditions under which the wastes can be treated completely, where the conditions for the second demonstration are not substantially the same as the conditions for the first demonstration	3
18. Changes in vegetative cover requirements for closure	2
L. Incinerators, Boilers, and Industrial Furnaces	
1. Changes to increase by more than 25% any of the following limits authorized in the permit: a thermal feed rate limit, a feedstream feed rate limit, a chlorine/chloride feed rate limit, a metal feed rate limit, or an ash feed rate limit. The Director will require a new trial burn to substantiate compliance with the regulatory performance standards unless this demonstration can be made through other means	3
2. Changes to increase by up to 25% any of the following limits authorized in the permit: A thermal feed rate limit, a feedstream feed rate limit, a chlorine/chloride feed rate limit, a metal feed rate limit, or an ash feed rate limit. The Director will require a new trial burn to substantiate compliance with the regulatory performance standards unless this demonstration can be made through other means	2

Modifications	Class
3. Modification of an incinerator, boiler, or industrial furnace unit by changing the internal size or geometry of the primary or secondary combustion units, by adding a primary or secondary combustion unit, by substantially changing the design of any component used to remove HCl/Cl_2 metals or particulate from the combustion gases, or by changing other features of the incinerator, boiler, or industrial furnace that could affect its capability to meet the regulatory performance standards. The Director will require a new trial burn to substantiate compliance with the regulatory performance standards unless this demonstration can be made through other means.	3
4. Modification of an incinerator, boiler, or industrial furnace unit in a manner that would not likely affect the capability of the unit to meet the regulatory performance standards but which would change the operating conditions or monitoring requirements specified in the permit. The Director may require a new trial burn to demonstrate compliance with the regulatory performance standards	2
5. Operating requirements:	
a. Modification of the limits specified in the permit for minimum or maximum combustion gas temperature, minimum combustion gas residence time, oxygen concentration in the secondary combustion chamber, flue gas carbon monoxide and hydrocarbon concentration, maximum temperature at the inlet to the particulate matter emission control system, or operating parameters for the air pollution control system. The Director will require a new trial burn to substantiate compliance with the regulatory performance standards unless this demonstration can be made through other means.	3
b. Modification of any stack gas emission limits specified in the permit, or modification of any conditions in the permit concerning emergency shutdown or automatic waste feed cutoff procedures or controls	3
c. Modification of any other operating condition or any inspection or record-keeping requirement specified in the permit	2
6. Burning different wastes:	
a. If the waste contains a POHC that is more difficult to burn than authorized by the permit or if burning of the waste requires compliance with regulatory performance standards different from those specified in the permit. The Director will require a new trial burn to substantiate compliance with the regulatory performance standards unless this demonstration can be made through other means.	3

Modifications	Class
b. If the waste does not contain a POHC that is more difficult to burn than authorized by the permit and if burning of the waste does not require compliance with different regulatory performance standards than specified in the permit	2

Note: See 40 CFR 270.42(g) for modification procedures to be used for the management of newly listed or identified wastes.

Modifications	Class
7. Shakedown and trial burn:	
a. Modification of the trial burn plan or any of the permit conditions applicable during the shakedown period for determining operational readiness after construction, the trial burn period, or the period immediately following the trial burn	2
b. Authorization of up to an additional 720 hours of waste burning during the shakedown period for determining operational readiness after construction, with the prior approval of the Director	*1
c. Changes in the operating requirements set in the permit for conducting a trial burn, provided the change is minor and has received the prior approval of the Director	*1
d. Changes in the ranges of the operating requirements set in the permit to reflect the results of the trial burn, provided the change is minor and has received the prior approval of the Director	*1
8. Substitution of an alternative type of nonhazardous waste fuel that is not specified in the permit	1
M. Containment Buildings	
1. Modification or addition of containment building units:	
a. Resulting in greater than 25% increase in the facility's containment building storage or treatment capacity	3
b. Resulting in up to 25% increase in the facility's containment building storage or treatment capacity	2
2. Modification of a containment building unit or secondary containment system without increasing the capacity of the unit	2
3. Replacement of a containment building with a containment building that meets the same design standards provided:	
a. The unit capacity is not increased	1
b. The replacement containment building meets the same conditions in the permit	1
4. Modification of a containment building management practice	2
5. Storage or treatment of different wastes in containment buildings:	
a. That require additional or different management practices	3

Modifications	Class
b. That do not require additional or different management practices	2
N. Corrective Action	
1. Approval of a corrective action management unit pursuant to 40 CFR 264.552	3
2. Approval of a temporary unit or time extension for a temporary unit pursuant to 40 CFR 264.553	2

APPENDIX E

INFORMATION SOURCES

This appendix contains various sources of further information concerning hazardous waste. The information is divided into five sections:

- General regulatory information sources
- Publication sources
- Electronic information sources
- Federal environmental agencies
- State environmental agencies

REGULATORY INFORMATION SOURCES

The Federal Register

The *Federal Register* is the most complete and helpful source of regulatory information. Proposed and final regulations, as well as selected notices, publications, interpretations, and other related federal government information, are published in the *Federal Register.* Each regulation (advanced notice of, proposed, and final) that is published contains a preamble, which is essentially a discussion of EPA's rationale for the regulation and background information and interpretative guidance.

Published daily, the annual $494 hardcover version (or the $433 michrofiche version) of the *Federal Register* is available at this address:

New Orders
Superintendent of Documents
P.O. Box 371954
Pittsburgh, PA 15250
202-512-1800

The *Federal Register* is also available free at the following Internet sites:

http://www.gpo.ucop.edu
http://www.access.gpo.gov
http://www.gpo.gov/nara/index.html

Code of Federal Regulations

Each year the *Federal Register* is codified into the *Code of Federal Regulations* (CFR). The CFR contains promulgated final regulations only, not preambles or proposed regulations. This annual document, issued by the Government Printing Office, is updated in July of each year, but is not published until January or February of the following year. Title 40 of the CFR contains EPA's regulations and Title 49 contains DOT's regulations.

The CFR is now available electronically via the Internet at the following addresses:

http://www.gpo.ucop.edu
http://www.gpo.gov/nara/index.html

Regulatory Information Dockets

Official files of rulemaking documents, including public hearing transcripts, litigation records, and pubic comments, can be obtained through the following phone numbers:

RCRA	703-603-9230
Superfund	703-603-9232
TSCA	202-260-3587

Written Regulatory Interpretation

A person can obtain an interpretation of a regulation or policy by writing to EPA and requesting it. However, at least six weeks is needed for one to obtain a response. To expedite a response, a letter could be routed through a congressional representative.

There is a set time for a response to a letter from a member of Congress requesting a regulatory interpretation, usually 15 working days.

Freedom of Information

The Freedom of Information Act (FOIA), enacted in 1966, established an effective statutory right of access to federal government records. Under the provisions of FOIA, all records under the custody and control of federal executive branch agencies (with specific exceptions) are covered. FOIA requests are required to be responded to within 10 working days. Additionally, the requester can be charged for reasonable research and copying fees.

For further information concerning the limitations and procedures of the Freedom of Information Act, one should contact the Government Printing Office for the publication *Your Right to Federal Records: Questions and Answers on the Freedom of Information Act and the Privacy Act* (Order No. 052-071-00752-1, $1.75).

Hotlines and Clearinghouses

The following hotlines may be consulted:

RCRA/Superfund/EPCRA Hotline	
(9:00 to 6:00 EST)	800-424-9346
	703-412-9810
TSCA/PCB Hotline	
(8:30 to 5:00 EST)	202-554-1404
DOT Hazardous Materials Information Hotline	202-366-4488

Miscellaneous Information Sources The following information sources are available to provide information on other federal regulatory programs:

Asbestos Ombudsman	800-368-5888
EPCRA	800-535-0202
Green Lights Program	202-775-6650
Hazardous Waste Test Methods	703-821-4789
Indoor Air Hotline	800-438-4318
Lead Hotline	800-532-3394
Radon Hotline	800-767-7236
Safe Drinking Water Act	800-426-4791
Solid Waste Assistance Program	800-677-9424
Wetlands Hotline	800-832-7823

PUBLICATION SOURCES

EPA Publications

National Center for Environmental Publications and Information The National Center for Environmental Publications and Information is the central repository for all EPA documents and links to other organizations distributing EPA documents. Documents can be searched and ordered on-line at the following Internet address:

http://www.epa.gov/ncepihom

Documents from the NCEPI also can be ordered by phone:

1-800-490-9198

RCRA Information Center EPA's Office of Solid Waste and Emergency Response, located in Washington, DC, distributes EPA publications dealing with solid waste, hazardous waste, and underground storage tanks through its RCRA Information Center. A complete listing of available documents and the ordering information are contained in EPA's *Catalog of Hazardous and Solid Waste Publications*. This document is available and can be searched on-line at the following Internet location:

http://www.epa.gov/epaoswer/osw/catalog.htm

The catalog can also be ordered directly from the RCRA/Superfund hotline:

RCRA/Superfund/EPCRA Hotline
(9:00 to 6:00 EST) 800-424-9346
703-412-9810

Superfund Publications EPA Superfund documents can be searched for and retrieved on-line or ordered through following Internet site:

http://www.epa.gov/superfund/oerr/techres

EPCRA Publications Documents and publications related to the Emergency Planning and Community Right-to-Know Act (EPCRA) can be searched for, retrieved from, and ordered from the following Internet address:

http://www.epa.gov/swercepp/pubs.html

Government Printing Office

The Government Printing Office (GPO) has many publications available concerning pollution control, including both EPA and other federal agency publications. Because GPO has thousands of documents, it will not send a list of publications. The following is GPO's telephone number:

U.S. GPO 202-512-11800

GPO publications, the *Federal Register*, and other resources can be accessed on the Internet at the following address:

http://www.gpo.gov

The National Technical Information Service

NTIS is a depository for federal government publications, and in recent years EPA has sent a majority of its publications to NTIS for distribution. A complete listing of all NTIS documents is contained in *U.S. Government Reports and Announcement Documents,* which is available at any U.S. Government Depository Library. NTIS also publishes its *Environmental Highlights* catalog, which contains government reports, studies, regulations, guidance, software, and data pertaining to the environment. This free catalog (No. PR-868) can be obtained from NTIS by calling or ordering through E-mail or the Internet.

NTIS may be contacted at

National Technical Information Service
U.S. Department of Commerce
5285 Port Royal Road
Springfield, VA 22161
703-487-4650

E-mail address: orders@ntis.fedworld.gov
Internet address: http://www.ntis.gov/catalogs.htm

The General Accounting Office

The General Accounting Office is an independent government agency that provides oversight functions on behalf of Congress. Periodically, Congress requests that GAO conduct an investigation of a particular federal program, and upon completion of the investigation GAO prepares and distributes its report. These reports provide in-depth information on the program of interest. You can telephone GAO at the following number to obtain a list of publications issued for the previous year:

The General Accounting Office 202-512-6000

GAO reports can be searched for and obtained electronically via the Internet at the following address:

http://www.gpo.gov

Miscellaneous Publication Sources

Additional information is available from the following sources:

Air	919-541-2777
Pesticides	703-557-4460
EPA's National Enforcement Investigations Center	303-236-5122
Drinking Water	202-260-5533
Water Planning and Standards	202-260-7115
Radiation	703-557-9351
EPA's Public Information Center	202-260-2080

ELECTRONIC INFORMATION SOURCES

With the explosion of the Internet, there are a number of sources of available information relating to hazardous waste and regulations. Listed below are a few of the major sources. Because this system is relatively new, constant changes are expected.

USEPA Office of Solid Waste and Emergency Response—the home page for EPA's Office of Solid Waste and Emergency Response, which administers Superfund, RCRA, and EPCRA. Detailed information, reports, announcements, limited copies of policy statements, and selected *Federal Register* notices can be obtained from this address.

http://www.epa.gov/oswer

USEPA's Office of Pollution Prevention and Toxics—the home page for the EPA office that implements the Toxic Substances Control Act. General information and publications can be obtained at this site:

http://www.epa.gov/opptintr

USDOT's Research and Special Programs Administration—the home page for the Research and Special Programs Administration, which is the office responsible for regulating hazardous materials transportation:

http://hazmat.dot.gov

Government Printing Office—a source for the *Code of Federal Regulations*, the *U.S. Code,* congressional bills, and the *Congressional Record:*

http://www.gpo.gov

U.S. House of Representatives Internet Law Library—a source of federal and state laws, the *U.S. Code,* the *Code of Federal Regulations*, and the *Congressional Record:*

http://law.house.gov

Enviroene—a multi-federal-agency-sponsored database that focuses on pollution prevention initiatives, reports, studies, and related information:

http://www.epa.gov/envirosense

National Institute of Standards and Technology—a home page where viewers can access information on research and technology, some of which relates to environmental control:

http://www.nist.gov or 301-975-2000

FEDERAL ENVIRONMENTAL AGENCIES

Pertinent addresses and/or telephone numbers of several agencies are listed below:

U.S. Environmental Protection Agency

Headquarters

401 M Street, SW
Washington, DC 20460

Personnel Locator:	202-260-2090
Personnel Locator/Directory Assistance via Internet:	http://www.epa.gov:6706/natlocdcd/owa/locator.home
EPA's Internet Home Page:	http://www.epa.gov

EPA Regional Offices

USEPA Region I Connecticut, Maine, Massachusetts, New Hampshire, Rhode Island, Vermont

JFK Federal Building
Room 2203
Boston, MA 02203

Internet	http://www.epa.gov/region01
General	617-565-3420
RCRA	617-565-3698
Superfund	617-565-3279
PCBs	617-565-3257
EPCRA	617-565-3240

USEPA Region II New Jersey, New York, Puerto Rico
26 Federal Plaza
New York, NY 10278

Internet	http://www.epa.gov/region2
General	212-264-9628
RCRA	212-264-8672
Superfund	212-264-6136
PCBs	908-906-6817
EPCRA	908-906-6890

USEPA Region III Delaware, District of Columbia, Maryland, Pennsylvania, Virginia, West Virginia

841 Chestnut Street
Philadelphia, PA 19107

Internet	http://www.epa.gov/region03
General	215-597-9800
RCRA	215-597-8132
Superfund	215-597-7668
PCBs	215-566-2145
EPCRA	215-597-9302

USEPA Region IV	Alabama, Florida, Georgia, Kentucky, Mississippi, North Carolina, South Carolina, Tennessee

345 Courtland Street, NE
Atlanta, GA 30365

Internet	http://www.epa.gov/region04
General	404-347-4727
RCRA	404-347-4097
Superfund	404-347-3864
PCBs	404-562-8977
EPCRA	404-347-2904

USEPA Region V	Illinois, Indiana, Michigan, Minnesota, Ohio, Wisconsin

230 South Dearborn Street
Chicago, IL 60604

Internet	http://www.epa.gov/region05
General	312-353-2000
RCRA	312-655-7570
Superfund	312-353-1428
PCBs	312-886-1332
EPCRA	312-886-6219

USEPA Region VI	Arkansas, Louisiana, New Mexico, Oklahoma, Texas

1445 Ross Avenue
Suite 1200
Dallas, TX 75270

Internet	http://www.epa.gov/region06
General	214-665-6444
RCRA	214-665-6705
Superfund	214-665-9899
PCBs	214-665-7579
EPCRA	214-655-8013

USEPA Region VII Iowa, Kansas, Missouri, Nebraska

726 Minnesota Avenue
Kansas City, MO 66101

Internet	http://www.epa.gov/region07
General	913-551-7000
RCRA	913-236-2855
Superfund	913-236-2835
PCBs	913-551-7395
EPCRA	913-551-7020

USEPA Region VIII Colorado, Montana, North Dakota, South Dakota, Utah, Wyoming

One Denver Place
999 18th Street
Suite 500
Denver, CO 80202

Internet	http://www.epa.gov/region08
General	303-312-6312
RCRA	303-293-1519
Superfund	303-293-1174
PCBs	303-293-1686
EPCRA	303-312-6419

USEPA Region IX Arizona, California, Guam, Hawaii, Nevada

75 Hawthorne Place
San Francisco, CA 94105

Internet	http://www.epa.gov/region09
General	415-744-1305
RCRA	415-744-1117
Superfund	415-556-8389
PCBs	415-556-8071
EPCRA	415-744-1128

USEPA Region X	Alaska, Idaho, Oregon, Washington

1200 Sixth Avenue
Seattle, WA 98101

Internet	http://www.epa.gov/region10
General	206-553-1200
RCRA	206-442-1987
Superfund	206-442-4153
PCBs	206-553-6693
EPCRA	206-553-4016

U.S. Department of Justice

Environment and Natural Resources Division This division is responsible for civil and criminal enforcement of environmental regulations, primarily on behalf of the Environmental Protection Agency.

Telephone:	202-514-2000
Internet:	http://www.usdoj.gov/enrd

U.S. Coast Guard/National Response Center

Telephone:	202-267-2229
Internet:	http://www.nrc.uscg.mil
Environmental Protection Office:	202-426-2200

STATE ENVIRONMENTAL AGENCIES

Alabama

Department of Environmental Management
1751 W. L. Dickinson Drive
Montgomery, AL 36109

Main	334-271-7700
Fax	334-279-3050
Internet	http://www.adem.state.al.us
EPCRA	205-260-2717
Hazardous waste	334-271-7730

Alaska

Department of Environmental Conservation
410 Willoughby Avenue
Suite 105
Juneau, AK 99801

Main	907-465-5065
Internet	http://www.state.ak.us/local/akpages/env.conserv/home.htm
EPCRA	907-465-5220
Hazardous waste	907-465-5168

Arizona

Department of Environmental Quality
3033 N. Central Avenue
Phoenix, AZ 85012

Main	602-207-2300
Internet	http://www.adeq.state.az.us
EPCRA	602-231-6346
Hazardous waste	602-207-4105

Arkansas

Department of Environmental Quality
8001 National Drive
Little Rock, AR 72219

Main	501-682-0744
Internet	http://www.adeq.state.ar.us
EPCRA	501-562-7444
Hazardous waste	501-682-0876

California

Department of Toxic Substances Control
400 P Street
Sacramento, CA 95814

Main	916-324-1826
Internet	http://www.calepa.cahwnet.gov
EPCRA	916-324-9924

Hazardous waste	916-324-1781
	800-618-6942

Colorado

Department of Health
4300 Cherry Creek Drive South
Denver, CO 80222

Main	303-692-3300
Internet	http://www.cdphe.state.co.us/cdphehom.htm
EPCRA	303-692-3017
Hazardous waste	303-692-3320

Connecticut

Bureau of Waste Management
Department of Environmental Protection
79 Elm Street
Hartford, CT 06106

Main	860-424-3000
Fax	860-424-4058
Internet	http://dep.state.ct.us
EPCRA	806-424-3024
Hazardous waste	806-424-3021
PCBs	806-424-3369

Delaware

Department of Natural Resources and Environmental Control
P.O. Box 1401
Dover, DE 19903

Main	302-739-6400
Internet	http://www.dnrec.state.de.us
EPCRA	302-739-4791
Hazardous waste	302-739-3689

District of Columbia

Department of Consumer and Regulatory Affairs
Environmental Regulation Administration
2100 Martin Luther King Jr. Avenue, SE
Suite 203
Washington, DC 20020

Main	202-645-6617
EPCRA	202-673-2101, ext. 3161
Hazardous waste	202-645-6080, ext. 3011

Florida

Department of Environmental Protection
2600 Blair Stone Road, Twin Towers
Tallahassee, FL 32399

Main	904-488-1554
Internet	http://www.dep.state.fl.us
EPCRA	800-535-7179
	904-413-9970
Hazardous waste	904-488-0300

Georgia

Department of Natural Resources
205 Butler Street, SE
Floyd Towers East
Atlanta, GA 230334

Main	404-656-4713
Internet	http://www.dnr.state.ga.us
EPCRA	404-656-6905
Hazardous waste	404-656-7802

Hawaii

Department of Health
919 Ala Moana Boulevard
2d Floor

Honolulu, HI 96814

Main	808-586-4850
EPCRA	808-586-4694
Hazardous waste	808-586-4235

Idaho

Division of Environmental Quality
Department of Health and Welfare
1401 N. Hilton Street
Boise, ID 83706

Main	208-334-0502
Internet	http://www2.state.id.us/deq
EPCRA	208-334-3263
Hazardous waste	208-334-5898

Illinois

Environmental Protection Agency
2200 Churchill Road
Springfield, IL 62794

Main	217-782-3397
Internet	http://www.epa.state.il.us
EPCRA	217-785-0830
Hazardous waste	217-785-8604

Indiana

Department of Environmental Management
105 N. Senate Avenue
Indianapolis, IN 46206

Main	317-232-8603
Internet	http://www.state.in.us/idem
EPCRA	317-232-8172
Hazardous waste	317-232-4417

Iowa

Environmental Protection Division
Department of Natural Resources
Henry Wallace State Office Building
900 E. Grand Avenue
Des Moines, IA 50319

Main	515-281-5145
Internet	http://www.state.ia.us/government/dnr/index.html
EPCRA	515-281-8852
Hazardous waste	913-551-7058 (U.S. EPA Region VII)

Kansas

Department of Health and Environment
Forbes Field, Bldg. 740
Topeka, KS 66620

Main	913-296-1606
Fax	913-296-8909
Internet	http://www.state.ks.us/public/kdhe
EPCRA	913-296-1690
Hazardous waste	913-296-1608

Kentucky

Department of Environmental Protection
14 Reilly Road
Frankfort Office Park
Frankfort, KY 40601

Main	502-564-2150
Internet	http://www.state.ky.us/agencies/nrepc/dep/dep2.htm
EPCRA	502-564-2150
Hazardous waste	502-564-6716

Louisiana

Department of Environmental Quality
7290 Bluebonnet Drive
Baton Rouge, LA 70884

Main	504-765-0353

Fax	504-765-0617
Internet	http://www.deq.state.la.us
EPCRA	504-765-0737
Hazardous waste	504-765-0272

Maine

Department of Environmental Protection
17 State House Station
Augusta, ME 04333

Main	207-287-7688
Internet	http://www.state.me.us/dep
EPCRA	800-452-8735
	207-289-4080
Hazardous waste	207-289-2651

Maryland

Department of the Environment
2500 Broening Hwy
Baltimore, MD 21224

Main	410-631-3000
Internet	http://www.mde.state.md.us
EPCRA	410-631-3800
Hazardous waste	410-631-3345

Massachusetts

Department of Environmental Protection
One Winter Street
7th Floor
Boston, MA 02108

Main	617-292-5856
Fax	617-574-6880
Internet	http://www.state.ma.us/dep/dephome.htm
EPCRA	617-292-5870
Hazardous waste	617-292-5574

Michigan

Department of Natural Resources
608 W. Allegan
1st Floor
Lansing, MI 48933

Main	800-662-9278
Internet	http://www.deq.state.mi.us
EPCRA	517-373-8481
Hazardous waste	517-335-8410

Minnesota

Pollution Control Agency
520 N. Lafayette Road
St. Paul, MN 55155

Main	800-657-3864
Internet	http://www.pca.state.mn.us/netscape.shtml
EPCRA	612-282-5396
Hazardous waste	612-297-8512

Mississippi

Office of Pollution Control
Department of Environmental Quality
2380 Highway 80 West
Jackson, MS 39204

Main	601-961-5171
Internet	http://www.deq.state.ms.us
EPCRA	601-960-9000
Hazardous waste	601-961-5171

Missouri

Division of Environmental Quality
Department of Natural Resources
205 Jefferson Street
Jefferson Building

Jefferson City, MO 65102

Main	573-751-6892
Internet	http://www.state.mo.us/dnr/deq
EPCRA	573-526-6627
Hazardous waste	573-751-3176

Montana

Department of Environmental Quality
P.O. Box 200901
Helena, MT 59620

Main	406-444-2544
Internet	http://www.deq.state.mt.us
EPCRA	406-444-1374
Hazardous waste	406-444-1430

Nebraska

Department of Environmental Quality
1200 N Street
The Atrium, Suite 400
Lincoln, NE 68509

Main	402-471-2186
Internet	http://www.deq.state.ne.us
EPCRA	402-471-4230
Hazardous waste	402-471-4217

Nevada

Division of Environmental Protection
Department of Conservation and Natural Resources
333 W. Nye Lane
Carson City, NV 89710

Main	702-687-5872
EPCRA	702-784-1717
Hazardous waste	800-882-3233

New Hampshire

Department of Environmental Services
6 Hazen Drive
Concord, NH 03301

Main	603-271-2900
Internet	http://www.state.nh.us/des
EPCRA	603-271-2231
Hazardous waste	603-271-2942

New Jersey

Department of Environmental Protection
401 E. State Street
Trenton, NJ 08625

Main	609-292-2885
Fax	609-292-7695
Internet	http://www.state.nj.us/dep
EPCRA	609-984-3219
Hazardous waste	609-292-8341

New Mexico

Environment Department
1190 S. St. Francis Drive
Santa Fe, NM 87505

Main	505-827-2855
Internet	http://www.nmenv.state.nm.us
EPCRA	505-827-9223
Hazardous waste	505-827-4308

New York

Department of Environmental Conservation
50 Wolfe Road
Albany, NY 12233

Main	518-457-6934
Fax	518-457-0629
Internet	http://unix2.nysed.gov/ils/executive/encon

EPCRA	518-457-4107
Hazardous waste	518-485-8988

North Carolina

Department of Environment, Health, and Natural Resources
P.O. Box 29603
Raleigh, NC 27611

Main	919-733-2178
Fax	919-715-3605
Internet	http://www.ehnr.state.nc.us/ehnr
EPCRA	919-733-3865
Hazardous waste	919-733-2178

North Dakota

Hazardous Waste Program
Division of Waste Management
Department of Health
1200 Missouri Avenue
P.O. Box 5520
Bismarck, ND 58506

Main	703-328-2372
Fax	701-328-5200
Internet	http://www.ehs.health.state.nd.us/ndhd/environ/wm
EPCRA	701-328-2111
Hazardous waste	701-328-5166

Ohio

Ohio Environmental Protection Agency
1800 Watermark Drive
Columbus, OH 43215

Main	614-644-3020
Internet	http://www.epa.ohio.gov
EPCRA	614-644-3606
Hazardous waste	614-644-2944

Oklahoma

Department of Environmental Quality
1000 N.E. Tenth Street
Oklahoma City, OK 73117

Main	405-271-1400
Fax	405-271-1317
Internet	http://www.deq.state.ok.us
EPCRA	405-271-1152
Hazardous waste	405-271-5338

Oregon

Department of Environmental Quality
811 S.W. Sixth Avenue
Portland, OR 97204

Main	503-229-5696
Fax	503-229-6124
Internet	http://www.deq.state.or.us
EPCRA	503-378-3473
Hazardous waste	503-229-5913

Pennsylvania

Department of Environmental Resources
400 Market Street
14th Floor
Harrisburg, PA 17101

Main	717-783-2300
Fax	717-783-8926
Internet	http://www.dep.state.pa.us
EPCRA	717-783-2071
Hazardous waste	717-787-6239

Rhode Island

Department of Environmental Management
291 Promenade Street

Providence, RI 02908

Main	401-277-2774
EPCRA	401-277-2808
Hazardous waste	401-277-2979

South Carolina

Department of Health and Environmental Control
2600 Bull Street
Columbia, SC 29201

Main	803-734-5383
Internet	http://www.state.sc.us/dhec/eqchome.htm
EPCRA	803-896-4117
Hazardous waste	803-896-4171

South Dakota

Department of Environment and Natural Resources
523 E. Capitol Avenue
Foss Building
Pierre, SD 57501

Main	605-773-3151
Fax	605-773-6035
Internet	http://www.state.sd.us/executive/denr
EPCRA	800-438-3367
Hazardous waste	605-773-3153

Tennessee

Department of Environment and Conservation
401 Church Street
L&C Tower, 21st Floor
Nashville, TN 37243

Main	888-891-8332
Internet	http://www.state.tn.us/environment
EPCRA	800-262-3300
	615-741-2986
Hazardous waste	615-532-0780

Texas

Industrial and Hazardous Waste Division
Natural Resources Conservation Commission
P.O. Box 13087
Austin, TX 78711

Main	512-239-1000
Internet	http://www.tnrcc.state.tx.us
EPCRA	512-239-3100
Hazardous waste	512-239-6592

Utah

Department of Environmental Quality
P.O. Box 144880
Salt Lake City, UT 84114

Main	801-536-4404
Internet	http://www.eq.state.ut.us
EPCRA	801-536-4100
Hazardous waste	801-538-6170

Vermont

Department of Environmental Conservation
Agency of Natural Resources
103 S. Main Street
West Building
Waterbury, VT 05671

Main	802-241-3800
Internet	http://www.anr.state.vt.us
EPCRA	802-241-3626
Hazardous waste	802-241-3868

Virginia

Department of Environmental Quality
P.O. Box 10009
Richmond, VA 23240

Main	804-698-4000

Internet	http://www.deq.state.va.us
EPCRA	804-698-4489
Hazardous waste	804-698-4115

Washington

Division of Hazardous Waste and Toxics Reduction
Department of Ecology
P.O. Box 47600
Olympia, WA 98504

Main	360-407-7000
Internet	http://www.wa.gov/ecology
EPCRA	360-407-6727
Hazardous waste	360-407-6703

West Virginia

Division of Environmental Protection
Bureau of Environment
State Complex Building
1356 Hansford Street
Charleston, WV 25301

Main	304-759-2506
Internet	http://charon.osmre.gov
EPCRA	304-558-5380
Hazardous waste	304-558-5929

Wisconsin

Division of Environmental Quality
Department of Natural Resources
P.O. Box 7921 SW/3
Madison, WI 53707

Main	608-266-1099
Internet	http://www.dnr.state.wi.us
EPCRA	608-266-9255
Hazardous waste	608-266-2111

Wyoming

Department of Environmental Quality
122 W. 25th Street
Herschler Building
Cheyenne, WY 82002

Main	307-777-7937
Internet	http://deq.stat.wy.us
EPCRA	307-777-4900
Hazardous waste	307-777-7752

APPENDIX F

ACRONYMS

AA	Assistant Administrator
ACL	Alternate Concentration Level
ADI	Acceptable Daily Intake
AEA	Atomic Energy Act
ALJ	Administrative Law Judge
ANPR	Advanced Notice of Proposed Rulemaking
AO	Administrative Order
APA	Administrative Procedures Act
AQCRs	Air Quality Control Regions
ARARs	Applicable or Relevant and Appropriate Requirements
ASHAA	Asbestos School Hazard Abatement Act
ASTM	American Society for Testing and Materials
ATA	American Trucking Association
ATSDR	Agency for Toxic Substances and Disease Registry
ATS	Action Tracking System
AX	Administrator's Office
BAT	Best Available Technology Economically Achievable
BDAT	Best Demonstrated Available Technology
BIF	Boilers and Industrial Furnaces
BOD	Biological Oxygen Demand
BRC	Below Regulatory Concern
BPT	Best Practicable Control Technology
BTU	British Thermal Unit

CA	Cooperative Agreement
CAA	Clean Air Act
CAAA	Clean Air Act Amendments
CAG	Carcinogen Assessment Group
CAP	Criteria Air Pollutants
CAS	Chemical Abstract Service
CASRN	Chemical Abstract Service Registry Number
CBA	Cost Benefit Analysis
CDC	Center for Disease Control
CERCLA	Comprehensive Environmental Response, Compensation, and Liability Act
CERCLIS	Comprehensive Environmental Response, Compensation, and Liability Information System
CERI	Center for Environmental Research Information
CESQG	Conditionally Exempt Small Quantity Generator
CFR	*Code of Federal Regulations*
CGL	Comprehensive General Liability
CHIP	Chemical Hazard Information Profile
CLP	Contract Laboratory Program
CM	Corrective Measures
CMI	Corrective Measures Implementation
CMS	Corrective Measures Study
CPF	Carcinogenic Potency Factor
CSF	Carcinogenic Slope Factor
CWA	Clean Water Act
DOC	Department of Commerce
DOD	Department of Defense
DOE	Department of Energy
DOJ	Department of Justice
DOL	Department of Labor
DOT	Department of Transportation
DQOs	Data Quality Objectives
DRE	Destruction and Removal Efficiency
EA	Environmental Assessment
EDD	Enforcement Decision Document
EDF	Environmental Defense Fund
EE/CA	Engineering Evaluation/Cost Analysis
EIL	Environmental Impairment Liability
EIR	Exposure Information Report
EIS	Environmental Impact Statement
EPA	Environmental Protection Agency
EPCRA	Emergency Planning and Community Right-to-Know Act
EPOW	Extraction Procedure for Oily Wastes

EP TOX Extraction Procedure Toxicity Test
ERCS Emergency Response Cleanup Services
ERL Environmental Research Laboratory
ERRIS Emergency and Remedial Response Information System
ESA Endangered Species Act

FDA Food and Drug Administration
FFA Federal Facility Agreement
FFDCA Federal Food, Drug, and Cosmetic Act
FHSA Federal Hazardous Substances Act
FIFRA Federal Insecticide, Fungicide, and Rodenticide Act
FIT Field Investigation Team
FML Flexible Membrane Liner
FOIA Freedom of Information Act
FONSI Finding of No Significant Impact
FR *Federal Register*
FS Feasibility Study
FWPCA Federal Water Pollution Control Act
FY Fiscal Year

GA Grant Agreement
GLP Good Laboratory Practices
GPO Government Printing Office
GWPS Groundwater Protection Standard

HA Health Advisory
HAP Hazardous Air Pollutant
HAZCOM Hazard Communication Standard
HAZWOPER Hazardous Waste Operations
HCA Hazardous Communications Act
HEEP Health and Environmental Effects Profile
HMT Hazardous Materials Table
HMTA Hazardous Materials Transportation Act
HMTR Hazardous Materials Transportation Regulations
HOC Halogenated Organic Compound
HRS Hazard Ranking System
HSL Hazardous Substance List
HSWA Hazardous and Solid Waste Amendments of 1984
HW Hazardous Waste

ICP Integrated Contingency Plan
IM Interim Measure
ISCL Interim Status Compliance Letter
ITC Interagency Testing Committee

KG	Kilogram
LAER	Lowest Achievable Emissions Rate
LC_{50}	Lethal Concentration—50 percent
LCRS	Leachate Control and Removal System
LD_{50}	Lethal Dose—50 percent
LLRW	Low-Level Radioactive Waste
LOIS	Loss of Interim Status
LQG	Large-Quantity Generator
LQHUW	Large-Quantity Handler of Universal Waste
MCL	Maximum Contaminant Level
MCLG	Maximum Contaminant Level Goal
MEK	Methyl Ethyl Ketone
MOA	Memorandum of Agreement
MOD	Memorandum of Decision
MOU	Memorandum of Understanding
MPRSA	Marine Protection, Research, and Sanctuaries Act
MSCA	Multi-Site Cooperative Agreement
MSDS	Material Safety Data Sheet
MTB	Materials Transportation Board
MTU	Mobile Treatment Unit
NA	North America
NAA	Non-Attainment Areas
NAAQS	National Ambient Air Quality Standards
NACE	National Association of Corrosion Engineers
NBAR	Nonbinding Preliminary Allocation of Responsibility
NCAPS	National Corrective Action Prioritization System
NCP	National Contingency Plan
NDD	Negotiation Decision Document
NEC	National Electric Code
NEIC	National Enforcement Investigations Center
NEPA	National Environmental Policy Act
NESHAP	National Emission Standards for Hazardous Air Pollutants
NFRAP	No Further Remedial Action Planned
NIEHS	National Institute of Environmental Health Sciences
NIMBY	Not In My Backyard
NIOSH	National Institute of Occupational Safety and Health
NLAP	National Lab Audit Program
NOD	Notice of Decision
NOD	Notice of Deficiency
NOEL	No Observed Effect Level
NOI	Not Otherwise Indexed
NOIBN	Not Otherwise Indexed By Name

NOID	Notice of Intent to Deny Permit
NON	Notice of Noncompliance
NOS	Not Otherwise Specified
NOV	Notice of Violation
NPDES	National Pollutant Discharge Elimination System
NPDWR	National Primary Drinking Water Regulations
NPL	National Priorities List
NRC	National Response Center
NRC	Nuclear Regulatory Commission
NRDC	Natural Resources Defense Council
NRT	National Response Team
NSPS	New Source Performance Standards
NTIS	National Technical Information Service
NTP	National Toxicological Program
OECD	Organization for Economic Cooperation and Development
OECM	Office of Enforcement and Compliance Monitoring
OERR	Office of Emergency and Remedial Response
O&M	Operation and Maintenance
OMB	Office of Management and Budget
O/O	Owner or Operator
OPP	Office of Pesticide Programs
ORC	Office of Regional Counsel
ORM	Other Regulated Material
ORNL	Oak Ridge National Laboratory
OSC	On-Scene Coordinator
OSHA	Occupational Safety and Health Administration
OSW	Office of Solid Waste
OSWER	Office of Solid Waste and Emergency Response
OTA	Office of Technology Assessment
OTS	Office of Toxic Substances
OUST	Office of Underground Storage Tanks
OWPE	Office of Waste Programs Enforcement
P2	Pollution Prevention
PA	Preliminary Assessment
PAT	Permit Assistance Team
PCB	Polychlorinated Biphenyl
PCP	Pentachlorophenol
PCT	Polychlorinated Terphenyl
PFLT	Paint Filter Liquids Test
PL	Public Law
PMN	Premanufacture Notification
POHC	Principal Organic Hazardous Constituent
POTW	Publicly Owned Treatment Works

PPB	Parts Per Billion
PPM	Parts Per Million
PPMW	Parts Per Million by Weight
PPT	Parts Per Trillion
PRP	Potentially Responsible Party
PSD	Prevention of Significant Deterioration
QA	Quality Assurance
QC	Quality Control
RA	Regional Administrator
RA	Remedial Action
RAMP	Remedial Action Master Plan
RCRA	Resource Conservation and Recovery Act
RD	Remedial Design
RD&D	Research, Demonstration, and Development
REM	Remedial Planning Contractor
RFA	RCRA Facility Assessment
RfD	Reference Dose
RFI	RCRA Facility Inspection
RI	Remedial Investigation
RIA	Regulatory Impact Analysis
RIM	Regulatory Information Memorandum
RIP	RCRA Implementation Plan
RMCL	Recommended Maximum Contaminant Level
ROD	Record of Decision
RPAR	Rebuttable Presumption Against Registration
RPM	Remedial Project Manager
RQ	Reportable Quantity
RS	Remedial Selection
RTECS	Registry of Toxic Effects of Chemical Substances
SAB	Science Advisory Board
SACM	Superfund Accelerated Cleanup Model
SARA	Superfund Amendments and Reauthorization Act
SBA	Small Business Administration
SCAP	Superfund Comprehensive Accomplishments Plan
SCRP	Superfund Community Relations Plan
SDWA	Safe Drinking Water Act
SI	Site Inspection
SIC	Standard Industrial Classification
SIP	State Implementation Plan
SITE	Superfund Innovative Technology Evaluation
SMOA	Superfund Memorandum of Agreement
SNARL	Significant No Adverse Reaction Level

SPCC	Spill Prevention, Control, and Countermeasure Plan
SPW	Supplemental Well
SQHUW	Small-Quantity Handler of Universal Waste
SQG	Small-Quantity Generator
SSC	State Superfund Contract
SWDA	Solid Waste Disposal Act
SWMU	Solid Waste Management Unit
TAG	Technical Assistance Grant
TAT	Technical Assistance Team
TBC	To Be Considered Guidelines and Controls
TC	Toxicity Characteristic
TCLP	Toxicity Characteristic Leaching Procedure
TOC	Total Organic Carbon
TSCA	Toxic Substances Control Act
TSD	Treatment, Storage, and/or Disposal
TSDF	Treatment, Storage, and/or Disposal Facility
TSS	Total Suspended Solids
UCR	Unit Cancer Risk
UIC	Underground Injection Control
UN	United Nations
USCG	United States Coast Guard
USDW	Underground Source of Drinking Water
USGS	United States Geological Survey
USPS	United States Postal Service
UST	Underground Storage Tank
UTS	Universal Treatment Standards
VOC	Volatile Organic Compound
WMA	Waste Management Area
WQC	Water Quality Criteria
ZHE	Zero Headspace Extraction Vessel

INDEX